1993

ECOLOGY AND EVOLUTION OF PLANT REPRODUCTION

ECOLOGY AND EVOLUTION OF PLANT REPRODUCTION

Edited by

Robert Wyatt

Chapman & Hall
New York   London

First published in 1992 by
Chapman & Hall
an imprint of
Routledge, Chapman & Hall, Inc.
29 West 35 Street
New York, NY 10001-2291

Published in Great Britain by
Chapman & Hall
2-6 Boundary Row
London SE1 8HN

© 1992 Routledge, Chapman & Hall, Inc.

Printed in the United States of America on acid free paper.

Library of Congress Cataloging-in-Publication Data

Ecology and evolution of plant reproduction / [edited by Robert
  Wyatt].
     p. cm.
  Includes bibliographical references and index.
  ISBN 0-412-03021-7
  1. Plants—Reproduction. 2. Plants, Sex in. 3. Botany—Ecology.
4. Plants—Evolution. I. Wyatt, Robert Edward, 1950–   .
QK827.E26 1992
581.1′66—dc20   92-30108
                                              CIP

British Library Cataloguing in Publication Data also available.

Contributors

W. T. Adams
Department of Forest Science
Oregon State University
Corvallis, OR 97331-5705
USA

Tia-Lynn Ashman
Department of Botany and Center for
    Population Biology
University of California
Davis, CA 95616
USA

Spencer C. H. Barrett
Department of Botany
University of Toronto
Toronto, Canada M5S 3B2
Canada

David S. Birkes
Department of Statistics
Oregon State University
Corvallis, OR 97331-4606
USA

Mitchell B. Cruzan
Department of Botany
University of Toronto
Toronto, Canada M5S 3B2
Canada

Pamela K. Diggle
Department of Environmental,
    Population, and Organismic Biology
University of Colorado
Boulder, CO 80303
USA

Michael J. Donoghue
Department of Ecology and
    Evolutionary Biology
University of Arizona
Tucson, AZ 85721
USA

V. J. Erickson
Umatilla National Forest
USDA Forest Service
Pendelton, OR 97801
USA

Michael W. Folsom
Department of Biology
University of New Mexico
Albuquerque, NM 87131
USA

Laura F. Galloway
Department of Botany and Center for
  Population Biology
University of California
Davis, CA 95616
USA

Kent E. Holsinger
Department of Ecology &
  Evolutionary Biology, U-43
University of Connecticut
Storrs, CT 06269-3043
USA

Terryn Hough
School of Botany
University of Melbourne
Parkville, Victoria 3052
Australia

Bruce Knox
School of Botany
University of Melbourne
Parkville, Victoria 3052
Australia

Joshua R. Kohn
Department of Botany
University of Toronto
Toronto, Canada M5S 3B2
Canada

Tak-Cheung Lau
Department of Biology
The Pennsylvania State University
University Park, PA 16802
USA

David G. Lloyd
Department of Plant and Microbial
  Sciences
University of Canterbury
Christchurch
New Zealand

Diane L. Marshall
Department of Biology
University of New Mexico
Albuquerque, NM 87131
USA

Susan J. Mazer
Department of Biological Sciences
University of California
Santa Barbara, CA 93106
USA

David L. Mulcahy
Department of Botany
University of Massachusetts
Amherst, MA 01003
USA

Gabriella B. Mulcahy
Department of Botany
University of Massachusetts
Amherst, MA 01003
USA

Mauricio Quesada
Department of Biology
The Pennsylvania State University
University Park, PA 16802
USA

Samuel M. Scheiner
Department of Biological Sciences
Northern Illinois University
DeKalb, IL 60115
USA

Karen B. Searcy
Department of Botany
University of Massachusetts
Amherst, MA 01003
USA

Mohan Singh
School of Botany
University of Melbourne
Parkville, Victoria 3052
Australia

Maureen Stanton
Department of Botany and Center for
  Population Biology
University of California
Davis, CA 95616
USA

Andrew G. Stephenson
Department of Biology
The Pennsylvania State University
University Park, PA 16802
USA

Cenk Suphioglu
School of Botany
University of Melbourne
Parkville, Victoria 3052
Australia

James D. Thomson
Department of Ecology and Evolution
State University of New York
Stony Brook, NY 11794-5245
USA
and
Rocky Mountain Biological
    Laboratory
Crested Butte, CO 81224-0519
USA

Barbara A. Thomson
Department of Ecology and Evolution
State University of New York
Stony Brook, NY 11794-5245
USA
and
Rocky Mountain Biological
    Laboratory
Crested Butte, CO 81224-0519
USA

Joseph Travis
Department of Biological Sciences B-
    142
Florida State University
Tallahassee, FL 32306-2043
USA

James A. Winsor
Department of Biology
The Pennsylvania State University
Altoona, PA 16601-3760
USA

Helen J. Young
Department of Biological Sciences
Barnard College
3009 Broadway
New York, NY 10027-6598

Contents

Preface

Studies of the ecology and evolution of plant reproduction have their roots in classical observations of pollination biology by nineteenth-century naturalists. The field, however, has been revitalized and redirected as a result of pathbreaking advances over the past 10–15 years. There has been an intellectual revolution, stimulated in part by a number of technical and quantitative breakthroughs. In the area of mating-system evolution, for example, workers have turned away from static descriptions of pollination and genetic self-incompatibility to dynamic models that explain evolutionary transitions in pollen dispersal modes and breeding structure. New discoveries have also forced us to consider ecological, as well as genetic, factors that may drive such transitions. Population geneticists and ecologists have developed explicit models whose assumptions can be cleanly tested by empirical observations and experiments. Moreover, using appropriate biochemical and molecular genetic markers, it is now possible to obtain reliable quantitative estimates of critical biological parameters such as outcrossing rates, gene flow, and the breeding structure of natural plant populations.

The purpose of this book is to make some of the exciting new discoveries in this field accessible to a wide audience. The contributions included here represent a diverse mix of the best current research on all aspects of the ecology and evolution of plant reproduction. The chapters attempt to summarize and review much of the new knowledge in the field, while simultaneously giving the flavor of ongoing research at the leading edge of the field. Each author has strived to describe the importance of his or her own research in a broader context. Thus, the literature cited section of each paper can serve as an efficient entry into the literature of that specific research area.

An unusual aspect of the book's genesis is that it is based on a two and one-half day conference held 12–14 April 1991 in Athens, Georgia. This gave the diverse range of scientists represented the opportunity to share viewpoints and exchange information—a rare occasion in a synthetic field, given the present-day penchant of biologists for fragmentation into specialist societies. The conference speakers represented morphology, systematics, genetics, and cell biology, in addition to the core of ecology and evolutionary biology. Collectively, the authors covered the full spectrum of approaches that contribute vigor to this active area of plant research.

In some sense this volume may mark a watershed for the field of plant reproductive biology. There are those who argue that interest in this subject has peaked and that its momentum will not be sustained into the next century. Judging, however, from the ever-growing number of young scientists entering this area and presenting exciting new discoveries at national and international meetings, it appears that interest in the ecology and evolution of plant reproduction is continuing to increase. It is my hope that this book will help to stimulate even more rapid advances in our knowledge of these aspects of the biology of plants and will encourage even more young scientists to direct their energies to this arena.

Acknowledgments

I thank the members of the Organizing Committee, Elizabeth G. Williams, James L. Hamrick, and William E. Friedman, for helping me to make the conference a reality. Primary support for the conference came from the Office of the Vice President for Academic Affairs under a state-of-the-art conference grants program started by William F. Prokasy. Some additional funding was provided by the Department of Botany. Finally, I thank a large number of conscientious reviewers who read the authors' manuscripts and offered constructive suggestions for improvement: Spencer C. H. Barrett, Steven B. Broyles, James S. Clark, Pamela S. Diggle, Michael J. Donoghue, Michelle R. Dudash, Michael W. Folsom, James L. Hamrick, John S. Heywood, Jeffrey P. Hill, Kent E. Holsinger, Suzanne Koptur, David G. Lloyd, Diane L. Marshall, Susan J. Mazer, David L. Mulcahy, David C. Queller, Tammy L. Sage, Andrew F. Schnabel, Allison A. Snow, Timothy P. Spira, Maureen L. Stanton, Andrew G. Stephenson, James D. Thomson, and Joseph A. Travis.

1

Pollen Presentation and Viability Schedules in Animal-Pollinated Plants: Consequences for Reproductive Success

James D. Thomson and Barbara A. Thomson

State University of New York and Rocky Mountain Biological Laboratory

Introduction: Pollen Presentation Schedules as an Object of Study

Our investigation begins with an observation so commonplace that the reasons for it have scarcely been sought: few plants produce only a single, big anther. In particular, the deployment of pollen accross many anthers and many flowers frequently results in a characteristic temporal schedule of pollen presentation to pollinators. Such schedules, we contend, are an important, but comparatively neglected, component of the floral phenotype.

Percival (1955) pioneered the study of pollen presentation schedules, but her survey of the British flora remained an isolated example for some years. Her work primarily concerned the food value of the plants to pollinators and, therefore, the food choices of the animals. She did point out that gradual presentation was very common. More recent interest in pollen presentation patterns has concentrated on the reproductive success of plants, especially on "male" success through pollen donation. Thomson and Barrett (1981) showed that in *Aralia hispida*, the timing of male-flower anthesis could substantially affect the plant's functional gender (i.e., the relative proportions of its genes that are passed on through pollen and ovules: Lloyd and Bawa, 1984). They argued that selection on male function should favor prolonged pollen presentation, because that would secure more mating opportunities (see also Thomson et al., 1989). Brantjes (1983) invoked weather-induced loss of viability as another selective agent with a similar effect.

Lloyd and Yates (1982; Lloyd, 1984) proposed a more general reason for temporal staggering of presentation: that the mechanics of pollen transfer would often result in a higher proportion of grains being delivered to stigmas if those grains were removed by numerous pollinators rather than by one or a few. Using empirical data on removal and deposition of *Er-*

ythronium spp. pollen by bumble bees, Harder and Thomson (1989) modeled the effects of various pollen presentation patterns on successful donation. This analysis confirmed the Lloyd–Yates conjecture for a generous region of the parameter space examined. Specifically, there will be an optimal presentation schedule for any expected pollinator visitation rate. Infrequent visits favor simultaneous presentation of all pollen, to avoid the pollen wastage entailed by presenting pollen after the last visit. With more frequent visits, however, plants can donate severalfold more of their grains to other stigmas if they package or dispense their pollen so that all visitors encounter some pollen.

The optimal presentation schedules of the Harder–Thomson models are based on a number of simplifying assumptions, including (1) all the grains that reach stigmas are of equal value for male reproductive success, and (2) all pollinator visits are equivalent. Here, we use simulation models to investigate the consequences of realistically relaxing these assumptions. First, we consider the loss of pollen viability that is known to occur in many plants following exposure of the grains. Some pollen will be dead on arrival. Second, we model the effects of different pollinators with different removal and delivery characteristics.

To illustrate the importance of pollen viability schedules, we first address a paradox arising from our observations of *E. grandiflorum*: although this plant provided most of the parameters for the Harder–Thomson model, it appears to contradict that model's principal conclusion. Specifically, casual observations suggest a very *low* visitation rate coupled with *gradual* pollen presentation. Here, we present data on pollen presentation and viability schedules, and on visitation rates, and demonstrate that adding pollen viability constraints to the model can resolve the contradiction.

We then use the new model to ask how presentation and viability schedules affect pollen donation when the suite of pollinators includes different species with different pollen transfer characteristics. Numerous studies have compared the pollination services provided by different flower-visiting animals (e.g., Primack and Silander, 1975; Motten et al., 1981; Tepedino, 1981; Herrera, 1987), but most of these have treated single visits. Only a few have treated multiple visits by individuals of one species or combinations of more than one species (Young, 1988; Young and Stanton, 1990a). Almost none has explicitly considered how temporal patterns of pollen presentation by the plant interact with particular temporal sequences of visits by various pollinators (Tepedino, 1981). Our "pollen depletion" model emphasizes that pollen removed by one visitor is no longer available to be picked up by subsequent visitors, and this essential fact generates some complex consequences for the plants when the pollinators differ in the amount of pollen they subsequently deposit. The value of a particular pollinator to a plant may depend very much on the other pollinators that are available, as well as on the presentation and viability schedule.

Methods

Pollen Presentation Schedules

In May–June 1990, we observed the timing of anthesis and the progress of anther dehiscence in inflorescences of *Erythronium grandiflorum* Pursh (Liliaceae), growing in large, dense stands in subalpine meadows at Irwin, Colorado (39°30'N, 107°6'W, 3275 m elevation). Flowering individuals produce only a few flowers at most; single-flowered plants ("singles") are most abundant, although in certain local areas "doubles" may be as common. Three- and four-flowered plants ("triples" and "quads") are uncommon. Each pendant flower bears six stamens. The anthers are long and thin, typically ca. 1.5 mm in width and 10–25 mm long. Anthers dehisce by "unzipping": a suture opens at the distal end, and the split gradually extends to the proximal end as the thecae turn inside out, exposing the pollen. The amount of pollen contained in an anther can be closely estimated by simply measuring its length (Harder et al., 1985; Thomson and Thomson, 1989). Stigmatic receptivity begins about when dehiscence begins, and extends approximately 1 day after dehiscence is complete, depending on weather and pollination status.

We located, in bud, 16 plants in each flower–number category, one through four. As each flower opened, we marked it and also individually marked the undehisced stamens with dots of ink on the filaments. We measured the length of the undehisced stamens with digital calipers reading to 0.01 mm. Then, at 2-hr intervals during daylight, and 4- or 6-hr intervals during the night, we repeatedly measured the *undehisced* portion of each anther. Observations continued until all anthers of all flowers were completely dehisced. Unfortunately, a number of these plants were lost to herbivores before the completion of anthesis, and many sequences were interrupted by rain.

To consider how pollen presentation schedules might change with an increased number of flowers, we calculated two summary indices for each of the 24 curves we obtained during fair weather. *Duration* is simply the length of time from first dehiscence until all pollen has been exposed. *Evenness* is a measure how evenly pollen is presented over time: the observed presentation curve is broken into hour-long segments, the proportion of grains newly presented (P_i) in each segment i is estimated by linear interpolation between the sequential observations, and the evenness index is calculated as $1/\Sigma(p_i^2)$ (Krebs, 1989).

Pollinator Visitation Rates

Although *E. grandiflorum* flowers are also visited by hummingbirds and small bees, most pollination is effected by large *Bombus* queens (Thomson, 1986). The activity of these insects—and, hence, the visitation rate received by flowers—varies greatly with the weather and with the season. We have noted in several seasons, for example, that the earliest flowers precede the emergence of most of the queens and thus are virtually unattended (Thomson, 1982; personal observation; D. Taneyhill, personal observation). We have not systematically studied variation in visitation rate. For the models in this paper, we required only a rough estimate of visitation under favorable circumstances.

In June 1990, during a sunny period about two-thirds of the way through the flowering season of *E. grandiflorum*, we marked 30 singles with green surveyor's stake flags. These plants were widely spread across ca. 1 ha of the population. All were recently opened, with fewer than three anthers dehisced; all had clean stigmas on inspection by hand lens. We harvested the styles 24 hr after the initial inspection, being careful to prevent contamination. Each stigma was examined microscopically and classified as "unvisited," "possibly visited," or "definitely visited." "Unvisited" stigmas were free of pollen. "Definitely visited" stigmas bore loads of 400 or more grains, deposited in patterns uncharacteristic of autogamous deposition (i.e., in the central cleft of the tripartite stigma and on the downward-facing portions of the papillose stigma lobes). "Possibly visited" flowers had numerous (often ca. 100) grains, but these were typically on the upward-facing and peripheral portions of the lobes. Pollen could have been deposited in these areas without an insect visit, although a visit by a "side-working" bee (Thomson, 1986) could also produce such a pattern. Although subjective, these judgments were informed by considerable experience in observing and counting stigmatic loads produced by *Bombus* visits.

We estimated the visitation rate (as visits flower^{-1} day^{-1}) from the fraction of flowers visited, by assuming a Poisson process (i.e., the fraction of unvisited flowers equals e^{-m}, where m is the mean number of visits per flower). Because the flowers were widely spaced, we expect that the independence assumption of the Poisson was adequately met. To circumvent the ambiguity caused by the "possibly visited" category, we calculated the visitation rate twice, once with these flowers counted as visited and once with them counted as unvisited. The resulting two estimates delimit a range that probably contains the true value.

Viability of Pollen

In a small experiment in June 1989, M. B. Cruzan and N. O'Connor used the fluorochromatic reaction (FCR) test (Heslop-Harrison et al., 1984;

Shivanna and Johri, 1985) to determine how long *E. grandiflorum* pollen grains remained viable after exposure, on the dehisced anther, under field conditions. Freshly cut flowers were kept in vases. Initially, a sample of ca. 200 grains was taken from the splitting cleft of one dehiscing anther from each flower; these were the freshest grains possible. At intervals, further samples of grains were removed from the same area of the anther. Because the cleft moves as the anther unzips and shrinks from desiccation, there is no reference point to assure that one is resampling from precisely the same spot, but this was attempted.

Each sample of grains was treated with a 10% sucrose solution to which fluorescein diacetate in acetone had been added dropwise until the mixture became cloudy. After culturing for a few minutes under a cover slip, at least 100 grains were examined under UV. Brightly fluorescing grains were scored as viable.

Pollen-Depletion Models

We wrote FORTRAN programs (available on request) to investigate the effects of pollen presentation schedules, viability schedules, and mixed pollinator faunas. These simulation models track pollen movement from a plant over a period of anthesis divided into 100 (200 in some runs) equal intervals.

Each model starts with a "pollen presentation curve," or a plot of cumulative number of pollen grains made available to pollinators through the 100 time intervals. During each interval, whether a pollinator visits or not is determined by a random number between 0 and 1, which is compared to a chosen probability. For example, if the probability is 0.1 per interval, 10 randomly spaced visits are expected during anthesis. For simplicity, only one visit is allowed per interval. During a visit, the pollinator moves some of the available pollen from the anthers, leaving the rest behind. Of the removed pollen, some is subsequently deposited on recipient stigmas. Removal and deposition are governed by deterministic functions whose parameters are simplified versions of those measured empirically by Thomson and Thomson (1989) and Harder and Thomson (1989) for the *Erythronium–Bombus* system. When the next pollinator comes, the pollen presented comprises the residuum left after the first visit, plus any newly presented pollen, as determined from the presentation curve. The total number of grains deposited on other flowers by all visitors is calculated from the (pollinator-specific) deposition function. This measure of pollen donation can serve as an index of expected plant reproductive success, in situations where pollen receipt does not limit female function. We ignore self-deposition in estimating male success because competitive pollinations indicate that self-pollen seldom fertilizes ovules when outcross pollen is also present

(Rigney et al., 1992). All results are based on 500 iterations. Parameter values and other details for specific simulations are given below, with the results, for ease of interpretation.

Results

Pollen Presentation

Pollen presentation is far from simultaneous in a plant of *E. grandiflorum* (Figs. 1.1 and 1.2). The six anthers are grouped into two whorls, and the outer whorl of three opens before the inner whorl. Often, dehiscence is complete in the outer whorl before it has begun in the inner; less commonly, there is some overlap. Within a whorl, the three anthers often show staggered openings.

Anthers unzip most quickly in warm, dry weather; dehiscence continues overnight, but at a slower pace, contributing a somewhat steplike character to the cumulative presentation curves (Figs. 1.1 and 1.2). Partial synchronization of anthers within whorls also contributes a step-like component. The gradual overnight presentation of pollen in the absence of visitors will presumably result in an accumulation of grains for the first visit of the morning. Otherwise, cumulative presentation is almost linear during the day.

Overcast weather with high relative humidity greatly slows anther dehiscence. If enough rain falls to wet the anthers, the unzipping process can actually be reversed, as the rehydrated anthers appear to close. Clearly, such reversals make it difficult to estimate cumulative pollen production. There were two rainy periods during our study, and, for simplicity, we have removed from the data set all of the plants that were caught in mid-presentation by either of these rains. Thus, the data represent dehiscence patterns under better-than-usual conditions.

Adding more flowers to the inflorescence could, in principle, change the duration or the evenness, or both, or neither. Duration would be increased by opening the flowers sequentially, rather than simultaneously; evenness could be affected if different flowers have staggered periods of most rapid presentation, thus averaging out presentation over time. Only a few of the multiple-flowered plants escaped the rainy periods, giving small sample sizes. Nevertheless, Figure 1.3 makes it clear that duration increases with flower number, and hints that evenness is maximized in three-flowered plants. Flowers generally open sequentially in *E. grandiflorum*, often with no overlap. Uppermost flowers are the largest, open first, have the most ovules, and are most likely to set fruit (Thomson, 1989). In triples, the lower two flowers usually follow the same pattern (i.e., the second to open is larger and more likely to set fruit than the third, but occasionally the

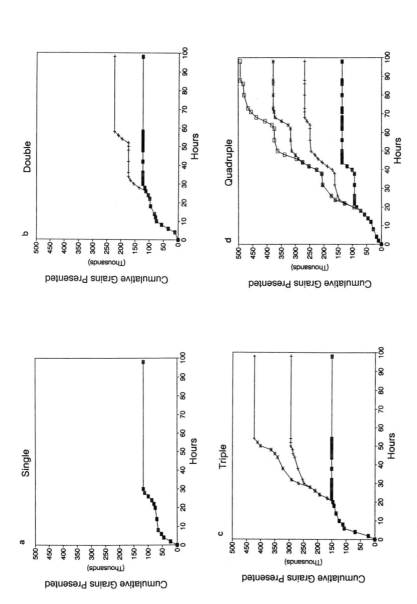

Figure 1.1. Observed pollen presentation curves for typical *Erythronium grandiflorum* plants with different numbers of flowers. (a) A single-flowered plant; (b) a double; (c) a triple; (d) a quadruple. In b–d, the graphs are stacked, so that the lowest line shows the pollen presented by the first flower, the next line shows the total amount presented by flowers 1 and 2, etc. Thus, the top line shows cumulative production by the whole plant.

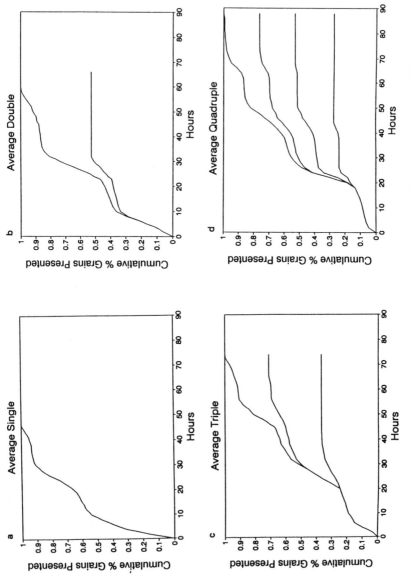

Figure 1.2. Average pollen presentation curves for *Erythronium grandiflorum* plants from different flower–number classes (a–d as in Fig. 1). Averages were computed by first interpolating grain numbers for each flower in a class so that all flowers had a number of grains presented for every hour. Sample sizes are 11, 7, 3, and 3 plants for panel a, b, c, and d, respectively.

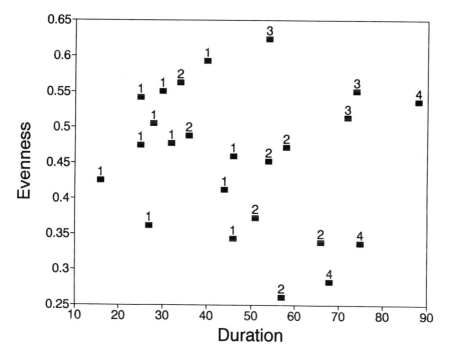

Figure 1.3. Temporal evenness of pollen presentation vs. duration of pollen presentation for *Erythronium grandiflorum* plants from different flower–number classes. Numerals by each point indicate the flower–number class of the plant.

lower two appear as "identical twins," borne at the same point on the scape, of the same size, and undergoing anthesis in parallel). We believe that the probability of such twins increases in quadruple inflorescences, possibly limiting the average evenness obtained by such plants.

Visitation Rate

The fraction of flowers "definitely visited" and "possibly visited" during the 24-hr test period were 0.57 and 0.80, respectively ($n = 30$), leading to minimum and maximum estimates of visitation rate of 0.65 and 0.82 visits flower^{-1} day^{-1}. As these estimates reflect the most favorable weather conditions, realized visitation rates are almost certainly lower, on the order of one visit per flower lifetime. Thomson (1982) reported visitation rates in this same general range at another site, using different methods.

Pollen Viability

FCR scores declined rapidly with time of exposure (Fig. 1.4). Desiccation is the most likely cause of the decline; drying of pollen is known to damage

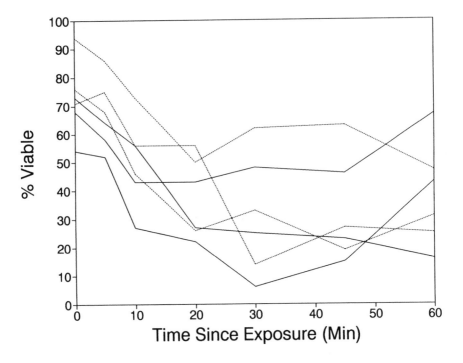

Figure 1.4. Time course of viability loss for *Erythronium grandiflorum* pollen exposed to field conditions for various lengths of time. Each line represents pollen taken from a single anther. The broken lines are data from 7 July 1989, a sunny day; the solid lines are from 8 July, a cloudy day.

membrane function (e.g., Shivanna and Johri, 1985; Hoekstra and van Roekel, 1988), and the FCR procedure assays membrane function (as well as enzyme activity). It is not certain that the nonfluorescing grains are truly dead; exposure to a humid atmosphere might well restore membrane integrity and, with it, pollen function (Shivanna and Johri, 1985; Linskens et al., 1989). Nevertheless, dehydrated grains will probably be slower to germinate on stigmas and thus be less likely to contribute to reproductive success than fresher grains. Additional experiments from 1991 have confirmed the general shape of the pollen survival curve and have suggested that the apparent "resurrections" seen in some curves in Figure 1.4 are probably artifacts due to againg of the fluorescein diacetate solution (Karoly et al., 1992).

Simultaneous vs. Gradual Presentation of Pollen

The data on visitation rate and on pollen presentation raise an apparent paradox. *Erythronium grandiflorum* appears to stagger pollen presentation nearly as much as its morphology allows, yet the models of Harder and Thomson (1989) suggest that rarely visited plants should *not* stagger presentation. To confirm that the observed presentation phenotype appears maladaptive under the Harder–Thomson assumptions, we modeled a situation in which the expected number of bee visits was either 1.0 or 10.0 during the life of a plant. Each visit removed 70% of the available pollen, and the number of grains subsequently deposited on other stigmas is estimated by the square root of the number of grains removed. These values approximate the mean values found for bumble bee visits by Thomson and Thomson (1989).

In addition to varying visitation rate, we used three different pollen presentation curves: one observed curve, taken from a typical two-flowered plant, and two hypothetical extremes. In one extreme, all grains were presented simultaneously (no staggering); in the other, equal numbers of grains were presented in each interval (complete staggering).

The total amount of pollen donated by the hypothetical plant with simultaneous pollen presentation was *greater* than that of either the observed phenotype or of the completely staggered plant when only a single visit is expected (Fig. 1.5a). The advantages shift more toward the staggered schedules when more visits are expected (Fig. 1.5b), in accordance with the expectation above.

If, however, the model is further modified so that pollen gradually loses viability after it is exposed in the anthers and if only viable grains are counted toward the total donated, the observed phenotype becomes superior to the simultaneous presenter (Fig. 1.5a), even at low visit numbers. The combination of low visitation and short-lived pollen probably best represents the conditions governing *E. grandiflorum* pollination. In these runs, pollen survival was modeled as an exponential decay process, with 10% mortality of the grains in each time interval. Formulating this loss rate in terms of typical *Erythronium* pollen presentation curves yields a pollen half-life of about 3 hr (6 intervals to 50% decay, 100 intervals per approximately 50-hr anthesis period). This is a conservative loss rate, in view of our preliminary data on viability schedules (Fig. 1.4).

Different Types of Pollinators

In this set of simulations, we consider the effects and interactions of three pollinator types that differ in their removal and delivery of pollen. For mnemonic convenience, we name the types *"good," "bad,"* and *"ugly,"* and for brevity we will consider them all to be bees, although they could represent any type of pollinator (Table 1.1). Good bees remove large amounts of pollen and redeposit it in relatively large amounts; their parameters are simplified versions of functions measured for bumble bees on

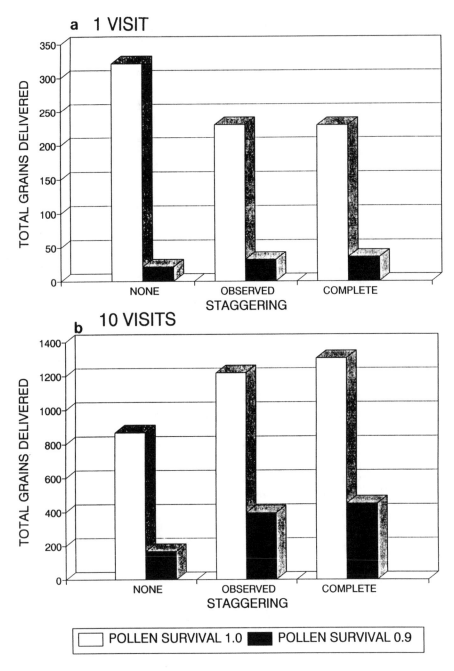

Figure 1.5. Results of simulation model of pollen delivery for three pollen presentation schedules, two viability schedules, and two pollinator visitation rates.

Table 1.1. The Deterministic Pollen Removal and Deposition Functions Used to Model the Three Different Types of Pollinators.[a]

	Pollinator type		
Function	Good	Bad	Ugly
Removal	$R = 0.7P$	$R = 0.2P$	$R = 0.7P$
Deposition	$D = R^{0.5}$	$D = R^{0.4}$	$D = R^{0.4}$

[a]P, number of grains presented in anthers at time of visit; R, number removed by pollinator; D, number of removed grains subsequently deposited on stigmas. The values used for the good pollinator approximate those observed for *Bombus occidentalis* queens on *Erythronium grandiflorum* (Thomson and Thomson, 1989; Harder and Thomson, 1989); the others are judged to be reasonable approximations for hypothetical pollinators that pick up and deposit less pollen.

Erythronium. We view these parameters as representing a situation of reasonable coadaptation of plant and pollinator. The other two categories are hypothetical variants. *Bad* bees remove less of the available pollen and deposit less of what they remove. They have relatively little impact on pollen flow. *Ugly* bees, in contrast, remove pollen as do good bees but deliver what they remove as do bad bees. They move more pollen than bads but also take a larger amount out of circulation. Other variants could be considered, of course, but these are sufficient to produce complex interactions.

A good bee always delivers more pollen than the others, but the relative values of the others to the plant depend on other parameters (Fig. 1.6). When presentation is staggered and pollen is immortal, all three bees deliver pollen as a slightly decelerating function of visit number (Fig. 1.6a). The bad and ugly bees both deliver about one-third as much per visit as good bees. If these long-lived grains are instead presented simultaneously, both good and ugly bees deplete the anthers of pollen fairly quickly, yielding saturating curves of delivery vs. visit number (Fig. 1.6b). Pollen delivery by ugly bees saturates at about 39% of the saturation level for good bees. The bad bees, less effective at removal, do not deplete the anthers so quickly, and their curve continues to rise, crossing that of the uglies at about four visits. Under these conditions, the relative values of bads and uglies as pollinators are not absolute, but depend on the number of visits, with the bad bees becoming more valuable as they are more common.

Having short-lived pollen changes things. With staggered presentation, none of the curves shows satuation over the range shown, and the uglies are almost twice as effective as the bads (Fig. 1.6c). This relatively better performance by the uglies shows how low pollen viability puts a premium on high removal, even if subsequent delivery to stigmas is not very effective. Unremoved grains will soon die in the anthers in any case, so there would

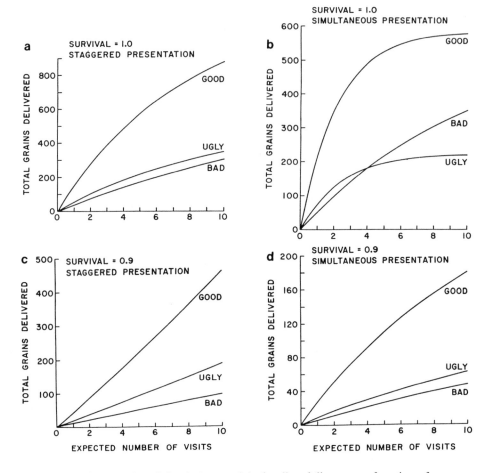

Figure 1.6. Results of simulation model of pollen delivery as a function of expected visit number by three hypothetical pollinator types (see Table 1.1 for the pollinator characteristics) under four combinations of pollen viability and presentation schedules. Note different vertical scales.

be little advantage in their "waiting" for a better pollinator to arrive. The curves do not saturate (1) because staggered production prevents depletion of the anthers and (2) because each of the many small cohorts of grains has a short life and thus needs a visit during that short life to achieve effective dispersal.

When short-lived pollen is presented simultaneously (Fig. 1.6d), the curve shapes are very similar to those for staggered, long-lived pollen (Fig. 1.6a), but total amounts of viable pollen delivered are very different. The agreement in shapes seems fortuitous, as the parameters of the model are so different.

We also explored the consequences of mixed groups of pollinators. In our simulations, we examined mixtures either of goods and bads or of goods and uglies. At each interval of the pollen presentation curve, a random number was chosen to determine, with a certain probability, whether or not a visit would occur. If a visit did occur, a second random draw determined whether the visitor was a good bee or of the other type (bad or ugly, depending on the run). Thus, both the expected total number of visits and the fraction of those visits made by good bees varied stochastically in these runs.

In addition to confirming that pollen delivery is increased by more visits and by a higher proportion of good visits, these simulations reveal some of the complexities and nonlinearities that govern how two different pollinators interact (Fig. 1.7). Some of these parameter combinations are explored in Table 1.2. Under the conditions outlined in Table 1.2, we see that four uglies alone deliver 1.41 times more grains than four bads alone. We might tend to conclude, therefore, that uglies are simply better than bads for plants with the indicated presentation and viability schedules. However, if four good bees are *also* visiting, the values of the bads and uglies reverse. Four uglies *add* only $528 - 492 = 36$ grains to the number that would have been delivered by the four good bees alone; four bads *add* 1.92 times that amount. In this situation, the ugly bees live up to their mnemonic; although the ugly bees are more effective than bads when by themselves, they do not combine well with better bees.

If we examine the same comparison with short-lived pollen, bad and ugly reverse again (Table 1.2). Four ugly bees now *add* 1.77 times more grains to the amount delivered by four goods alone than do four bads. Once again, short-lived pollen puts a premium on removal; in this simulation, four ugly bees *add* more grains when in the company of goods than they deliver alone.

The final comparison—long-lived pollen, simultaneously presented–reveals ugly bees at their ugliest (Table 1.2). Here, *adding* four visits by ugly bees to four visits by good bees actually *reduces* total pollen delivery by 21%. Adding bads *increases* delivery by 7%. That adding uglies reduces total delivery is doubly striking: first, the ugly bees are good pollinators when they alone visit. In fact, they are as good as the bads. Second, eight visits, all by effective pollinators, deliver less pollen than four of those visits would alone. The peculiar interaction of the two bees with the presentation and viability schedules turns the less effective pollinator into a functional parasite. Its "crime" is that it removes pollen that, if left behind, would be delivered better by the other visitor.

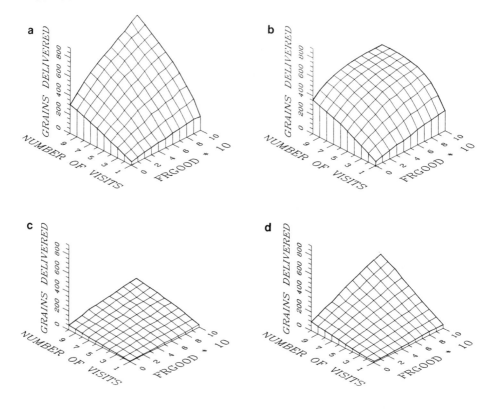

Figure 1.7. Results of simulation models of pollen delivery as a function of the abundances of different hypothetical pollinator types. (a–d) Mixtures of good and bad pollinators; (e–h) mixtures of good and ugly pollinators. The axis labeled "FRGOOD * 10" indicates the expected proportion of visits that are by good pollinators (multiplied by 10 for the convenience of the graphing package used). (a and e) Long-lived grains and staggered presentation; (b and f) long-lived grains and simultaneous presentation; (c and g) short-lived grains and staggered presentation; (d and h) short-lived grains and simultaneous presentation.

Discussion

We began this work with the idea—derived from our measurements of pollen removal and delivery—that pollen presentation schedules should become objects of quantitative study for evolutionary ecologists. We ended by concluding that pollen viability schedules deserve equal attention. To paraphrase George Williams on life history attributes: *schedules*, no less than teeth or chromosomes, evolve. However, measuring these schedules

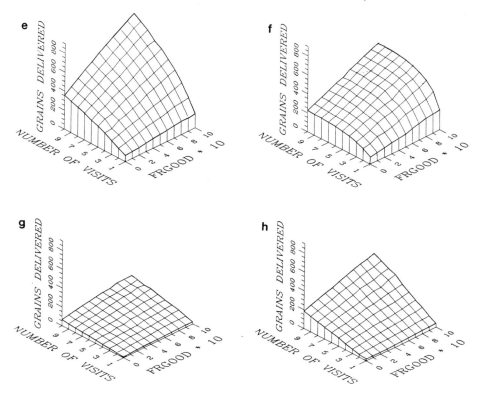

Figure 1.7. (*Continued*)

Table 1.2. Interaction of Different Pollinators for Some Selected Parameter Combinations Extracted from Figure 1.7.[a]

Expected set of visits	Total viable grains delivered		
	Long-stag	Short-stag	Long-simul
4 goods only	492	173	492
4 bads only	133	36	179
4 uglies only	187	67	181
4 goods + 4 bads	561	218	525
4 goods + 4 uglies	528	252	390

[a]Long, long-lived pollen; short, short-lived pollen; stag, staggered presentation; simul, simultaneous presentation.

poses practical difficulties, which we will discuss briefly before taking up more subtle points of interpretation.

Measuring pollen presentation schedules seems straightforward in principle. Although few plants have large, linear, "unzipping" anthers like those that allowed us to follow gradual dehiscence at the individual stamen level in *Erythronium*, such detail may be unnecessary in plants with smaller anthers. But harder problems come in deciding which aspects of the measured presentation curves are relevant. Our analyses included whole curves, measured on a 24-hr basis. Our simulations send in visitors at random times, regardless of day or night. Although suitable for heuristic purposes, such simplifications would compromise any applied study. To do better, however, would require additional details on the diel patterning of pollinator visits, which requires a particularly demanding and tedious variety of field work.

Equally worrisome is the effect of weather. We achieved simplicity by ignoring rainy periods, but this cost us two-thirds of our data. More importantly, natural selection on pollen presentation schedules does not, presumably, ignore rainy periods. Even mild overcasts and temperature fluctuations influence dehiscence; they influence animal pollinators as well. A complete analysis of a real system must tie weather variation to both supply and demand aspects of pollen–pollinator interactions. Furthermore, desiccation (as cited by Brantjes, 1983) is not the only cause of pollen death; grains that stay too long in undehisced anthers due to humid weather may lose viability as well (Linskens et al., 1989), an additional factor to measure and model. We need a more integrative view of weather in floral biology; humidity and temperature will certainly be important driving variables (see Corbet et al., 1979).

A more enjoyably solved challenge of this pollen–demographic approach will lie in establishing more communication between evolutionary ecologists concerned with the fitness consequences of floral phenotypic variation and physiologists, who have developed most of our understanding of the basic processes of dehiscence, viability, and germination. Both groups need to understand each other's literature before some of the most basic questions can be answered satisfactorily. One of the most important of these is: "How should viability be measured?" It is clear that none of the traditional "stainability" methods gives the same information as the FCR test itself, and it is equally clear that the FCR test does not always predict germinability (Shivanna and Johri, 1958). For models like ours, we need to know more about the prospects for *partially* desiccated pollen to sire seeds after landing on a stigma along with fresher grains. Pollen variability tends to be treated as a " +/- " condition, but there may be considerable variation in vigor among the viable grains (Young and Stanton, 1990b).

We are beginning such studies on *Erythronium*, using electrophoretic markers to determine the success of stressed pollen in competitive pollinations.

Pollen Removal and Deposition by Different Visitors

Detailed studies of removal and deposition are still rare and mostly recent (e.g., Strickler, 1979; Snow and Roubik, 1987; Wolfe and Barrett, 1989), but seem to be increasing. The advent of electronic particle counters has made pollen counting so quick that new types of study are practical. Because directly estimating a plant's male reproductive success is so difficult, several workers have proposed using the amount of pollen removed from a plant's anthers as an index of male success (e.g., Cruzan et al., 1988; Young and Standon, 1990a; Murcia, 1990). Although these indirect estimates are better than more primitive estimates, such as the number of visits (e.g., Thomson, 1988), pollen removal can be a very poor index of subsequent pollen delivery, especially if pollinator faunas are heterogeneous. "Ugly" pollinators may contribute to high removal rates, even as they decrease successful delivery. Such outcomes are not purely hypothetical; Wilson and Thomson (1991) have demonstrated that pollen-foraging honey bees do play the role of "ugly" pollinators in *Impatiens capensis*, where nectar-feeding bumble bees are the "good" pollinators. Patches dominated by the honey bees show higher removal rates but lower stigmatic deposition. One cannot draw valid inferences from removal data without knowing something about the pollinators.

The necessary information is virtually unavailable in the literature. Few existing studies of the differences among pollinators consider multiple visits (Young, 1988; Young and Stanton, 1990a; Murcia, 1990; and Harder, 1990a provide exceptions) or simultaneously quantify both removal and deposition. Thus, most studies imply that a particular visitor has a fixed pollination value. In contrast, our pollen-depletion models show that the value of any single visit depends greatly on what other visits have previously occurred or will occur in the future.

There are two especially weak links in our model; both concern the removal and delivery functions. First, all our knowledge of removal in *Erythronium* comes from first visits. The models assume that freshly presented pollen and pollen remaining after a visit have identical removal characteristics. In fact, the pollen remaining may have remained precisely because it resisted removal (e.g., it may have been stuck to tapetal tissue or wedged in fissures). This issue—the equivalency of fresh pollen and remaining pollen—is hard to approach experimentally.

Second, we use deterministic functions for removal and deposition, even though these processes are actually highly stochastic. The empirical relationship between pollen removal and subsequent deposition (Thomson and

Thomson, 1989; Harder and Thomson, 1989) shows so much scatter that fitting *any* function through the points is partly an act of faith and needs to be bolstered by other knowledge (e.g., the positive relation between pollen load on the bee and the bee's rate of grooming (Thomson, 1986; Harder, 1990b). We use these data because better data do not exist, and we use deterministic functions to clarify the contrasts between the different pollinator types in a model with heuristic intent. Any attempts to measure significant differences in pollen transfer by different pollinators will require very large sample sizes, due to the highly stochastic nature of anther and stigma contact (P. Wilson, personal communication; see Thomson, 1986).

It will also be important to measure deposition as completely as possible. If deposition is measured only on the first few recipient flowers, for example, a pollinator with very extensive pollen carryover might appear to be "ugly," whereas in fact it might provide superior pollen delivery (M. Stanton, personal communication).

Furthermore, natural selection on floral phenotypes, including scheduling aspects, will also reflect, and will be impeded by, the great stochasticity of pollen transfer. Our models present an unusually "clean" view of the linkage between phenotype and fitness.

Visit Sequence

Our models assume a randomized order of arrival of the two pollinator types. This assumption is crucial; ugly bees, for example, would have no detrimental effects if their visits occurred after the visits of the good bees. In some real situations (e.g., honey bees vs. bumble bees), random arrivals may be an acceptable assumption, but in others [e.g., those involving early-morning or "matinal" bees such as *Peponapis* (squash bees) (Tepedino, 1981)] sequential arrivals of different visitors may be predictable and important. In such cases, pollen depletion by the early visitors may render later visitors completely ineffectual and therefore of no evolutionary importance to the plants. It would be interesting to see whether matinal solitary bees are more likely to show oligolecty than nonmatinal species.

Practical Implications

In numerous agricultural crops, harvests are typically limited by insufficient pollinations (Free, 1979; Robinson et al., 1989). The near-universal "remedy" is to bring in hives of honey bees. Even though honey bees are known, in some cases, to be less effective pollinators than certain wild bees (Parker et al., 1987), there appears to be a common feeling that one can always bring in enough of them to "saturate the system." Our models suggest two caveats. First, honey bees are often very active pollen collectors. Although Free (1970) implies that pollen-foraging bees are likely to be better pollinators than nectar collectors, this is hardly necessary; an efficient pollen collector that often misses the stigma will still be an "ugly" bee. This is true of honey bees on *Impatiens capensis* (Wilson and Thomson, 1991), and it may be true in some agricultural systems. If so, adding honey bees may be ineffective or even harmful. Furthermore, even if there are no negative effects, "saturating the system" may involve pollen *delivery* saturating as a function of visit number as in Figure 1.6B. Under the conditions depicted, one could add an infinite number of ugly bee visits without equaling the pollen delivery provided by only three good ones. It would be worthwhile to know more about bee-specific pollen transfer dynamics before embarking on a managed system of pollination by honey bees or before calculating the cash value of pollination services provided by honey bees (Robinson et al., 1989).

Competition

When we have spoken of adding bees to a system, we have assumed no interaction between pollinators (e.g., we "add" four uglies by comparing pollen transfer by four goods alone to that by four goods plus four uglies). In real systems, competition among pollinators might occur, such that adding new pollinators would entail the loss of some old ones. The surfaces in Figure 1.7 can be used to make simple comparisons of this sort. Generally, competition makes bad pollinators worse, but the effects vary greatly depending on other parameters. Competition effects should also be considered when supplementing (supplanting?) wild bees with managed hives.

Conclusion

The male reproductive success of a plant depends on the amount of pollen it donates to stigmas. In animal-pollinated plants, this amount is influenced by the schedules of pollen presentation, pollen survivorship, and pollinator visits. Although each of these factors acts in a straightforward, comprehensible way when only a single pollinator type is attracted, heterogeneous pollinator faunas produce complicated interactions with the scheduling variables. Consequently, the value of a given pollinator to a plant may vary from positive to negative depending on context. Also, two pollinators may have very different values to plants that are identical in all respects except scheduling. So far, scheduling has received little attention from workers interested in male reproductive success, but the relevant parameters are generally accessible through direct observation. We encourage more studies.

Acknowledgments

Field assistance and ideas were contributed by Mitch Cruzan, Lisa Rigney, Paul Wilson, Dale Taneyhill, Johanne Brunet, Don Stratton, Helen Young, Lawrence Harder, Bob Unnasch, Amy Seidl, Andrea Lowrance, George Weiblen, Rita Hurault, and Teri Anderson. More ideas have come from Bruce Knox, Jo Kendrick, K. R. Shivanna, Kent Holsinger, David Lloyd, Chris Plowright, Nick Waser, Mary Price, and others. Lawrence Harder, Maureen Stanton, Helen Young, and an anonymous reviewer provided comments on an early draft. *Erythronium* work has been supported by the American Philosophical Society, the E. N. Huyck Preserve Foundation, and NSF Grants BSR 8614207, 8601104, 9001065, and 9006380. Susan Allen and the staff of the Rocky Mountain Biological Lab provided superb support, especially in an emergency. Contribution No. 798 from Ecology and Evolution, SUNY-Stony Brook.

Literature Cited

Brantjes, N.B.M. 1983. Regulated pollen issue in *Isotoma*, Campanulaceae, and evolution of secondary pollen presentation. Acta Bot. Neerl. 32:213–222.

Corbet, S.A., D.M. Unwin and O.E. Prys-Jones. 1979. Humidity, nectar and insect visits to flowers, with special reference to *Crataegus, Tilia* and *Echium*. Ecol. Ent. 4:9–22.

Cruzan, M.B., P.R. Neal, and M.F. Wilson. 1988. Floral display in *Phyla incisa:* Consequences for male and female reproductive success. Evolution 42:505–515.

Free, J.B. 1970. Insect Pollination of Crops. Academic Press, London.

Harder, L.D. 1990a. Pollen removal by bumble bees and its implications for pollen dispersal. Ecology 71:1110–1125.

Harder, L.B. 1990b. Behavioral responses by bumble bees to variation in pollen availability. Oecologia 85:41–47.

Harder, L.D. and J.D. Thomson. 1989. Evolutionary options for maximizing pollen dispersal of animal-pollinated plants. Am. Nat. 133:323–344.

Harder, L.D., J.D. Thomson, M.B. Cruzan and R.S. Unnasch. 1985. Sexual reproduction and variation in floral morphology in an ephemeral vernal lily, *Erythronium americanum*. Oecologia 61:286–291.

Herrera, C.M. 1987. Components of pollinator "quality": Comparative analysis of a diverse insect assemblage. Oikos 50:79–90.

Heslop-Harrison, J., Y. Heslop-Harrison and K.R. Shivanna. 1984. The evaluation of pollen quality and a further appraisal of the fluorochromatic (FCR) test procedure. Theor. Appl. Genet. 67:367–375.

Hoekstra, F.A., and T. van Rocrel. 1988. Desiccation tolerance of *Papaver dubium* pollen during development in anther: Possible role of phospholipid composition and sucrose content. Plant Physiol. 88:626–632.

Karoly, K., L. Rigney, and J. Thomson. 1992. In preparation.

Krebs, C.J. 1989. Ecological Methodology. Harper & Row, New York.

Linskens, H.F., F. Ciampolini, and M. Cresti. 1989. Restrained dehiscence results in stressed pollen. Proc. Kon. Ned. Acad. Wet. Ser. C 92:465–475.

Lloyd, D.G. 1984. Gender allocations in outcrossing cosexual plants. *In* R. Dirzo and J. Sarukhan, eds., Perspectives on Plant Population Ecology. Sinauer, Sunderland, MA, pp. 277–300.

Lloyd, D.G., and K.S. Bawa. 1984. Modification of the gender of seed plants in varying conditions. Evol. Biol. 17:255–338.

Lloyd, D.G., and J.M.A. Yates. 1982. Intrasexual selection and the segregation of pollen and stigmas in hermaphrodite plants, exemplified by *Wahlenbergia albomarginata* (Campanulaceae). Evolution 36:903–913.

Motten, A.F., D.R. Campbell, D.E. Alexander, and H.L. Miller. 1981. Pollination effectiveness of specialist and generalist visitors to a North Carolina population of *Claytonia virginica*. Ecology 62:1278–1287.

Murcia, C. 1990. Effect of floral morphology and temperature on pollen receipt and removal in *Ipomoea trichocarpa.* Ecology 71:1098–1109.

Parker, F.D., S.W.T. Batra, and V.J. Tepedino. 1987. New pollinators for our crops. Agri. Zoo. Rev. 2:279–304.

Percival, M.S. 1955. The presentation of pollen in certain angiosperms and its collection by *Apis mellifera*. New Phytol. 54:353–368.

Primack, R.B., and J.A. Silander. 1975. Measuring the relative importance of different pollinators to plants. Nature (London) 255:143–144.

Rigney, L.P., J. Thomson, M.B. Cruzan, and J. Brunet. 1992. Differential donor success in *Erythronium grandiflorum*, a self-compatible lily. Evolution. (Submitted).

Robinson, W.S., R. Nowogrodzki, and R.A. Morse. 1989. The value of honey bees as pollinators of U.S. crops. Am. Bee J. 129:411–423, 477–487.

Shivana, K.R., and B.M. Johri. 1985. The Angiosperm Pollen: Structure and Function. Wiley Eastern Limited, New Delhi.

Snow, A.A., and D.W. Roubik. 1987. Pollen deposition and removal by bees visiting two tree species in Panama. Biotropica 19:57–63.

Strickler, K. 1979. Specialization and foraging efficiency of solitary bees. Ecology 60:998–1009.

Tepedino, V.J. 1981. The pollination efficiency of the squash bee (*Peponapis pruinosa*) and the honey bee (*Apis mellifera*) on summer squash (*Cucurbita pepo*). J. Kansas Entomol. Soc. 54:359–377.

Thomson, J.D. 1982. Patterns of visitation by animal pollinators. Oikos 39:241–250.

Thomson, J.D. 1986. Pollen transport and deposition by bumble bees in *Erythronium:* Influences of floral nectar and bee grooming. J. Ecol. 74:329–341.

Thomson, J.D. 1988. Effects of variation in inflorescence size and floral rewards on the visitation rates of traplining pollinators of *Aralia hispida*. Evol. Ecol. 2:65–76.

Thomson, J.D. 1989. Deployment of ovules and pollen among flowers within inflorescences. Evol. Trends Plants 3:65–68.

Thomson, J.D., and S.C.H. Barrett. 1981. Temporal variation of gender in *Aralia hispida* Vent. (Araliaceae). Evolution 35:1094–1107.

Thomson, J.D. and B.A. Thomson. 1989. Dispersal of *Erythronium grandiflorum* pollen by bumblebees: Implications for gene flow and reproductive success. Evolution 43:657–661.

Thomson, J.D., M. McKenna, and M. Cruzan. 1989. Temporal patterns of nectar and pollen production in *Aralia hispida:* Implications for reproductive success. Ecology 70:1061–1068.

Wilson, P. and J.D. Thomson. 1991. Heterogeneity among floral visitors leads to discordance between removal and deposition of pollen. Ecology 72:1503–1507.

Wolfe, L.M., and S.C.H. Barrett. 1989. Patterns of pollen removal and deposition in tristylous *Pontederia cordata* L. (Pontederiaceae). Biol. J. Linn. Soc. 36:317–329.

Young, H.J. 1988. Differential importance of beetle species pollinating *Dieffenbachia longispatha* (Araceae). Ecology 69:832–844.

Young, H.J., and M.L. Stanton. 1990a. Influences of floral variation on pollen removal and seed production in wild radish. Ecology 71:536–547.

Young, H.J., and M.L. Stanton. 1990b. Influence of environmental quality on pollen competitive ability in wild radish. Science 248:1631–1633.

2

Evolutionary Genetics of Pollen Competition

David L. Mulcahy, Gabriella B. Mulcahy, and Karen B. Searcy
University of Massachusetts

Introduction

Approximately 60% of the genes expressed in the sporophytic portion of the life cycle are expressed also in the microgametophyte (Willing et al., 1988; Ottaviano and Mulcahy, 1989). Thus, these genes are subject to natural selection in the haploid condition. Pollen competition has been suggested to be a potentially important phenomenon in flowering plants (Jones, 1928; Mulcahy, 1979). In this chapter we consider some of the factors that might influence pollen competition. Our interest in this subject has been stimulated by the resemblance between pollen and microorganisms: both are haploid, allowing the expression of rare recessive alleles, and both are available in extremely large numbers, allowing reliable production of otherwise rare genetic combinations. It is important to understand that, to the extent possible, we have limited our review to phenomena that relate to these characteristics of pollen. The quality of a pollen grain will be influenced not only by its own haploid genotype but also by the phenotype of the sporophyte that produces it. Thus, if pollen from separate sporophytes is compared, an additional source of variation is introduced, one that lacks the features of haploidy and large population sizes that make pollen unusual. For that reason, we have not reviewed an extensive collection of interesting and important studies that compare the performance of pollen grains from different plants (e.g., Jones, 1928; Murakami et al., 1972; Marshall and Whittaker, 1989; Karron et al., 1990). Exceptions include cases in which phenomena have been investigated only through the use of pollen mixtures. We deal also with topics, such as selective fertilization and interference between pollen grains, that appear to be emerging as interesting topics for further investigation.

Nongenetic Influences on Pollen Quality

It is axiomatic in genetic studies that the phenotype is determined by both environment and genotype. For pollen, the environment includes not only the sporophyte that produces the pollen but also the external environment, which impinges on both the sporophyte and the pollen itself. Young and Stanton (1990) recently demonstrated the influence of environmental quality on pollen competitive ability in wild radish (see also Schoch-Bodmer, 1940; Bell, 1959). Our own experience is that when the plant producing pollen is in poor condition or when a plant that normally produces mostly good pollen exhibits low pollen viability, then any viable pollen grains are extremely sensitive to stress. Conversely, there is clear evidence of acclimation in pollen. In soybeans, pollen tubes will continue to grow at 41°C if the incubation temperature is raised from 29 to 41°C incrementally by 4°C per 15 min. In contrast, pollen germinated at 29°C and suddenly transferred to 41°C is immediately inhibited (Altschuler and Mascarenhas, 1985; Xiao and Mascarenhas, 1985) In *Juglans regia*, the optimum temperature for pollen germination is closely correlated ($r = 0.998$) with the mean number of degree-days experienced by the pollen during maturation (Polito et al., 1991). These demonstrations that pollen can effectively acclimate to changes in temperature may explain unsuccessful attempts to use pollen to select for cold tolerance in the cultivated tomato (den Nijs et al., 1986). In that study, plants were raised under low temperatures and, very likely, the failure of pollen to respond to selection for cold resulted from the ability to acclimate to the cold stress before selection was applied. In the case of acclimation to cold, the mechanism is well known: unsaturated double bonds in lipids cause kinks in their hydrophobic tails, reducing their ability to interlock into a frozen lattice. (An unresolved issue is the extent to which the effects of environmental influence persist if, during pollen tube growth, the environment changes.) Other cases of acclimation by pollen are known (Xiao and Mascarenhas, 1985); thus, it is necessary to control for environmental effects on pollen performance to the fullest extent possible.

Competition between Pollen Grains during Pollen Ontogeny

Some evidence indicates that microspores begin to compete with each other for access to resources within the anther. For example, a mutant for small pollen (*sp*) in corn (*Zea mays* L.) is expressed postmeiotically so that a heterozygous plant (+ /*sp*) produces pollen of two sizes (Singleton and Mangelsdorf, 1940). The smaller pollen grains carry the *sp* allele and larger grains, the + allele. Moreover, the largest individual grains from the bimodal distributions of heterozygotes are larger than the largest individual grains from the +/+ plants. This suggests that pollen carrying the + allele is able to capture parental resources that otherwise would have been invested in pollen carrying the *sp* allele.

Recently, we have measured pollen sizes within the ornamental *Salpiglossis sinuata* Ruis and Pav. (Solanaceae), a species that sheds pollen in loosely joined tetrads. As is often the case with cultivars, pollen quality, and even viability, is variable. Here this variability serves the useful function of indicating if viable (and perhaps even particularly competitive) microspores are able to utilize, or perhaps preempt, resources of nonviable or less competitive microspores. We found that the smallest pollen grains were obtained when all four members of a tetrad were viable (as indicated by aniline blue staining for the presence of cytoplasm). The largest grains were those in which only one viable grain occurred within a tetrad (Fig. 2.1). Grains from tetrads containing two or three viable grains were intermediate in size. These data indicate that viable microspores are indeed able to utilize resources relinquished by nonviable microspores. Thus, we conclude that pollen competition for resources begins when the microspores are still contained within the tetrad.

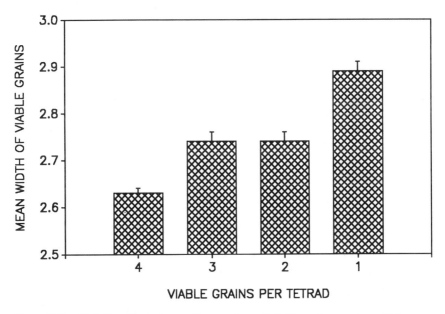

Figure 2.1. Relationship between the number of viable pollen grains within a tetrad and the mean widths of those viable grains. Error bars extend one standard error beyond the means.

Competition within tetrads may also affect pollen germination and pollen tube growth rate. Williams and Rouse (1990) reported that in one semi-sterile plant of *Rhododendron* that produced only two viable pollen grains per tetrad, pollen tubes exhibited an unusually high growth rate. In *Salpiglossis sinuata*, one pollen tube often emerges from a tetrad well before others. Consequently, even when all members of a tetrad are viable, there are measurable differences in the speed of pollen tube germination. These differences could result from both genetic and environmental differences.

Apparently some of the effects of competition within tetrads can be overcome once microspores separate from the tetrad. Shivanna et al. (1991) indicated that storage can be used as a selection pressure on pollen, and we have found that storing pollen of *Nicotiana alata* and *Lycopersicon esculentum* at room temperature for 1–3 days sometimes results in a significant increase in both percentage germination and speed of pollen tube growth. Storage may allow pollen that is immature, because of intratetrad competition or other factors, to complete development before being exposed to germination conditions. In contrast, microspores of barley, *Hordeum vulgare*, exhibit a dimorphism that probably will persist. There, microspores migrate to the tapetum immediately after release from the callosic tetrad, with the single germination pore facing the tapetum. Not all microspores succeed in attaining this position, but rather remain within the lumen of the anther locule. At anthesis, pollen from a single anther consists of two cytoplasmically distinct populations: that derived from microspores directly in contact with the tapetum and that developing within the anther locule (Rose, 1987; Roberts-Oehlschlager and Dunwell, 1991). Deprived of tapetal contact before anthesis, those pollen grains are unlikely to recover.

At least in some cases, differences among members of a tetrad very likely have a genetic basis. Differences within the dimorphic pollen populations of barley, however, are probably due largely to chance. In *Petunia hybrida*, storing pollen (or ovules) at room temperature had a significant effect on the vigor of the resulting progeny and may thus have a genetic basis (Mulcahy et al., 1982). Storage reduced *in vitro* pollen germination to 29.2 ± 0.7% of the value in fresh pollen. Plants resulting from fertilizations by gametes of pollen which had aged 5 days before pollination flowered significantly earlier [$\chi^2(4) = 78.27$, $p < 0.01$] than progeny resulting from fertilizations by fresh pollen. This observation may relate to the competition observed between members of a tetrad. Because there is a positive correlation between the vigor of a pollen genotype and the vigor of the resulting sporophyte (McKenna and Mulcahy, 1983; Ottaviano and Mulcahy, 1989), competition within a tetrad could endow the most vigorous microspores with disproportionate quantities of resources. These might therefore be better able to survive storage, and the selected survivors could give rise to unusually vigorous offspring. It is not known, however, to what extent larger pollen grains from within a tetrad are better able to tolerate storage than are small grains. Nor do we know if larger grains give rise to more vigorous progeny, or what the correlation is between *intraplant* pollen size variation and pollen tube growth rate (but see Williams and Rouse, 1990).

Kumar and Sarkar (1980) examined the correlation between pollen grain diameter and pollen tube growth rate between different lines of *Zea mays*, but such studies do not bear on the present question, as the sporophyte has a great effect on average pollen diameter. Comparison of pollen from different plants is largely a comparison of sporophytic, rather than gametophytic, qualities. The mean pollen diameter declines from F_1 to F_7 in *Zea mays*, and this is a sporophytic effect. Concurrently, the coefficient of variation in pollen diameter decreases, a gametophytic effect that reflects the convergence of haploid genotypes (Johnson and Mulcahy, 1978).

Evidence indicates that microspheres do compete within the anther. It is unknown, however, to what extent this will be expressed in different pollen tube growth rates or in greater longevity of highly competitive pollen genotypes.

The Angiosperm Stigma and Style as Experimental Plots

Competition has been intensively studied, and the classical approaches to such studies (e.g., Harpker, 1977; Firbank and Watkinson, 1990) can be applied to pollen, considering the stigma and style as experimental plots. Stigmas and styles have long been known to have a significant effect on pollen growth rate (Jones, 1928; Pfahler, 1967; Mulcahy and Mulcahy, 1986). [It is for this reason that although there are some exceptions (Sari Gorla et al., 1989; Zamir et al., 1982; Zamir and Gadish, 1987), pollen behavior *in vitro* is usually an inaccurate predictor of pollen behavior *in vivo* (Mazer, 1987).] This is not too surprising, since simple artificial germination medium is certainly a poor substitute for the angiosperm style.

With the miniature experimental plot, we can vary density, placement, and genetic composition of the competing population. The density of a pollen population often has a dramatic influence on pollen behavior, both *in vitro* and *in vivo* (e.g., Brink, 1924; Brewbaker and Majumder, 1961; Iwanami, 1970). Thomson (1989) found that more dispersed clumps of pollen grains on stigmatic lobes of *Erythronium grandiflorum* germinated faster than very dense clumps. Cruzan (1986) sectioned styles of *Nicotiana glauca* and found that the frequency of pollen germination on the stigma increased with increasing pollen density. Pollen tube penetration of the stigma was unchanged as density increased from low to moderate levels, but it was reduced at high densities. Pollen tube penetration of the style was enhanced by increasing density. This might be due to facilitation, as there is extensive interaction between pollen and pistil (Mulcahy and Mulcahy, 1985) and between pollen and pollen (Sari Gorla and Rovida, 1980). It might be due also to a larger pollen population and the greater probability of having some pollen tubes which grow very rapidly.

A particularly elegant use of the style as an experimental plot is Landi and Frascaroli's (1988) study of *Zea mays*. Pollen from two sources was mixed with 20–80% (by weight) of a standard aleurone marker pollen, in effect creating a classical replacement series except that it lacks the unmixed (100%) parental treatments. The resulting ears were divided into five segments of equal size, and the kernels were separated by color. The proportions of each type in the apical segment indicate the proportion of pollen types present in the mixture. The slope of the changing proportion from one segment to another within an ear indicates the relative speed of pollen tube growth for the two pollen types. This analytical system is possible because pollen grains attach to various positions of the style (silk), and styles are of unequal lengths. Longer styles provide more distance for faster tubes to overtake slower tubes. Thus, with short styles, fertilizations will be accomplished by gametes of both fast- and slow-growing pollen tubes. With long styles, slower pollen tubes will be less successful in fertilizations. The difference between success of fast and slow pollen tubes will depend on both the lengths of the style and the differences in pollen tube growth rates. The apical kernels in *Zea mays* possess the shortest styles in the inflorescence, and style length increases steadily toward the base of the inflorescence. Thus, the inflorescence is divided into five transverse segments, and the proportion of fertilizations by each pollen type, in each segment, can be determined. To measure the relative pollen tube growth rates for two pollen types, one need only calculate the regression slope of fertilization success by either pollen type across the five transverse segments. The faster pollen tube type will increase (giving a positive slope) in direct proportion to the speed with which it surpasses the slower pollen tube type.

Comparing the contents of the five segments (Fig. 2.2), the relative speed of pollen tube growth rate showed a curvilinear relationship in proportion to the percentage of the tested pollen type included in the mixture (i.e., the probability of a grain's success was highest when it was very rare or very abundant within the mixture). This fits the mutual antagonism model (III) of deWit (Harper, 1977) and might indicate that each pollen type "damaged the environment of the other more than it damaged its own environment." Harper (1977) further suggests that luxury consumption of a limiting factor by one competitor could have this effect. If both competitors followed this pattern, the results observed by Landi and Frascaroli (1988) would be produced. The damaging effect of pollen on the success

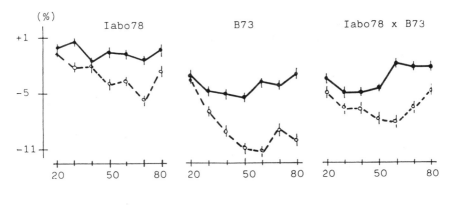

COMPOSITION OF THE POLLEN MIXTURE (% TESTED POLLEN)

Figure 2.2. Regression coefficients of the pollen sources as affected by the pollen mixture. Solid lines indicate that the style was the same genotype as the pollen being tested. Broken lines indicate an unrelated style.

of its neighbors is shown by the fact that, even at very low densities, dispersing pollen grains on the stigma increases the probability of pollen success (Thomson, 1989).

In studying pollen–pollen interactions on the stigma, it is important to recognize that a critical number of pollen grains may be necessary to stimulate pollen germination. In *Leucaena leucocephala* (Leguminosae), Ganeshaiah and Shaanker (1988) found that pollen reduced the pH of the stigma and thereby destroyed an acid-sensitive pollen germination inhibitor. In contrast, Murdy and Carter (1987) found that timing of pollen germination in *Talinum mengesii* (Portulacaceae) is independent of pollen density and under control of the stigma. Presumably, by delaying germination, this mechanism allows the accumulation of a higher pollen density. Each of these mechanisms, one requiring a minimum number of pollen grains, the other delaying time of pollen germination, should accentuate pollen competition in natural populations. These mechanisms could exist in other species and should be considered in studying the effect of pollen population density.

Effect of the Stylar Genotype on Pollen Competition

At lease three genetic components must be considered in regard to pollen performance *in vivo:* the stylar genotype, the pollen genotype(s), and the relative genetic similarities between the style and the pollen. Here too, the data of Landi and Frascaroli (1988) are informative. In each of the three mixtures they examined, pollen tubes were better competitors, relative to the standard marker pollen, when the style was genetically related (self or F_1). This observation is similar to one made by Jones (1928). He found that, when mixtures of self- and nonself-pollen were applied to stigmas of *Zea mays*, self-pollen was relatively more competitive than nonself-pollen. This observation may indicate the operation of two contributing influences: the extent to which pollen genotypes are adapted to that particular stylar environment and the relationship between the stylar and pollen cytoplasm.

Progressing from F_1 through F_7, self-pollen became increasingly effective as a competitor against standard pollen testers (Johnson and Mulcahy, 1978). This suggested that each generation of selfing selected for pollen genotypes that grew more rapidly through that stylar environment. This selective process more than balances the loss of vigor resulting from inbreeding, (which has been demonstrated to have an effect on pollen competition (Yamada and Murakami, 1983). In addition, Jones (1928) found that the relative competitive ability of self-pollen was strongly influenced by the cytoplasm of the style. This phenomenon was more clearly demonstrated in a series of experiments designed to maximize fruit-set in tomato by pollinating with mixtures of pollen (Hornby and Li, 1975). In each case, the pistillate parent was an F_1 hybrid between cv. Bonnie Best (BB) and cv. Immur Prior Beta (IPB). Following Jones (1928), pollen grains of the two types were combined and then placed on the stigmas of the F_1 hybrid BB × IPB or IPB × BB. In each case, the performance of BB pollen was significantly better when the pistillate parent was BB × IPB, rather than IPB × BB. Similarly, the IPB pollen was more successful when the pistillate parent contained IPB cytoplasm. Clearly, there are significant interactions between the stylar cytoplasm and the pollen.

From Hornby and Li's (1975) study, it was not possible to determine how important the pollen cytoplasm is in pollen–style interactions. In *Plantago coronopus*, however, nuclear genes that restore pollen fertility in cytoplasmically determined male-sterile plants are known to have an effect on pollen tube growth rate (Van Damme, 1991). This suggests that pollen tube growth rate is influenced by the effect of pollen nuclear genes on pollen cytoplasm. In conjunction with Hornby and Li's (1975) data, this suggests that pollen tube growth rate could be influenced by the cytoplasmic genotypes of the style and of the pollen and by the effect of pollen nuclei on the pollen cytoplasm. Presumably, the nuclei in the style will have a comparable effect.

Selective Fertilization

Selective fertilization is the genetically determined attraction between specific pollen tube types and specific embryo sac types. Normally, heterozygous plants produce thousands of different pollen and embryo sac genotypes, making it difficult to determine if pollen of one genotype is attracted by embryo sacs with specific genotypes. In many species of *Oenothera*, however, a series of reciprocal translocations within the seven pairs of chromosomes unites the nuclear genetic material into two, largely non-recombining, linkage groups, each termed a "Renner complex" (Cleland, 1972). As a consequence, both embryo sacs and pollen grains exist as one of two genetic configurations. This provides an opportunity to examine selective fertilization. Schwemmle (1968) pollinated *O. berteriana*, which contains the Renner complexes called B and l, with a trisomic plant l.II. This trisomic produces only one type of pollen, that containing the l complex. The progeny should contain either complexes B and l (= B.l) or l and l (= l.l), but each fruit contained an average of 26.2 B.l seeds and no l.l seeds. There are two possible explanations for this observation: abortion or selective fertilization. Perhaps the l complex carries a recessive zygotic lethal that causes all l.l zygotes to abort. Alternatively, l pollen tubes might selectively enter B ovules. To test this possibility, Schwemmle (1968) first pollinated B.l pistils with l.II and then, 17 hr later, with *O. scabra*, a homozygous species producing only one type of pollen, termed *hsc*. The *hsc* pollen, by itself, fertilizes equal frequencies of B and the l ovules, forming B.hsc and l.hsc zygotes. However, if pollination of B.l pistils by l pollen fertilizes all the ovules, followed by the abortion of l.l., neither B.hsc nor l.hsc zygotes will result from the second pollination. Schwemmle (1968) found that the average pollinated fruit contained 15.1 B.l zygotes, 7.6 B.hsc zygotes, and 18.1 l.hsc zygotes, indicating that the *hsc* pollen is able to overtake some of the first pollen, even though it was applied 17 hr after the l pollen. Thus, the frequency of B.l zygotes was reduced by competition from *hsc* pollen for B zygotes. Significantly, for every 7.6 B.hsc zygotes formed, there were 18.1 l.hsc zygotes. This indicated that the l pollen tubes of the first pollination fertilized a substantially greater portion of the B ovules than of the l ovules. Schwemmle (1968) termed this phenomenon "selective fertilization."

Lamprecht (1954) reported selective fertilization in species of *Pisum*, and there is ancedotal evidence that it exists in other species. Certainly the Renner complexes make it relatively easy to demonstrate this phenomenon in *Oenothera*. Perhaps a limited number of ovule or pollen types makes it possible for unambiguous signals at the time of ovule–pollen contact, thus facilitating selective fertilization. Alternatively, perhaps selective fertilization operates in many other taxa. Knowing that it exists at all suggests that the search seems worth undertaking.

Literature Cited

Altschuler, M., and J.P. Mascarenhas. 1985. Transcription and translation of heat shock proteins in seedlings and developing seeds of soybean exposed to a gradual temperature rise. Plant Mol. Biol. 5:291–297.

Bell, C.R. 1959. Mineral nutrition and flower to flower pollen size variation. Am. J. Bot. 46:621–624.

Brewbaker, J.L., and S.K. Majumder. 1961. Cultural studies of the pollen population effect and the self incompatibility inhibition. Am. J. Bot. 48:457–464.

Brink, R.A. 1924. The physiology of pollen. IV. Chemotropism; effects on growth of grouping grains; formation and function of callose plugs; summary and conclusions. Am. J. Bot. 11:417–436.

Cleland, R.E. 1972. *Oenothera* Cytogenetics and Evolution. Academic Press, New York.

Cruzan, M.B. 1986. Pollen tube distribution on *Nicotiana glauca:* Evidence for density dependent growth. Am. J. Bot. 73:902–908.

Den Nijs, A.P.M., B. Maisonneuve, and N.G. Hogenboom. 1986. Pollen selection in breeding glasshouse tomatoes for low energy conditions. *In* D. L. Mulcahy, G. Bergamini Mulcahy, and E. Ottaviano, eds. Biotechnology and Ecology of Pollen. Springer-Verlag, New York. pp. 125–130.

Firbank, L.G., and A.R. Watkinson. 1990. On the effects of competition: From monocultures to mixtures. *In* J. B. Grace and D. Tilman, eds. Perspectives on Plant Competition. Academic Press, New York, pp. 165–192.

Ganeshaiah, K.N., and R. Uma Shaanker. 1988. Regulation of seed number and female incitation of mate competition by a pH-dependent proteinaceous inhibitor of pollen grain germination in *Leucaena leucocephala.* Oecologia 75:110–113.

Harper, J.L. 1977. Population Biology of Plants. Academic Press, London.

Hornby, C.A., and Shin-Chai Li. 1975. Some effects of multiparental pollination in tomato plants. Can. J. Plant Sci. 55:127–132.

Iwanami, Y. 1970. Physiological researches of pollen. XX. On population effect and mixture effect in pollen culture. Bot. Mag. (Tokyo) 83:364–373.

Johnson C.M., and Mulcahy D.L. 1978. Male gametophyte in maize: II. Pollen vigor in inbred plants. Theor. Appl. Genet. 51:211–215.

Jones, D.F. 1928. Selective Fertilization. University of Chicago Press, Chicago.

Karron, J.D., D.L. Marshall, and D.M. Oliveras. 1990. Numbers of sporophytic self-incompatibility alleles in populations of wild radish. Theor. Appl. Genet. 79:457–460.

Kumar, D., and K.R. Sarkar. 1980. Correlation between pollen diameter and rate of pollen tube growth in maize (*Zea mays* L.). Ind. J. Exp. Bot. 18:1242–1244.

Lamprecht, H. 1954. Selektive Befruchtung. Agric. Hort. Genet. 12:1–37.

Landi, P.A., and E. Frascaroli. 1988. Pollen-style interactions in *Zea mays* L. *In* M. Cresti, P. Gori, and E. Pacini, eds., Sexual Reproduction in Higher Plants. Springer-Verlag, Heidelberg, pp. 315–320.

Marshall, D.L., and K.L. Whittaker. 1989. Effects of pollen donor identity on offspring quality in wild radish, *Raphanus sativus.* Am. J. Bot. 76:1081–1088.

Mazer, S.J. 1987. Parental effects on seed development and seed yield in *Raphanus raphanistrum:* Implications for natural and sexual selection. Evolution 41:355–371.

McKenna, M., and D.L. Mulcahy. 1983. Ecological aspects of gametophytic competition in *Dianthus chinensis. In* D.L. Mulcahy and E. Ottaviano, eds., Pollen: Biology and Implications in Plant Breeding. Elsevier, New York. pp. 419–424.

Mulcahy, D.L. 1979. The rise of the angiosperms: A genecological factor. Science 206:20–23.

Mulcahy, G.B., and D.L. Mulcahy. 1985. Ovarian influence on pollen tube growth, as indicated by the semivitro method. Am. J. Bot. 72:1078–1080.

Mulcahy, G.B., and D.L. Mulcahy. 1986. More evidence on the preponderant influence of the pistil on pollen tube growth. *In* M. Cresti and R. Dallai, eds., Biology of Reproduction and Cell Motility in Plants and Animals. University of Siena, Siena, Italy. pp. 139–144.

Mulcahy, G.B., D.L. Mulcahy, and P.L. Pfahler. 1982. The effect of delayed pollination in *Petunia hybrida.* Acta Bot. Neerl. 31:97–103.

Murakami, K.I., M. Yamada, and K. Takayanagi. 1972. Selective fertilization in maize, *Zea mays.* I. Advantage of pollen from F_1 plants. Jpn. J. Breeding 22:202–208.

Murdy, W.H., and M. Carter. 1987. Regulation of timing of pollen germination by pistil in *Talinum mengesii* (Portulacaceae). Am. J. Bot. 74:1888–1892.

Ottaviano E., and D.L. Mulcahy. 1989. Genetics of angiosperm pollen. Adv. Genet. 26:1–64.

Pfahler, P.L. 1967. Fertilization ability of maize pollen grains. II. Pollen genotype, female sporophyte and pollen storage interactions. Genetics 57:513–521.

Polito, V.S., S.A. Weinbaum, and T.T. Muraoka. 1991. Adaptive responses of walnut pollen germination to temperature during pollen development. J. Am. Soc. Hort. Sci. 116:552–554.

Roberts-Oehlschlager, S.L., and J.M. Dunwell. 1991. Barley anther culture: The effect of position on pollen development *in vivo* and *in vitro.* Plant Cell Rep. 9:631–634.

Rose, J.B. 1987. Anther culture of *Lolium temulentum, Festuca pratensis,* and *Lolium* × *Festuca* hybrids. II. Anther and pollen development *in vivo* and *in vitro.* Ann. Bot. 60:203–214.

Sari Gorla M., and Rovida E. 1980. Competitive ability of maize pollen. Intergametophytic effects. Theor. Appl. Genet. 57:37–41.

Sari-Gorla, M., E. Ottaviano, E. Frascaroli, and P. Landi. 1989. Herbicide-tolerant corn by pollen selection. Sex. Plant Reprod. 2:65–69.

Schoch-Bodmer, H. 1940. The influence of nutrition upon pollen grain size in *Lythrum salicaria.* J. Genet. 40:393–402.

Schwemmle, J. 1968. Selective fertilization in *Oenothera.* Adv. Genet. 14:225–324.

Shivanna, K.R., H.F. Linskens, and M. Cresti. 1991. Pollen viability and pollen vigor. Theor. Appl. Genet. 81:38–42.

Singleton, W.R., and P.C. Mangelsdorf. 1940. Gametic lethals on the fourth chromosome of maize. Genetics 25:366–390.

Thomson, J.D. 1989. Germination schedules of pollen grains: Implications for pollen selection. Evolution 43:220–223.

Van Damme, J.M.M. 1991. A restorer gene in gynodioecious *Plantago coronopus* subject to selection in the gametophytic and seeling stage. Heredity 66:19–28.

Williams, E.G. and J.L. Rouse. 1990. Relationships of pollen size, pistil length and pollen tube growth rates in *Rhododendron* and their influence on hybridization. Sex. Plant Reprod. 3:7–17.

Willing, P.R., D. Bashe, and J.P. Mascarenhas. 1988. Analysis of the quantity of diversity of messenger RNAs from pollen and shoots of *Zea mays*. Theor. Appl. Genet. 75:75–3.

Xiao C.M., and J.P. Mascarenhas. 1985. High temperature-induced thermotolerance in pollen tubes of *Tradescantia* and heat-shock proteins. Plant Physiol. 78:887–890.

Yamada M., and K. Murakami. 1983. Superiority in gamete competition of pollen derived from F_1 plant in maize. *In* D.L. Mulcahy and E. Ottaviano, eds. Pollen: Biology and Implications for Plant Breeding. Elsevier, New York.

Young, H.J., and M.L. Stanton. 1990. Influence of environmental quality on pollen competitive ability in wild radish. Science 248:1631–1633.

Zamir, D., and I. Gadish. 1987. Pollen selection for low temperature adaptation in tomato. Theor. Appl. Genet. 74:545–548.

Zamir, D., S. Tanksley, and R.A. Jones. 1982. Haploid selection for low temperature tolerance of tomato pollen. Genetics 101:129–137.

3

Using Genetic Markers to Measure Gene Flow and Pollen Dispersal in Forest Tree Seed Orchards*

W. T. Adams and David S. Birkes
Oregon State University   and

V. J. Erickson   USDA Forest Service

Introduction

Mating patterns, including effective pollen dispersal within populations and gene flow between populations, are important determinants of the genetic structure of plants (Levin and Kerster, 1974; Hamrick and Schnabel, 1985). Mating patterns affect levels of inbreeding and maintenance of diversity within populations, as well as effective population size and the potential for subdivision as the result of selection or drift. In addition to their importance to population genetics and evolutionary biology, patterns of gene movement via pollen are of great practical significance in forestry. Knowledge of mating patterns in natural populations is important in assessing the genetic composition of seed collected from wild stands for purposes of reforestation or progeny testing of parent trees (Ledig, 1974; Sorenson and White, 1988; Kitzmiller, 1990). This knowledge is also necessary for formulating gene conservation strategies, such as the number and distribution of parent trees for *ex situ* seed collections or distances required to isolate *in situ* preserves (Ledig, 1986).

One type of artificial forest-tree population in which mating patterns are of particular interest is the seed orchard. Seed orchards are important because they are the primary source of genetically improved seeds used in artificial reforestation (Zobel and Talbert, 1984). In clonal seed orchards, cuttings of genetically superior trees are grafted onto rootstocks. The number of superior trees (clones) represented in each orchard varies from as few as 20 to several hundred, with each clone replicated many times (each replicate of a clone is called a ramet). To promote cross-fertilization between clones, ramets are systematically assigned positions in the orchard so that each has a different set of neighboring clones. For convenience and

---

*Paper 2760 of the Forest Research Laboratory, Oregon State University.

economy of management, orchard sites often contain more than one orchard (block), the clones in each block coming from a different geographical source than clones in other blocks. Each block produces seeds that are adapted to a different geographical region.

Seed orchards are expected to produce seeds that reflect both the genetic quality (superiority) and diversity of clones in the orchard. Three elements of the breeding structure of orchards are particularly relevant to meeting this expectation: the extent to which seeds are the result of (1) self-fertilization, or (2) immigrant pollen from outside the orchard (gene flow or pollen contamination), and (3) the extent of cross-fertilization among clones within the orchard (Woessner and Franklin, 1973; Adams and Joly, 1980). The proportion of seeds due to selfing is important because of the usual poor survival and growth of selfed offspring (Muona, 1989). Because of the low seed-set that follows self-fertilization in most forest trees, the proportion of viable progeny resulting from selfing is expected to be low. Pollen contamination is of concern because seeds resulting from nonorchard (unimproved) pollen sources will have lower genetic quality than seeds resulting from improved parents within the orchard block (Smith and Adams, 1983). If seeds are fertilized by pollen from trees that are poorly adapted to the intended planting region of the block, a less well-adapted seed crop will result. Finally, patterns of cross-fertilization among individuals within orchards are important in determining the amount of genetic diversity in seed crops.

Despite the evolutionary and economic significance of mating patterns in plant populations, our knowledge on this subject is remarkably limited (Hamrick and Schnabel, 1985; Brown, 1989; Adams, 1992). In recent years, however, a number of statistical procedures have been described for investigating mating patterns based on the segregation of genetic markers (primarily isozymes) in the offspring of mother plants of known or inferred genotype. The most widely used of these procedures is the application of the mixed-mating model to estimate the proportion of progeny due to selfing versus outcrossing (Brown, 1989; Adams and Birkes, 1991). Accumulated evidence from a relatively large number of studies in forest trees, especially conifers, indicates that levels of outcrossed progeny are typically $> 90\%$ in both natural stands and seed orchards (Muona, 1989; Adams and Birkes, 1991). Occasionally, however, individual trees display levels of outcrossing as low as 50% (Erickson and Adams, 1990). Nevertheless, because self-fertilization normally accounts for only a small proportion of viable offspring, the major challenge to understanding breeding structure in most forest trees is to describe patterns of outcrossing. In this chapter, we briefly describe and compare statistical procedures that have been developed for examining mating patterns in plants and review their application in conifer seed orchards. Particular emphasis is placed on the most recent techniques used to investigate effective pollen dispersal within populations.

Seed orchards have a number of features that make them useful model populations for investigating breeding structure in wind-pollinated trees. Variation in tree spacing and size is much lower in seed orchards than in natural stands, reducing these factors as variables in mating studies. Site uniformity is also much greater in seed orchards. The limited number of genotypes (clones) in orchards makes it relatively easy to describe all genotypes in a population (block) and facilitates assessment of gene flow between populations. Replication of clones in different parts of blocks makes it possible to separate genetic and environmental factors influencing mating success in replicated trials. Finally, by assaying both the haploid megagametophyte and the diploid embryo in gymnosperm seeds, isozyme genotypes of the pollen gametes that fertilized the seeds can be determined directly (Adams, 1983). In angiosperms, the pollen contribution to the embryo genotype cannot be inferred by this straightforward approach; thus, gymnosperms have a powerful advantage for mating studies. Although we specifically use conifers in our examples, the principles employed are applicable to angiosperms as well. The mechanics of mating pattern analyses of angiosperms, however, are more complicated because of the difficulty in inferring pollen gamete genotypes.

Two observations of pollen distribution in conifers have important implications for breeding structure in seed orchards. The first is that the density of pollen dispersed from individual trees decreases rapidly with increasing distance, suggesting that most mating within local populations occurs between neighboring individuals (Wang et al., 1960; Levin and Kerster, 1974). The second is that accumulated long-distance dispersal of small amounts of pollen from many trees can result in considerable pollen distribution over long distances when whole stands are considered as pollen sources (Silen, 1962; Lanner, 1966). Thus the potential for gene flow between nearby populations is great. Based on these observations, Sorensen (1972) developed a model apportioning the seed from an individual mother tree to various pollen sources. If pollen production in surrounding stands is large relative to that in the orchard, the model predicts that a substantial proportion of outcrossed progeny will be due to pollen contamination. Mating among trees within the orchard, however, will be restricted primarily to near neighbors (i.e., females will mate almost exclusively with males in the first two surrounding ranks of trees). In a situation where the density of pollen around orchard clones from nonorchard (background) sources is about 25% of that produced within the orchard, Sorensen (1972) predicted that 30% of outcrossed progeny would be due to pollen contamination, 65% to near-neighbor mating, and only 5% to other males within the orchard. This model does not take into account differences among orchard trees in floral phenology or pollen fecundity, factors that could strongly influence differential mating success of males and that have been observed to vary substantially among orchard clones (Schoen et al., 1986; Erickson and Adams, 1989).

Statistical Procedures for Estimating Mating Patterns

Gene Flow

Use of single-gene markers to identify seeds fertilized by immigrant pollen is generally not feasible in forest trees because allele frequency differences among local populations are usually small (Hamrick and Godt, 1989; Muona, 1989). This is true even when comparing seed orchard blocks that have clones from different geographical regions (Adams, 1983). A multilocus procedure (paternity exclusion) for estimating gene flow into a local (recipient) population can be applied, however, in cases where the recipient population contains a limited number of genotypes (< 300), which is usually the case in seed orchards (Smith and Adams, 1983). This method takes advantage of the fact that the recipient population has a limited number of multilocus genotypes relative to the much larger, surrounding (background) populations. Thus, the background pollen sources produce a greater variety of multilocus gametes, which collectively can be used to identify immigrant pollen.

The first step in the analysis is to compare the multilocus genotype of each pollen gamete sampled in the offspring to those that are possible based on the parents in the recipient population. All pollen gametes that could not have been produced by males within the recipient population are detected immigrants. The proportion of detected immigrants in the sample is a minimum estimate of gene flow, because some immigrants are likely to have multilocus genotypes indistinguishable from those that can be produced in the recipient population (i.e., will be undetected immigrants). To estimate the true proportion of immigrants (m), we note that the probability of observing a detected immigrant (b) is equal to md, where d is the probability that an immigrant pollen grain has a detectable genotype (detection probability); thus, $\hat{m} = \hat{b}/\hat{d}$.

In order to estimate d, we need to assume (1) that the background populations can be considered a single large population, (2) that pollen gametes received by the recipient population are randomly drawn from pollen parents in the background, and (3) that loci used in the analysis are in linkage equilibrium (Smith and Adams, 1983; Devlin and Ellstrand, 1990). Given these assumptions, d is estimated as $1 - \Sigma^z h_j$, where h_j is the frequency in the background pollen source of the jth multilocus gamete that can be produced by trees in the recipient population, of which there are a total of z different possible gamete genotypes. Each h_j is calculated as the product of the frequencies of the component alleles in a representative sample of adults in the background stands. The first two assumptions seem reasonable, given the large distances that wind-dispersed pollen can travel in forest trees and the limited genetic differentiation observed among populations at marker loci. The third assumption is not likely to be violated in large outcrossing populations of forest trees, unless linkage between marker loci is tight (Epperson and Allard, 1987).

When the genotypes of pollen gametes can be scored directly, the detection probability is simply the combined frequency in the background source of multilocus pollen gametes that cannot be produced in the recipient population. It is constant for all mother plants. In angiosperms, however, application of the paternity exclusion procedure to estimate gene flow is complicated by the fact that the detection probability varies depending on the genotype of the mother. The more heterozygous the mother, the lower the ability to detect which alleles in diploid progeny are the result of the father (i.e., the lower the detection probability). A Monte Carlo procedure for estimating detection probabilities in angiosperm species is described by Devlin and Ellstrand (1990).

Effective Pollen Dispersal within Populations

The simplest and most direct procedure to investigate mating patterns is to track pollen from individual males having rare marker alleles (*rare marker approach*). By scoring pollen gametes in the seed progeny of mother plants at varying distances and directions away from a marker male, patterns of cross-fertilization can be inferred. The application of this technique is limited because of the small number of individuals with rare markers. Moreover, if more than one individual in a population carries the rare marker, interpretation of results is confounded (Muller-Starck, 1982; Yazdani et al., 1989).

A second approach to investigate mating patterns within populations is *paternity analysis*. In this approach, the paternity of pollen gametes sampled in the progeny of individual mother trees is inferred by comparing the multilocus genotypes of the pollen gametes to those of potential males in the population. Thus, one requirement of this technique is that the genotypes of all potential males in the population must be known. Once the inferred pollen parents of a large number of offspring have been determined, mating patterns can be assessed from the inferred parental pairs. The major advantage of this approach over the use of rare markers is that it can be applied in virtually any population, as long as sufficient polymorphic marker loci are available. Paternity analysis is ideal if all potential males, except one, can be excluded on the basis of genetic incompatibility with the observed pollen gamete genotype. In practice, however, complete genetic exclusion is possible for only a portion of the offspring, even in situations where the number of potential males is few and the number of polymorphic loci used in the analysis is relatively large (Chakraborty et al., 1988; Adams, 1992).

Because complete genetic exclusion is usually not possible, a number of likelihood methods for inferring paternity among nonexcluded males have been proposed. One procedure (*most-likely method*) is to assign paternity to the male with the highest conditional likelihood of producing the multilocus offspring genotype (Meagher, 1986; called the transition probability by Devlin et al., 1988). When potential males share the highest likelihood, the observation must be thrown out. A problem with this method is that paternity assignment is biased toward homozygotes, as males with the highest number of loci homozygous for alleles compatible with the offspring genotype will always have the highest transition probabilities (Devlin et al., 1988; Brown, 1989). In the *fractional paternity method*, paternity is assigned in ambiguous cases not to single individuals, but is allotted fractionally to all nonexcluded males in proportion to their transition probabilities (Devlin et al., 1988). All offspring are assigned paternity, although some offspring are not assigned to a single male parent. Under the assumption of equal male fertility, the fractional method removes biases due to over-assignment of homozygotes. A difficulty with both the most-likely and fractional methods is that likelihoods are based on genetic compatibility alone, without consideration for location of males relative to the mother plant or to other factors that might influence mating success. Thus, unless the proportion of potential males excluded on the basis of transition probabilities (i.e., the exclusion fraction) is very high, estimates of gene dispersal parameters based on these methods can be substantially biased (Brown, 1989; Adams, 1992). For example, when the fractional likelihood procedure was applied to simulated samples of offspring in a population of 48 parents, exclusion fractions of at least 0.90 were necessary to obtain reasonable estimates of male fertilities, when actual male fertilities were not constant (Devlin et al., 1988). In addition, even at high levels of exclusion, the fractional paternity procedure underestimated fertilities in the high fertility males. In the case where male mating success decreases with distance from the mother tree, estimates of mean distance between mates is likely to be biased upwards using the most-likely or fractional methods (Adams, 1992).

A method to increase accuracy of paternity inference and, thus, to reduce bias in estimating mating parameters is to weight transition probability by probabilities of mating to derive a combined likelihood measure (*modified most-likely method*) (Ritland, 1983; Neal, 1984; Adams et al., 1992). Information on factors such as distance between males, relative pollen fecundity, and degree of phenological overlap in flowering could be used in estimating probabilities of mating. A problem with this approach is that if paternity analysis is to be used to estimate mating parameters, the use of estimated mating probabilities in calculating likelihoods of paternity is circular in reasoning (Devlin et al., 1988; Brown, 1989). Thus, the modified most-likely method should not be used unless the information on mating probabilities is reliable; otherwise, significant bias in parameter estimation could result.

Rather than derive mating parameters from inferred parent–offspring pairs, they can be estimated directly using maximum likelihood or similar estimation procedures from probability models designed to account for observed frequencies of multilocus genotypes in pollen gametes of offspring samples. This procedure (*mating model approach*) was used by Schoen and Stewart (1986) to investigate male fertilities among clones in a white spruce [*Picea glauca* (Moench) Voss] seed orchard. A similar model for estimating male fertilities in angiosperm species is described by Roeder et al. (1989). Both of these models require that populations be isolated from background pollen sources. A model that can be used to describe mating patterns within populations that are not isolated from immigrant pollen is the *neighborhood model* proposed by Adams and Birkes (1991). This model is described later in this chapter.

The advantage of the mating model approach over paternity analysis is that bias in estimating mating parameters can be eliminated when the probability model fitted to the data is a reasonable approximation of reality (Roeder et al., 1989; Adams and Birkes, 1991; Adams, 1992). The validity of the mating model approach, therefore, is dependent on a priori information regarding which model or models are appropriate. Paternity analysis is free of assumptions about mating patterns when likelihoods of mating are based solely on transition probabilities. The inferences derived from paternity analysis, however, may be a poor reflection of the true mating patterns within populations if exclusion fractions are not high. Exclusion fractions can be increased above those generally possible with isozymes by employing molecular markers to augment the number of polymorphic loci [e.g., restriction fragment length polymorphisms (RFLPs) or randomly amplified polymorphic DNA (RAPD)]. Adding loci, however, does not guarantee unambiguous assignment of paternity unless the number of potential males is small. Moreover, it increases the likelihood of linkage disequilibrium, which complicates paternity analysis (Smouse and Chakraborty, 1986). Hypervariable molecular markers have great potential for paternity analysis (Jeffreys et al., 1985; Nybom and Schaal, 1990), but inexpensive methods will be needed to assay large samples of offspring and parents.

Mating Patterns in Conifer Seed Orchards

Pollen Contamination

According to Sorensen's (1972) model, pollen contamination is expected to be relatively high in wind-pollinated seed orchards, because orchard blocks are normally separated by only a few hundred meters, at best, from large natural stands of the same species, with even less separation between blocks on the same site. Estimates of pollen contamination (m) in a variety of species and orchard situations support this expectation. When orchards are not subjected to pollen management regimes to limit contamination, estimates of m are often > 40%, even in older orchards where pollen production is considered to be heavy (Adams and Birkes, 1989).

On orchard sites with more than one block, pollen contamination can be partitioned into that due to immigrants from the surrounding natural stands and that due to adjacent blocks (Smith and Adams, 1983). In an Oregon Douglas-fir [*Pseudotsuga menziesii* (Mirb.) Franco] orchard with 10 blocks, pollen contamination from all sources was estimated, on average, to be 0.50, with 80% of the contamination in any one block due to pollen from the surrounding natural stands (Table 3.1). This result is not surprising since there was no isolation between the orchard and the natural stands

Table 3.1. Estimates of Pollen Contamination ($\hat{m}$) from Natural Stands, or from Other Orchard Blocks, in Seed Orchard Blocks Not Subjected to Pollen Management Regimes to Limit Contamination.

	Orchard			$\hat{m}$		
Species/location	Age	Total blocks	Mean block size (ha)	Natural stands	Other blocks	Combined (SE)
Douglas-fir/ Oregon	14	10	1.8	0.39[a]	0.11[a]	0.50[a](0.02)
Douglas-fir/ Washington	15	4	5.0	0.11[b]	0.33[b]	0.44[b](0.07)
Loblolly pine/ South Carolina	15	2	2.0	0.34[c]	0.11[c]	0.45[c](0.03)

[a]Unweighted means for 10 blocks in one cone crop year (Smith and Adams, 1983; Adams, unpublished).

[b]Unweighted means for two orchard blocks in one cone crop year and one orchard block in a second crop year (Wheeler and Jech, 1986).

[c]Unweighted means for two orchard blocks in three crop years (Friedman and Adams, 1985).

and pollen production within the orchard was light. A similar contamination pattern was observed in the two blocks of a loblolly pine (*Pinus taeda* L.) orchard (Table 3.1). Pollen contamination from both sources might have been expected to be lower, however, because pollen production was relatively heavy within the blocks and the blocks were separated from each other by 100 m and from natural stands by at least 122 m. The pattern of contamination observed in blocks of a Douglas-fir orchard in Washington is quite different: most (76%) of the immigrant pollen came from the adjacent blocks rather than from the surrounding stands. A number of factors are responsible for this different pattern. First, this orchard is probably the most isolated of the three, with only scattered Douglas-fir trees or small natural stands within 0.5 km of the orchard and the nearest continuous stand 2 km away. In addition, pollen production was heavy in adjacent blocks because of flower (i.e., male and female cone) stimulation treatments. The blocks for which estimates of *m* are reported in Table 3.1, however, did not receive flower stimulation treatments in the crop years investigated.

Pollen contamination is primarily a function of the relative proportion of within-block versus immigrant pollen available during the period that female flowers are receptive. Distance to background pollen sources, pollen production within blocks, and floral phenology within blocks relative to surrounding stands are all factors influencing this ratio. As observed in the loblolly pine orchard, limited physical isolation may have little effect on contamination, especially in years when pollen production in background stands is heavy. Pollen production within blocks, however, appears to influence levels of contamination significantly. Pollen production can be increased substantially by applying flower stimulation treatments such as nitrogen fertilization and partial stem girdling. Wheeler and Jech (1986) found that blocks receiving stimulation in the Douglas-fir orchard in Washington had two to three times the pollen production and less than half the contamination observed in untreated blocks. Orchard blocks that flower substantially earlier or later than background pollen sources will be temporally isolated from immigrant pollen. This is the basis for the "bloom delay" techniques employed in some orchards, whereby evaporative cooling (by cold water mists) in late winter and early spring is used to delay flowering in orchard blocks relative to background stands (Wheeler and Jech, 1986; Fashler and El-Kassaby, 1987).

Another example of the influence of floral phenology on pollen contamination comes from data in two blocks (5 and 10) of the Douglas-fir orchard in Oregon. Estimated contamination was nearly 30% greater in block 10 (*m* = 0.62) than in block 5 (*m* = 0.48) (Smith and Adams, 1983). Although this difference in contamination might be attributed to the greater pollen production in block 5, differences in floral phenology between the blocks are probably more important (Fig. 3.1). While the periods of male and female flower production were nearly coincident in block 5, peak female production lagged several days behind peak male production in block 10.

Patterns of Cross-Fertilization within Blocks

Some of the best information currently available on patterns of cross-fertilization within orchard blocks comes from rare marker studies. These studies have revealed that the mating success of individual males is negatively related to distance from mother trees, but not necessarily in the steadily decreasing fashion predicted by Sorensen's (1972) model. In an investigation conducted in one of the blocks of the Douglas-fir orchard in Washington, offspring of 38 mother trees ranging from 6 to 41 m away from two marker males were assayed for the rare marker alleles (Erickson and Adams, 1989). Few offspring in mother trees > 30 m away were fertilized by the marker males. Within 30 m, however, no relationship was found between the frequency of offspring sired by marker males and distance to mother trees. Similar lack of a consistent relationship between mating success of individual males and distance between mates (within 20–30 m) was found in two Scots pine orchards (*Pinus sylvestris* L.) (Shen et al., 1981; Muller-Starck, 1982). The primary factor modifying the relationship between mating success and distance between mates is large differences among clones in floral phenology and pollen fecundity (Shen et al., 1981; Erickson and Adams, 1989). For example, the 111 clones in the Douglas-fir orchard in Washington ranged 26 days in average date of peak female receptivity and 28 days in average date of peak pollen shed (Erickson and Adams, 1989). Thus, it is not surprising that degree of floral synchrony was a major factor in determining mating success of the marker males in this orchard. When floral synchrony was high, near neighbors accounted for large proportions of offspring; one marker male sired > 70% of the progeny of an adjacent mother tree (7.3 m away) with whom it coincided completely in flowering. However, even though the marker males were heavy pollen producers, on average they accounted for only a small percentage of the offspring in any one neighboring female; only 6.8% in mother trees within 30 m, and 11% within 20 m. This indicates that a large number of males normally contribute to the offspring of individual females, an observation consistent with the high levels of pollen contamination observed in orchards.

Large differences among orchard clones in pollen production suggest that male fertility might vary substantially among clones (Schoen et al., 1986). Fractional paternity analysis was used to estimate male fertilities in a 9-year-old loblolly pine orchard (Wheeler et al., 1992). Although estimated male fertilities ranged between 0.4 and 8.4% among the 42 clones

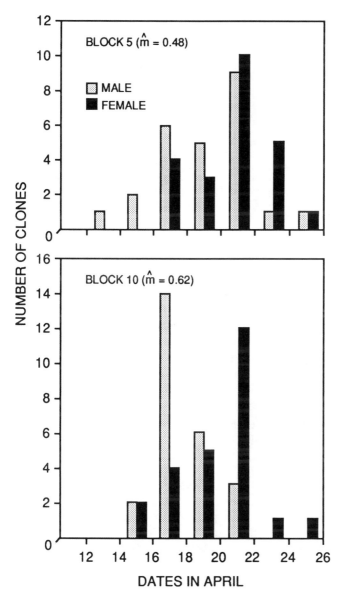

Figure 3.1. Histograms of mean dates of initiation of pollen shedding and female receptivity among the 25 clones in each of two blocks of a Douglas-fir seed orchard in Oregon. The estimated proportion of progeny due to pollen contamination ($\hat{m}$) is also given for each block.

in the orchard, most estimates were near the 2.4% predicted for uniform clonal contributions. As indicated earlier, however, fractional paternity analysis probably underestimates fertility differences. Using the mating model approach, Schoen and Stewart (1986) estimated male fertilities in an orchard of 33 white spruce clones. Twelve widespread clones were estimated to sire > 60% of the offspring, with four of these clones accounting for nearly 50%. Among the five clones for which individual estimates of male fertility could be obtained, fertilities ranged from 0 to 15%. In addition, male fertility was found to be significantly and positively related to male cone production. Similar large differences in male fertility were observed in a Scots pine (*Pinus sylvestris* L.) orchard (Muller-Starck and Ziehe, 1984).

Neighborhood Model

In the previously cited investigations of mating patterns within orchard blocks, the blocks were assumed either to be completely isolated from immigrant pollen sources or genetic exclusion was used to remove detected immigrants from the data set prior to analysis. The extent to which these procedures are suitable depends on the actual amount of gene flow, if any, and the degree to which immigrants can be detected by genetic exclusion. Because the probability of detecting immigrants is likely to be low (Smith and Adams, 1983; Friedman and Adams, 1985), the proportion of actual immigrants remaining undetected and contributing error to mating parameter estimation is expected to be high. Thus, whenever, more than limited gene flow is suspected, the validity of paternity analyses, or estimation models that ignore gene flow, for estimating pollen dispersal patterns within populations, must be seriously questioned (Adams, 1992; also see below). Another approach is to use a mating model that accounts for gene flow as well as mating patterns among individuals within populations. This is the rationale for the neighborhood model (Adams and Birkes, 1991).

In the neighborhood model, a specified area around a mother tree is called a neighborhood. Pollen successful in fertilizing viable offspring of the mother tree is ascribed to three sources: self-fertilization (with probability s), immigrant pollen from outside the neighborhood (w), and cross-fertilization with males inside the neighborhood ($1 - w - s$), with the relative mating success of the jth outcross male being ϕ_j. The probability of observing multilocus genotype g_i in pollen gametes of offspring from the mother tree is

$$p(g_i) = s\,p(g_i|M) + (1 - w - s)\,\Sigma^r\phi_j\,p(g_i|F_j) + wp(g_i|B),$$

where $p(g_i|M)$ is the probability that the mother tree produces pollen gametes with genotype g_i (transition probability for the mother tree), $p(g_i|F_j)$

is the transition probability for the jth outcross male in the neighborhood, and $p(g_i|B)$ is the probability that pollen gametes from immigrant pollen sources (background) have genotype g_i.

As we have seen, the relative mating success of each outcross male in the neighborhood (ϕ_j) can be determined by a number of factors, including distance from the mother tree, relative pollen fecundity, and degree of floral overlap with the mother tree. To investigate the relationship between mating success and factors expected to influence mating success, ϕ_j can be expressed as a function of these factors. Thus, the parameters in the model to be estimated are s, w, and one or more additional terms related to mating success inside the neighborhood. Maximum likelihood procedures can be used to estimate these parameters when the model is applied to offspring data from individual mother trees (Adams and Birkes, 1991). We have used for this purpose, the module for maximization of likelihood functions in the GAUSS Mathematical and Statistical System (Aptech Systems, Inc., Kent, Washington). Examples of computer programs written in GAUSS can be obtained by contacting one of us (D.S. Birkes). We are also working on the development of estimation programs using a more readily available statistical package such as SAS (SAS Institute Inc., Cary, North Carolina).

When the model is applied to seed orchards, it would seem logical to treat the entire orchard block as a neighborhood, but since mating among near neighbors [i.e., among trees within two or three ranks (20–30 m) of each other] is probably of greatest interest, neighborhoods do not need to be large. In addition, the greater the number of individuals within the neighborhood, the less will be the ability to discriminate genetically among neighboring males and between multilocus pollen gametes produced within and outside the neighborhood. This is particularly true in clonal orchards, where the number of ramets of individual clones increases with increasing neighborhood size.

We now apply the neighborhood model to two Douglas-fir data sets available from previous studies (Omi and Adams, 1986; Erickson and Adams, 1990). Both data sets were originally collected to investigate proportions of outcrossed progeny in orchard clones and provide minimal sample sizes for the neighborhood analysis; nevertheless, they are useful for illustrating a number of points. Our main interest in both cases is in the relationship between mating success of males (ϕ) and distance (d) to the mother tree. To relate ϕ to d, we use an exponential function with the parameter β (Fig. 3.2). When the value of β is 0, ϕ is unrelated to d, and as β increases, nearer males are increasingly favored in mating. For an appropriate value of β, the function seems sufficient to approximate any relationship we might reasonably expect to observe. Thus, with this form of the model, there are three parameters to estimate: s, w, and β.

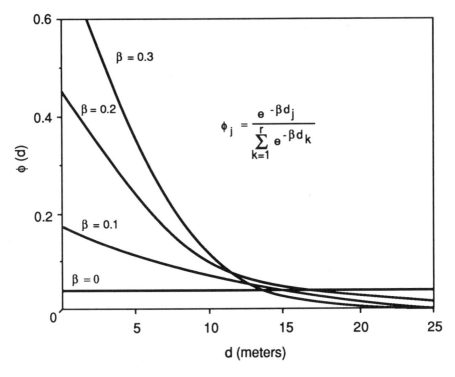

Figure 3.2.. Relationship between mating success of a male (ϕ) and distance to mother tree (d) for various values of the distance parameter (β). The formula for computing ϕ for the jth male (ϕ_j) is also given, where d_j is the distance between the jth male and the mother tree, and r is the number of males in the neighborhood.

The first data set comes from a 20-year-old orchard block with 25 clones (Omi and Adams, 1986). Genotypes of all trees in this block were known at 11 isozyme loci and estimates of pollen gamete frequencies in background sources were available for the same loci. All pollen-producing trees within a 25 m radius (25–26 trees) were considered the neighborhood around a particular mother tree. Using computer-simulated data sets from three mother trees in this orchard, we have tested the validity of maximum likelihood estimation of the neighborhood parameters under a variety of parameter conditions (Adams and Birkes, 1991; Adams 1992; Adams and Birkes, unpublished). The estimation procedure appears to give reasonably efficient estimates of all parameters, especially when compared to parameters estimated using paternity analysis. For example, offspring data samples consisting of 200, 10-locus pollen gametes were simulated for each of the three mother trees 500 times, using each of three sets of model parameters: $w = 0$, 0.10, or 0.40; $s = 0.05$; $\beta = 0.15$. For each sample of offspring, the mean distance between the mother tree and its outcross mates within the neighborhood (μ_d) was estimated using four methods: the most-likely and fractional paternity methods, and by estimating β using the neighborhood model approach, with w either included or not included in the estimation model, and calculating $\hat{\mu}_d$ from the $\hat{\phi}_j$ (i.e., $\hat{\mu}_d = \Sigma^r d_j \hat{\phi}_j$). Pollen gametes identified by genetic exclusion as immigrants were deleted from the data before calculating $\hat{\mu}_d$ in the most-likely and fractional analyses, and in the model approach where w was ignored. Mean values of $\hat{\mu}_d$ and its standard deviation over the 500 simulations were averaged over the three mother trees (Table 3.2). Given the low exclusion fractions for the three mother trees (mean values ranged from 0.57 to 0.74) and the relatively strong negative association assumed between male mating success and distance between mates (i.e., $\beta = 0.15$), we would not expect either the most-likely or fractional paternity methods to give reliable estimates of μ_d, regardless of the assumed amount of gene flow into the neighborhood, and this is what the simulations show. When w was included in the estimation model, the model approach gave nearly unbiased estimates of μ_d, although the standard deviation of $\hat{\mu}_d$ increased as w increased. When w was left out of the estimation model, μ_d was progressively overestimated as the assumed level of w was increased.

Using the model approach without w in the estimation model led to only minor bias in μ_d when the assumed level of gene flow was low ($w = 0.10$: Table 3.2). In other computer simulations, the fractional paternity approach gave reliable estimates of μ_d when $w = 0.10$, but only when ϕ was assumed to be uniform among males in the neighborhoods (i.e., when $\beta = 0$) (Adams, 1992). As expected, with $\beta > 0$, the fractional approach

Table 3.2. Comparison of Means and Standard Deviations (in Parentheses) of Estimates of the Mean Distance (m) between Mother Trees and Their Outcross Mates ($\hat{\mu}_d$) Obtained by Applying Each of Four Estimation Procedures to Computer-Simulated Samples of Pollen Gametes.[a]

Assumed level to gene flow (w)	Most-likely	Fractional likelihood	Neighborhood model	
			w ignored	w included
0.0	13.39(0.60)	15.63(0.20)	12.04(0.80)	12.04(0.80)
0.1	13.43(0.62)	15.67(0.20)	12.30(0.87)	12.06(0.90)
0.4	13.49(0.65)	15.83(0.22)	13.42(1.02)	12.13(1.37)

[a]The results are averages for three mother trees, with the mean value of true μ_d being 12.07 m. Based on 500 samples of 200 pollen gametes for each set of parameter conditions (see text for details). Exclusion fractions for pollen gametes ranged from 0.57 to 0.74 over the three mother trees.

increasingly overestimated μ_d as β increased, whereas the model approach (w included the estimation model) continued to provide reliable estimates. The most-likely paternity procedure gave poor estimates under all values of assumed β. From these observations, it appears that low levels of gene flow may have little or no negative effect on the ability to estimate parameters describing cross-fertilization patterns within populations. This idea, however, needs to be tested under a broader range of parameter conditions, including more values of w.

The computer simulations we have completed so far provide only limited information on the effects of sample size on the precision of parameter estimation using the neighborhood model. As expected, estimation efficiency improves with both increasing numbers of polymorphic loci (i.e., with increasing exclusion fractions) and offspring included in the sample. The precision of individual parameter estimates, however, decreases as the number of parameters in the estimation model increases. Given the above estimation model with three parameters (i.e., s, w, β) and 10 polymorphic loci, our data suggest offspring sample sizes (n) of around 100 should be considered a minimum, but $n \geq 200$ is recommended. In general, standard deviations (s) of estimates appear to be roughly proportional to $1/\sqrt{n}$; thus, a 4-fold increase in n would be required to halve s. Although the neighborhood model applies to individual mother trees, offspring data from several trees could be analyzed under a single model when homogeneity of mating parameters among mother trees can be assumed. The combined n in this case should also exceed 200. The number of mother plants needed to sample a population adequately depends on the heterogeneity of site (e.g., soil fertility, microtopography) and stand (e.g., plant density, distribution, and size) conditions. With relatively uniform site and stand conditions in seed orchards, we feel that samples of 75–100 offspring from each of 4–5 mother trees would be adequate for assessing mating patterns in individual orchard blocks.

Application of the neighborhood model to real offspring data from five ramets of four clones in the orchard block used for the computer simulations gave results consistent with earlier observations of mating patterns in seed orchards. Because a previous study had indicated that the proportions of selfed progeny were < 5% in these five ramets and sample size per ramet was only 30–40 offspring, samples from all trees (total = 173 offspring) were combined under a single model (Adams and Birkes, 1991). Resulting estimates were $\hat{s} = 0.00$ (SE = 0.03), $\hat{w} = 0.69$ (0.06), and $\hat{\beta} = 0.02$ (0.07), with only $\hat{w}$ being significantly ($p < 0.05$) different from zero. We conclude that most of the pollen that was effective in fertilizing viable offspring came from outside the neighborhoods, that the proportion of viable selfed offspring in these mother trees was near zero, and that mating success of individual males within the neighborhoods is unrelated to distance from mother trees. Given that pollen contamination into this block was previously estimated at 40% (Adams, unpublished), we would expect $\hat{w}$ to exceed this value, as the background of each neighborhood includes trees within the block, as well as pollen sources outside the block. The fact that $\hat{\beta}$ is nearly zero supports the lack of relationship between mating success and distance between mates among near neighbors that was observed in the rare marker studies.

The second Douglas-fir data set comes from the same orchard block in Washington in which cross-fertilization patterns were investigated using rare markers (Erickson and Adams, 1989). In a separate study, variation among individual trees in the proportion of progeny due to outcrossing (t_i) was investigated in five ramets, each of a different clone, using 11 polymorphic loci and a genetic exclusion estimation procedure based on the mixed-mating model (Erickson and Adams, 1990). These ramets were chosen for study because they had widely different floral phenology and pollen production, which was expected to result in differences in $\hat{t}_i$. In fact, $\hat{t}_i$ ranged from 0.50 to 1.07 and corresponded closely to levels of outcrossing expected based on the floral biology of the ramets. For example, the ramet with the unusually low $\hat{t}_i$ estimate of 0.50 was considerably out of phase with the other trees in the orchard, having peak female receptivity 11 days earlier than the mean date of peak receptivity for the block. Because of the wide variation in mating systems, the neighborhood model was applied to the offspring of each ramet separately ($n = 86–97$, mean $= 90.2$). In addition, because genotype data were not complete for all orchard trees, only 5–6 loci could be used per mother tree, and the neighborhoods had to be limited in size to 16 m [7–17 (mean $= 11$) potential males].

Estimates of s in the neighborhood model could not be calculated for three of the five mother trees because of failure of the numerical maximum likelihood procedure to converge (Table 3.3). We have found in computer simulations that, when true s is low and sample size is small, failure to converge almost always occurs when iterations are moving toward $s = 0$ (Adams and Birkes, unpublished). When there is failure and it is verified that the iterations were going toward zero prior to the failure, our procedure is to refit the model with only parameters w and β. Simulations show that when true s is low (< 0.15), ignoring s in the model has little effect on estimates of the other parameters (Adams and Birkes, unpublished). Neither estimates of β nor w varied significantly ($p < 0.05$) among the five ramets, so their weighted means are presented (Table 3.3). Consistent with the earlier application of the neighborhood model, the majority of offspring ($\hat{w} = 0.60$) are fertilized by males outside the neighborhoods. Nevertheless, neighborhoods are much smaller in this case, such that 40% of the offspring are estimated to result from males within the first two ranks around the mother tree. In agreement with the rare marker study

Table 3.3. Estimates of the Distance Parameter (β) and Proportions of Offspring due to Gene Flow (w) and Self-Fertilization (s) from the Neighborhood Model, and Proportions of Offspring due to Self-Fertilization from the Mixed-Mating Model ($1 - t_i$), for Each of Five Mother Trees in a Douglas-Fir Seed Orchard in Washington[a].

Mother tree	Sample		Outcross[b] males in neighborhood	$\hat{\beta}$	$\hat{w}$	$\hat{s}$	$(1 - \hat{t_i})$[c]
	Pollen gametes	Loci					
634-5-28	86	6	11	0.11(0.21)	0.57(0.10)	0.33(0.08)	0.50(0.06)
643-7-19	87	6	7	−0.06(0.02)	0.54(0.08)	—[d]	0.24(0.09)
639-8-14	95	6	10	−0.03(0.04)	0.74(0.14)	0.00(0.021)	0.12(0.06)
695-5-13	97	5	17	0.07(0.04)	0.59(0.10)	—[d]	0.16(0.06)
673-6-20	86	5	10	−0.04(0.03)	0.68(0.15)	—[d]	0 (0.04)
			Mean[e]	−0.03(0.01)	0.60(0.05)		

[a]See text for description of parameters. Standard errors are given in parentheses.
[b]Pollen-producing males within 15.8 m of the mother tree.
[c]From Erickson and Adams (1990).
[d]Estimates of s not obtained because the numerical maximum likelihood procedure failed to converge.
[e]Weighted by inverse of variances.

in the same orchard block (but using different mother trees), no relationship was found between mating success of near-neighbor males and distance to mates (i.e., $\hat{\beta} \cong 0$).

Assuming that lack of convergence means $\hat{s}$ is near 0, estimates of the proportion of selfed progeny based on the neighborhood model were lower than those based on the mixed-mating model $(1 - \hat{t}_i)$ in four of the five ramets (Table 3.3). One explanation is that different sets of loci (5–6 versus 11) were used in the two analyses. Another is that the difference in estimates is a function of the different models. In the mixed-mating model, fertilizations are the result of either selfing *or* random outcrossing, where the outcross pollen pool is derived from all sources of outcross pollen. In the neighborhood model, the outcross pollen pool is further subdivided into that due to males within the neighborhood (neighborhood pollen pool) and that due to gene flow (immigrant pollen pool). If multilocus pollen gamete frequencies differ between the neighborhood and immigrant pools, expectations for observed frequencies of pollen gametes due to selfing versus outcrossing will differ for the two models. Presumably, the neighborhood model is a more accurate reflection of mating patterns because it does not require the simplifying assumption that all outcrosses occur at random. Furthermore, in this case, nearby males appear to have a much greater chance of mating with the mother tree than expected if all outcrosses were completely at random. In addition, computer simulations have not revealed any tendency for maximum likelihood estimates of s to be downwardly biased when mating conforms to the neighborhood model (Adams and Birkes, unpublished). Thus, $\hat{s}$ based on the neighborhood model may be a more accurate reflection of proportions of selfed progeny in these ramets than estimates derived from the mixed-mating model. This conjecture needs to be tested by computer simulations. It is also of interest to determine whether subdivision in the outcross pollen pool causes a consistent direction of bias in $\hat{t}_i$ when the mixed-mating model is employed.

Conclusions

Although the data are limited, observations of outcross mating patterns in conifer seed orchards conform fairly closely to the predictions of Sorensen's (1972) model. Under natural wind pollination, pollen contamination accounts for a large proportion of viable offspring, especially in young orchards where pollen production is limited relative to that produced in surrounding stands. Mating success of individual males diminishes with distance from the mother tree, but not in the rapidly decreasing manner predicted from pollen distribution curves. Within orchard blocks, mother trees appear to mate primarily with males within the first two to three ranks of trees around them (20–30 m). Among these neighboring males, however, mating success is more a function of relative pollen production and degree of overlap in flowering with the mother tree, rather than distance to the mother tree. When flowering periods of mother trees and pollen parents are synchronous, the closest males will be most successful.

Because of detrimental effects of pollen contamination, manipulation of pollen production and timing of female receptivity have become important management tools in many seed orchards (Wheeler et al., 1992; Fashler and El-Kassaby, 1987). These treatments clearly reduce contamination, but little is known about their effects on other aspects of the mating system. For example, does increasing pollen production influence levels of selfed progeny or have any effects on patterns of cross-fertilization within orchard blocks? Treatments that delay receptivity in orchards relative to pollen shed in background stands also increase floral synchrony among clones within treated blocks (Fashler and El-Kassaby, 1987). What is the effect of more uniform phenology on mating patterns? Will mating success become primarily a function of proximity, such that mother trees now mate nearly exclusively with their nearest neighbors?

The extent to which mating patterns in seed orchards reflect those that occur in natural stands of wind-pollinated trees is unclear. Relative to the surrounding natural stands, seed orchards are typically small in both spatial extent and average tree size. Thus, the high amounts of gene flow observed in seed orchards may be peculiar to small populations surrounded by large ones. Gene flow, however, can be substantial even among orchard blocks of similar area and tree size. In addition, the limited isozyme diversity typically observed among natural populations of forest trees within geographical regions suggests that gene flow is a powerful force influencing natural population structure (Hamrick and Godt, 1989; Muona, 1989), but this needs to be tested more directly.

The strong influence of differences among males in floral phenology and pollen production on mating success may also be an artifact of clonal seed orchards. Because the clones in any one orchard block come from parent trees that are widely distributed over a geographical region, large differences among clones in phenology might be expected. The extent to which individuals within natural stands of conifers vary in floral phenology and pollen production is largely unknown (Shea, 1987; Smith et al., 1988). In the few cases where effective pollen dispersal within natural conifer stands has been studied, mating success of individual males appears to drop off steadily and rapidly with increasing distance between mates. This suggests that differences among trees in floral phenology and pollen production may have less influence on mating patterns in natural populations (Adams, 1992). When rare markers have been employed, marker males sire only small proportions of offspring, even on neighboring mother trees (Muller, 1977; Yazdani et al., 1989). This indicates, as found in seed orchards, that distant pollen sources may be primarily responsible for cross-fertilization in natural stands and that each female effectively mates with large numbers of males. Obviously, there is still much to be learned about cross-fertilization patterns in natural stands. In addition to the effects of distance between mates and differential flowering on mating success, it is important to determine how patterns of outcrossing are influenced by stand structure, including density, size, and age of individuals.

Additional work is also needed to develop and test statistical procedures for estimating patterns of cross-fertilization in plant populations. More information is required on how the efficiency of estimates based on paternity analysis is influenced by the magnitude of the exclusion fraction, number of potential male parents in the study population, and presence of unidentified immigrants. Even given a large number of polymorphic marker loci, paternity analysis is not likely to lead to precise estimates of parameters describing within-population cross-fertilization patterns, unless the population is isolated and the number of potential males is small [e.g., Broyles and Wyatt's (1990) study of the perennial herb *Asclepias exaltata*). In most forest trees, however, small isolated populations are rare except at the margins of the range, and the breeding structure of these populations may not be representative of typical populations of the species. With the limited exclusion fractions possible with isozymes, the model approach appears to be the most reliable means of estimating cross-fertilization patterns in most outcrossing plants. The neighborhood model seems particularly useful, because it allows for gene flow and can be employed to investigate cross-fertilization patterns among neighboring individuals, even when they are part of a large continuous stand. Variants of the model can be developed to account for a variety of breeding structures, although large samples of offspring will be necessary for efficient parameter estimation, especially when the models are complex.

Literature Cited

Adams, W.T. 1983. Application of isozymes in tree breeding. *In* S.D. Tanksley and T.J. Orton, eds., Isozymes in Plant Genetics and Breeding, Part A. Elsevier, New York, pp. 381–400.

Adams, W.T. 1992. Gene dispersal in forest tree populations. New Forests (in press).

Adams, W.T., and D.S. Birkes. 1989. Mating patterns in seed orchards. Proc. 20th South. Forest Tree Improv. Conf., Charleston, South Carolina, pp. 75–86.

Adams, W.T., and D.S. Birkes. 1991. Estimating mating patterns in forest tree populations. *In* S. Fineschi, M.E. Malvolti, F. Cannata, and H.H. Hattemer, eds., Biochemical Markers in the Population Genetics of Forest Trees. SPB Academic Publishing, The Hague, Netherlands, pp. 157–172.

Adams, W.T., and R.J. Joly. 1980. Allozyme studies in loblolly pine seed orchards: Clonal variation and frequency of progeny due to self-fertilization. Silvae Genet. 29:1–4.

Adams, W.T., A.R. Griffin, and G.F. Moran. 1992. Using paternity analysis to measure effective pollen dispersal in plant populations. Am. Nat. (in press).

Brown, A.H.D. 1989. Genetic characterization of plant mating systems. *In* A.H.D. Brown, M.T. Clegg, A.L. Kahler, and B.S. Weir, eds., Plant Population Genetics, Breeding, and Genetic Resources. Sinauer, Sunderland, MA, pp. 145–162.

Broyles, S.B., and R. Wyatt. 1990. Paternity analysis in a natural population of *Asclepias exaltata*: Multiple paternity, functional gender, and the "pollen-donation hypothesis." Evolution 44:1454–1468.

Chakraborty, R., T.R. Meagher, and P.E. Smouse. 1988. Parentage analysis with genetic markers in natural populations. 1. The expected proportion of offspring with unambiguous paternity. Genetics 118:527–536.

Devlin, B., and N.C. Ellstrand. 1990. The development and application of a refined method for estimating gene flow from angiosperm paternity analysis. Evolution 44:248–259.

Devlin, B., K. Roeder, and N.C. Ellstrand. 1988. Fractional paternity assignment: Theoretical development and comparison to other methods. Theor. Appl. Genet. 76:369–380.

Epperson, B.K., and R.W. Allard. 1987. Linkage disequilibrium between allozymes in natural populations of lodgepole pine. Genetics 115:341–352.

Erickson, V.J., and W.T. Adams. 1989. Mating success in a coastal Douglas-fir seed orchard as affected by distance and floral phenology. Can. J. Forest Res. 19:1248–1255.

Erickson, V.J., and W.T. Adams. 1990. Mating system variation among individual ramets in a Douglas-fir seed orchard. Can. J. Forest Res. 20:1672–1675.

Fashler, A.M.K., and Y.A. El-Kassaby. 1987. The effect of water spray cooling treatment on reproductive phenology in a Douglas-fir seed orchard. Silvae Genet. 36:245–249.

Friedman, S.T., and W.T. Adams. 1985. Estimation of gene flow into two seed orchards of loblolly pine (*Pinus taeda* L.). Theor. Appl. Genet. 69:609–615.

Hamrick, J.L., and M.J.W. Godt. 1989. Allozyme diversity in plant species. *In* A.H.D. Brown, M.T. Clegg, A.L. Kahler, and B.S. Weir, eds., Plant Population Genetics, Breeding, and Genetic Resources. Sinauer, Sunderland, MA, pp. 43–63.

Hamrick, J.L. and A. Schnabel. 1985. Understanding the genetic structure of plant populations: Some old problems and a new approach. *In* H-R Gregorius, ed., Population Genetics in Forestry. Springer-Verlag, New York, pp. 50–70.

Jeffreys, A.L., V. Wilson, and S.L. Thein. 1985. Individual-specific 'fingerprints' of human DNA. Nature (London) 316:76–79.

Kitzmiller, J.H. 1990. Managing genetic diversity in a tree improvement program. Forest Ecol. Manage. 35:131–149.

Lanner, R.M. 1966. Needed: A new approach to the study of pollen dispersion. Silvae Genet. 15:50–52.

Ledig, F.T. 1974. An analysis of methods for the selection of trees from wild stands. Forest Sci. 20:2–16.

Ledig, F.T. 1986. Conservation strategies for forest gene resources. Forest Ecol. Manage. 14:77–90.

Levin, D.A., and H.W. Kerster. 1974. Gene flow in seed plants. Evol. Biol. 7:139–220.

Meagher, T.R. 1986. Analysis of paternity within a natural population of *Chamaelirium luteum*. 1. Identification of most-likely parents. Am. Nat. 128:199–215.

Muller, G. 1977. Short note: Cross-fertilization in a conifer stand inferred from enzyme gene-markers in seeds. Silvae Genet. 26:223–226.

Muller-Starck, G. 1982. Reproductive systems in conifer seed orchards. I. Mating probabilities in a seed orchard of *Pinus sylvestris* L. Silvae Genet. 31:188–197.

Muller-Starck, G., and M. Ziehe. 1984. Reproductive systems in conifer seed orchards. 3. Female and male fitnesses of individual clones realized in seeds of *Pinus sylvestris* L. Theor. Appl. Genet. 69:173–177.

Muona, O. 1989. Population genetics in forest tree improvement. *In* A.H.D. Brown, M.T. Clegg, A.L. Kahler, and B.S. Weir, eds., Plant Population Genetics, Breeding, and Genetic Resources. Sinauer, Sunderland, MA, pp. 282–298.

Neale, D.B. 1984. Population genetic structure of the Douglas-fir shelterwood regeneration system in southwest Oregon. Ph.D. thesis, Oregon State Univ., Corvallis, Oregon.

Nybom, H., and B.A. Schaal. 1990. DNA "fingerprints" applied to paternity analysis in apples (*Malus × domestica*). Theor. Appl. Genet. 79:763–768.

Omi, S.K., and W.T. Adams. 1986. Variation in seed set and proportions of outcrossed progeny with clones, crown position, and top running in a Douglas-fir seed orchard. Can. J. Forest Res. 16:502–507.

Ritland, K. 1983. Estimation of mating systems. *In* S.D. Tanksley, and T.J. Orton, eds., Isozymes in Plant Genetics and Breeding, Part A. Elsevier, New York, pp. 289–302.

Roeder, K., B. Devlin, and B.G. Lindsay. 1989. Application of maximum likelihood methods to population genetic data for the estimation of individual fertilities. Biometrics 45:363–379.

Schoen, D.J. and S.C. Stewart. 1986. Variation in male reproductive investment and male reproductive success in white spruce. Evolution 40:1109–1120.

Schoen, D.J., D. Denti, and S.C. Stewart. 1986. Strobilus production in a clonal white spruce seed orchard: Evidence for unbalanced mating. Silvae Genet. 35:201–205.

Shea, K.L. 1987. Effects of population structure and cone production on outcrossing rates in Engelmann spruce and subalpine fir. Evolution 41:124–136.

Shen, H.-H., D. Rudin, and D. Lindgren. 1981. Study of the pollination pattern in a Scots pine seed orchard by means of isozyme analysis. Silvae Genet. 30:7–15.

Silen, R.R. 1962. Pollen dispersal considerations for Douglas-fir. J. Forest. 60:790–795.

Smith, C.C., J.L. Hamrick, and C.L. Kramer. 1988. The effects of stand density on frequency of filled seeds and fecundity in lodgepole pine (Pinus contorta Dougl.). Can. J. Forest Res. 18:453–460.

Smith, D.B. and W.T. Adams. 1983. Measuring pollen contamination in clonal seed orchards with the aid of genetic markers. Proc. 17th South. Forest Tree Improv. Conf., Athens, Georgia, pp. 69–77.

Smouse, P.E., and R. Chakraborty. 1986. The use of restriction fragment length polymorphisms in paternity analysis. Am. J. Human. Genet. 38:918–939.

Sorensen, F.C. 1972. The seed orchard as a pollen sampler: A model and example. USDA For. Serv. Res. Note PNW-175.

Sorensen, F.C., and T.L. White. 1988. Effect of natural inbreeding on variance structure in tests of wind-pollination Douglas-fir progenies. Forest Sci. 34:102–118.

Wang, C-W., T.O. Perry, and A.G. Johnson. 1960. Pollen dispersion of slash pine (Pinus elliotii Engelm.) with special reference to seed orchard management. Silvae Genet. 4:78–86.

Wheeler, N., and K. Jech. 1986. Pollen contamination in a mature Douglas-fir seed orchard. In A.V. Hatcher and R.J. Weir, eds., Proc. IUFRO Conf. on Breeding Theory, Progeny Testing and Seed Orchards, Williamsburg, Virginia, pp. 160–171.

Wheedler, N.C., W.T. Adams, and J.L. Hamrick. 1992. Pollen distribution in wind-pollinated seed orchards. In D.L. Bramlett, G.R. Askew, T.D. Blush, F.E. Bridgwater, and J.B. Jett, eds., Pollen Management Handbook, Vol. 2. USDA Forest Service Agric. Bulletin No. ____(in press).

Woessner, R.A., and E.C. Franklin. 1973. Continued reliance on wind-pollinated seed orchards, is it reasonable? Proc. 12th South. Forest Tree Improv. Conf., Baton Rouge, Louisiana, pp. 64–73.

Yazdani, R., D. Lindgren, and S. Steward. 1989. Gene dispersion within a population of *Pinus sylvestris*. Scand. J. Forest. 4:295–306.

Zobel, B.J., and J.T. Talbert. 1984. Applied Forest Tree Improvement. John Wiley, New York.

4

Estimating Male Fitness of Plants in Natural Populations

Maureen L. Stanton, Tia-Lynn Ashman, and Laura F. Galloway, University of California
and

Helen J. Young Barnard College

Introduction

In the past 15 years, evolutionary ecologists have dramatically changed their views on floral function and plant reproductive success. Previously, reproductive success was estimated directly from female function (i.e., the successful capture of pollen and maturation of seeds). From this perspective, the relative fitness of two individuals could readily be assessed by comparing their seed production. Recently, however, reproductive biologists have recognized that understanding plant reproduction requires knowledge of genetic transmission through pollen, as well as through seeds (Horovitz, 1978; Bell, 1985; Bertin, 1988). A plant that produces relatively few seeds may, in fact, be highly successful as a pollen donor (Horovitz and Harding, 1972; Bertin, 1982; Devlin and Ellstrand, 1990a). Because several studies suggest that selection on plant reproductive traits may act differently through male and female function (Stanton et al., 1986, 1989, 1991; Campbell, 1989; Galen and Stanton, 1989), efforts to document patterns of plant paternity have intensified. A deeper understanding of male fitness variation and its evolutionary importance will require increasing refinement of methods for determining or estimating male reproductive success and a body of work measuring associations between plant phenotype and male fitness components.

Despite a concerted empirical effort to study the movements and fates of genes carried by pollen grains, studies of male fitness variation in plants have encountered a series of stumbling blocks. Pollen does not "stand still and wait to be counted" as does the sporophyte generation (Harper, 1977). Rather, pollen is a mobile, often microscopic life stage that engages in a sequence of ecological interactions that is difficult to observe.

Genetic Markers: Direct Measurements of Plant Paternity

Measuring male reproductive success directly requires tracing genes of paternal origin into the next generation. Large plant populations contain many potential pollen donors, however, and the diffuse nature of pollen dispersal makes it difficult to identify the male parents of seeds or seedlings unambiguously. The use of rare genetic markers provides one technique for such analysis. Schaal (1980) introduced pollen-bearing plants homozygous for rare electrophoretic alleles into a population of *Lupinus texensis* and then genotyped seeds from surrounding individuals to characterize the resulting spatial distribution of genes from the pollen parents. Movement of bees in the population was highly leptokurtic, but gene flow by pollen was greater than that expected from single movements, presumably due to pollen carryover. Handel (1983) and Handel and Le Vie Mishkin (1984) employed a similar technique in artificial populations of cucumber (*Cucumis sativus*), by assaying seedlings for the presence of a dominant "bitter" taste allele derived from a central set of plants homozygous for that marker. The spatial distribution of the bitter allele was highly asymmetrical, suggesting that gene flow dynamics were more complex than those suggested by simple diffusion models (Morris, 1992).

Although useful for measurements of average gene flow, the "rare allele" technique is of limited use in comparative studies of male fitness. Almost by definition, only very few rare markers are available, so it is difficult to compare the reproductive success of pollen parents possessing an assortment of phenotypes. In the future, it may be possible to associate alternative phenotypes with unique markers via genetic transformation or via chemical markers transmitted through the pollen grain to the next generation (e.g., spermatophore labeling with rubidium in insects: Hayes, 1988).

An alternative method for directly measuring male reproductive success takes advantage of naturally occurring phenotypic and genetic variation among potential pollen parents. In these analyses, a genetic profile is constructed of each potential pollen donor, typically by assaying plants for variation at multiple electrophoretic loci. Genetic profiles are constructed similarly for seeds (Meagher, 1986; Devlin and Ellstrand, 1990a; Broyles and Wyatt, 1990) or seedlings (Meagher and Thompson, 1987). In very small and genetically polymorphic populations, many male parents can be excluded simply by identifying impossible genetic combinations (Ellstrand, 1984; Ellstrand and Marshall, 1985). When more than one male parent is possible, however, a variety of statistical estimation techniques must be employed to identify the most likely father (Devlin et al., 1988; Roeder et al., 1989; Devlin and Ellstrand, 1990b). This approach has yielded entirely new information about variance in male reproductive success in plant populations. For example, Devlin and Ellstrand (1990a) presented evidence that, contrary to expectation, male fitness variance was considerably less than female fitness variance in a wild population of *Raphanus sativus*.

Characterizing mating patterns in many natural plant populations may be impossible using analysis of protein electrophoretic markers. At present, multilocus paternity analyses can be applied only in relatively small populations (typically fewer than 60 potential male parents). Moreover, recent evidence for "correlated matings" between genetically related individuals in *Mimulus guttatus* (Ritland, 1989) and *Eichhornia paniculata* (Morgan and Barrett, 1990) suggests difficulties with using likelihood methods to discern mating patterns. The resolution of these methods may be enhanced by developing additional genetic markers via DNA fingerprinting techniques (Burke, 1989; Nybom and Schaal, 1990; Williams et al., 1990), but these techniques are presently impractical for population samples including thousands of seeds and hundreds of potential parents.

Because paternity analysis remains impractical in most natural populations, a number of researchers have turned to the use of synthetic population arrays in which the reproductive success of a limited number of male parental genotypes can be compared. Conifer seed orchards, with their randomized arrangements of a relatively small number of genotypes, have proven especially useful for these studies (Muller-Starck and Ziehe, 1984; Schoen and Stewart, 1986; Adams, 1992). Paternity analyses in seed orchards have demonstrated substantial variation in male fitness among clones, and in one case (Schoen and Stewart, 1986), male success was positively correlated with pollen production. Artificial genotypic arrays have also been used to measure male fitness variation associated with conspicuous floral variation in wild radish (*Raphanus* spp.: Stanton et al., 1986, 1989, 1991), morning glory (*Ipomoea purpurea*: Schoen and Clegg, 1985; Clegg and Epperson, 1988), and water hyacinth (*Eichhornia paniculata*: Barrett et al., 1992). As for multilocus studies conducted in natural populations, paternity analysis in genotypic arrays is limited to a relatively small number of genotypes and phenotypes.

Although not widely recognized, the dynamics of pollination and the consequences of phenotypic variation may differ between small, isolated arrays and larger natural populations. Long-distance dispersal of pollen (e.g., Ellstrand and Marshall, 1985) may cause underestimates of paternity in the context of small seed orchards or experimental arrays, compared with more extensive natural stands (Adams, 1992). Pollinator visitation may also be influenced by population size and degree of isolation. In large populations of California wild radish (*Raphanus sativus*), honeybees are the predominant visitor, whereas a wide variety of insects visit flowers in small populations and experimental arrays. The visitation patterns and fitness effects of these visitors differ, making it difficult to generalize results obtained in small arrays to large populations (Stanton et al., 1991; Thomson and Thomson, 1992).

In summary, plant reproductive biologists now find themselves in a challenging and somewhat frustrating position. Natural selection is almost certainly acting on male fitness variation in ways that cannot be ascertained by merely counting and measuring seed production. Comparative studies of paternal success using genetic markers are presently practical only in small populations, and it is not clear how well the patterns observed there will apply to larger populations. To make direct measurements of male fitness, investigators should select species and populations for which such difficulties are minimized. Comparative studies of male success in other systems will need to focus on measurable components of male fitness.

Measuring Components of Male Fitness Variation

A pollen-borne gene is passed to the next generation only after a sequence of improbable events takes place. First, the pollen grain must be removed from the anther and transferred to a receptive and unsaturated stigma. Second, a successful pollen grain must germinate and fertilize an ovule, possibly under competitive conditions. Third, the fertilized ovule must develop into a seed and be dispersed to a location suitable for seedling growth. Directly measuring male fitness requires knowledge of success across all stages in this reproductive process, whereas estimates of male fitness focus on single components of reproduction such as pollen removal or pollen competition. The postpollination events leading to production of mature seeds by various pollen parents are presently receiving intense scrutiny (Marshall and Folsom, 1992; Stephenson et al., 1992; Knox et al., 1992; Mulcahy and Mulcahy, 1992). In this chapter we discuss how rates of pollen removal from flowers and movements of pollen grains could contribute to male fitness variation (also see Thomson and Thomson, 1992).

Pollen flow, the movement of pollen from anthers to stigmas, is a major component of gene flow, but has proven difficult to measure in most situations. Unmarked pollen grains are inconspicuous, and pollen from different individuals can only rarely be distinguished on the basis of morphological characteristics (Thomson and Plowright, 1980; Thomson and Thomson, 1989). Accordingly, investigators have used various means to make pollen grains more visible and distinctive, so that their movements can be traced. Handel (1976) used neutron-activation analysis (NAA) to detect pollen grains labeled with rare earth elements, and other investigators have explored techniques for labeling pollen grains with radioactive markers (Schlising and Turpin, 1971; Reinke and Bloom, 1979). Unfortunately, the cost of NAA is prohibitive for population-level analyses, and radiolabeling of pollen for field studies presents many technical challenges.

The most commonly used method for marking pollen grains employs highly visible dyes. Fluorescent dye powders are dusted onto anthers and then detected on stigmas of surrounding plants under near-ultraviolet light. For species with very large pollen grains, dye particles adhering to the pollen surface make it possible to count the marked grains that have been transferred to other flowers (Hodges, 1990). For plants in which pollen grain size and dye particle size are more similar, the dye particles themselves are typically counted and treated as pollen analogs (Price and Waser, 1982; Waser and Price, 1982; Murawski, 1987; Campbell, 1989). Although very useful when employed carefully, the use of fluorescent dyes to estimate male fitness does present some problems. It is difficult to apply fluorescent dyes evenly to all grains and flowers, requiring many replications of experimental treatments to overcome this source of error. Carrying out many replications in one area may be hindered, however, by accumulation of dye particles on stigmas and pollinators. For pollinator species that groom between flower visits, it is not clear how significantly the movements of pollen grains and dye particles differ due to their contrasting sizes or electrostatic properties (Thomson et al., 1986). Finally, some pollinators may respond to changes in anther appearance due to the presence of conspicuous fluorescent powders (C. Galen and M. Stanton, personal observation). Studies using fluorescent powders to estimate pollen flow should carefully control for color-specific effects on dye movement.

Measurements of pollen removal from flowers have been used for some time as comparative estimates of male fitness in natural plant populations (Willson and Rathcke, 1974; Willson and Price, 1977; Schemske, 1980; Queller, 1983; Cruzan et al., 1988; Galen and Stanton, 1989; Young and Stanton, 1990; Harder, 1990). The earliest studies were on milkweeds or orchids, in which production and removal of pollinaria can be documented visually. More recently, the use of automated particle counting systems has made it possible to extend these analyses to species in which pollen grains are transferred individually. In these studies, the amount of pollen removed from a flower is calculated by subtracting the number of pollen grains remaining in anthers from the flower's estimated pollen production.

Several arguments have been advanced for using pollen removal from flowers as a comparative estimate of male fitness. First, although the fates of pollen grains removed from the anthers are unknown, pollen grains remaining in the anthers generally do not contribute to reproductive success. The amount of pollen removed from a flower thus sets an upper limit on the numbers of seeds that could have been sired. Second, although pollen export measures are tedious, they can be performed in many situations, and one can compare rates of pollen removal for a large number of phenotypes simultaneously. This latter attribute contrasts to marker studies, in which the number of markers (rare alleles, dye colors, etc.)

defines the maximum number of potential donors among which fitness components can be compared.

Despite these logistical advantages, male fitness estimates based on pollen export require a number of heroic assumptions, many of which may be violated in natural plant populations. Most importantly, one must assume that there are positive (perhaps even linear) relationships between pollen grains removed, viable pollen grains deposited on receptive stigmas, and male reproductive success. Despite the highly stochastic nature of pollen movement, some studies lend support to this assumption. Schoen and Stewart (1986) found a positive (but apparently nonlinear) relationship between pollen production and realized paternity in wind-pollinated white spruce. Broyles and Wyatt (1990) noted a significant positive correlation between pollinarium removal and both male and female reproductive success in an insect-pollinated milkweed. Galen (1992) has recently reported a significant correlation between pollen removal and paternal success in bumblebee-pollinated *Polemonium viscosum*. Still, Thomson and Thomson (1992) point out that the expected relationship between pollen removal and pollen deposition can become uncoupled by pollen wastage, rapid loss of pollen viability, and variation among pollinator species with respect to pollen transfer efficiencies. Their observations and models make it clear that pollen export must be used as a male fitness estimator with great caution. This is especially true for plant populations visited by a number of different pollinators. In such cases, flowers preferentially visited by inefficient pollinator species could have more pollen removed, but might actually transfer less of that pollen to receptive stigmas.

We have attempted to evaluate the use of pollen removal as an estimate of male fitness using wild radish (*Raphanus sativus*) as a model system. In central California, large populations of this weedy species are visited predominately by honeybees. First, we present one case study of pollen removal in a natural population of wild radish, where flowers of different phenotypes were contrasted. Our analysis of the case study discriminates between the *total amount* of pollen exported over the lifetime of a flower and the *rate* of pollen export during hours when fertilization opportunities are greatest. Second, we describe a study in which radioactive labeling was used to trace the movements of pollen grains in a controlled outdoor setting. Our results corroborate those for *Erythronium grandiflorum* (Thomson and Thomson, 1989; Harder and Thomson, 1989; Thomson and Thomson, 1992), showing that only a tiny fraction of the pollen exported from a flower arrives on stigmas of subsequently visited flowers.

Rates of Pollen Removal in a Field Population

Background

Successful pollen donation by a flower should be a function of both visitation by pollinators and its schedule of pollen presentation. To determine the effects of floral attractiveness, pollen production, and pollinator activity on rates of pollen removal, we conducted experiments in an extensive natural stand of *Raphanus sativus*. Because large populations are visited predominately by a single pollinator species (Stanton, 1987; Young and Stanton, 1990), we expect pollen removal to serve as a rough predictor of male reproductive success. Honeybees dominate such populations, so pollen grains not removed by honeybees are unlikely to reach a receptive stigma. Rates of visitation by honeybees should result in greater rates of pollen export and pollen delivery to other flowers, although these relationships are very likely not to be linear (Lloyd, 1984; Harder and Thomson, 1989; Young and Stanton, 1990).

The experiment described here differs in three principal ways from a previous study of pollen removal in *R. sativus* (Young and Stanton, 1990). First, in our earlier analysis, pollinator visitation was measured in small arrays of four flowering stems, within which extremes of naturally occurring petal sizes were compared. Although we could document the qualitative effects of petal size on honeybee visitation using this design, a strong phenotypic correlation between petal size and pollen production in naturally occurring plants made it difficult to discriminate between the effects of "floral attractiveness" (petal size) and male allocation on pollen removal. In the present study, we modified this phenotypic correlation in flowers by experimentally manipulating petal size to reduce its correlation with pollen production. Second, the focus of the previous analysis was on the relationship between visit number and pollen export; we controlled for day-to-day differences in pollinator abundance by allowing individual flowers to be visited a specified number of times. In the study described here, we presented 60 experimental flowers to pollinators simultaneously, intermixed within the natural stand instead of being placed in discrete arrays. Flowers were harvested after fixed periods of time, rather than after fixed numbers of visits, so that rates of pollen removal could be compared among floral phenotypes as a function of pollinator activity levels in the population as a whole. Finally, pollinators were previously presented with flowers of *R. sativus* in which anthers had fully dehisced (also see Harder and Thomson, 1989; Galen and Stanton, 1989). Natural pollen presentation is staggered over an hour or more, however, as buds of *R. sativus* open. The analyses of Harder and Thomson (1989) demonstrate that such "pollen packaging" can decrease rates of pollen removal while increasing the efficiency of pollen delivery. In the study presented here, we allowed pollen-donating flowers to open during the experiment and monitored rates of pollen removal during natural pollen dehiscence.

Methods

To measure the separate effects of petal size and pollen production on pollen removal in *R. sativus*, we attempted to uncouple their typically strong phenotypic association in two ways. Some of the plants used in the experiments were produced by two generations of artificial selection in the greenhouse, yielding flowers with unusually high or unusually low ratios of petal size to pollen production (Stanton and Young, unpublished). In addition, we carefully trimmed petals on another set of flowers to make them resemble much smaller flowers, while leaving their large anthers intact. For the experiment as a whole, petal area ranged from 44 to 300 mm^2 and pollen grain number per flower ranged from 33,400 to 122,600. Although the range for neither character was outside the normal phenotypic range for *R. sativus* (Stanton and Preston, 1988), our manipulations allowed us to reduce the phenotypic correlation between petal size and pollen production from "typical" values of 0.4–0.6 (Stanton and Preston, 1988) to a value of 0.2 for this experiment. Log-transformed floral characters were normally distributed; thus, we treated petal size and pollen production as continuous variates in regression analyses.

We established a circular transect 60 m in circumference through a dense stand of *R. sativus* in which pollen removal and pollinator visitation analyses had been conducted previously (Young and Stanton, 1990). Green bamboo stakes were placed at 1-m intervals along the transect so that the top of each stake was at the same height as the local flower canopy. Flowering stems of *R. sativus* were placed in florist's water pics (small, rubber-capped vials) and attached to each stake, making experimental flowers appear as though they were part of the natural floral display.

On four mornings over a 37-week period, 60 stems were set up along the transect between 0830 and 0900 hr PDST. Pollinator activity was just beginning at that time. Four stems from each of 15 different greenhouse-grown plants were placed into the transect at random locations, and each stem was assigned at random to be harvested 2, 4, 7, or 27 hr later. On each stem, one opening flower was marked for later harvest. Because we could not watch 60 stems simultaneously to note exact times of anthesis, the developmental stage of each marked flower was coded as a value between 0 and 2.0. Stage 0 was assigned to closed buds with petals beginning to emerge but still tightly closed. Stage 2 corresponded to freshly open flowers in which anthers were still undehisced and pressed together. Our observations in this population show that flowers typically make the transition from stage 0 to stage 2 within 1.5 hr. As dictated by the experimental design, petals of some flowers were trimmed as they began to open. Flowers were not used in the study if their anthers had begun to dehisce.

Additional stems from each plant were taken into the field and placed under insect-proof netting. At each of the four sampling intervals, anthers from a subset of unvisited "control" flowers and "experimental" flowers open to insect visitation were harvested into microcentrifuge tubes using fine-tipped forceps. Pollen grains within each tube were counted using an automated particle counting system (Elzone Model 180XY; Particle Data, Inc., Elmhurst, IL.; see Young and Stanton, 1990 for methods). Pollen removed from each experimental flower exposed to pollinators was estimated by subtracting the number of pollen grains remaining in its harvested anthers from that in anthers on a matched, unvisited control flower.

Pollinator activity was monitored throughout the experiment. In the center of the circular transect, we marked a 0.5 × 0.5 m area containing approximately 100 flowers. Each morning, the numbers of flowers within the observation area were counted. Numbers of insect visits made to those flowers were recorded for 10 min during each hour between 0800 and 1700 hr. From these data we calculated the mean number of visits per flower per hour for the time period over which each set of experimental flowers was exposed to pollinators.

We used multiple regression to determine how floral traits and pollinator activity levels influenced pollen removal from experimental flowers. Analyses were conducted separately for each time interval (2, 4, 7, or 27 hr) because pollinator observations were not made throughout the 27-hr treatment. Multiple regressions were used to detect effects of pollen production, petal size, flower developmental stage at the start of the experiment, and pollinator activity levels on pollen exported during successive intervals. To estimate the individual contribution made by each predictor variable to variation in pollen removal, separate regressions were conducted in which predictor variables were excluded from the model, one at a time. The decrease in the coefficient of determination relative to the complete regression model was used as an estimate of the partial r^2 for each predictor.

Results

In this heavily visited population, flowers exported most of their pollen within several hours after opening. On average, 34% of the pollen (over 25,000 pollen grains) was removed by insect visitors within 2 hr (Fig. 4.1). Over 60% of the pollen was removed within the first 4 hr. The rate of pollen export subsequently decreased, and after 27 hr an average of 69% of the available pollen had been exported. After 27 hr, flowers were entirely wilted and seemed unlikely to export additional pollen. Pollen removal varied considerably among flowers harvested at each time interval, but only rarely was more than 90% removed.

Pollinator activity in the observation area varied markedly among experimental trials conducted on different days (Fig. 4.2A–D). Activity levels depended strongly on weather: overcast, cool conditions during the first two trials resulted in rates of visitation 5–25% that seen in later trials on sunny, warm days. Honeybees were the dominant pollinator in the first three trials, with visitation rates to flowers reaching a maximum of 12 per

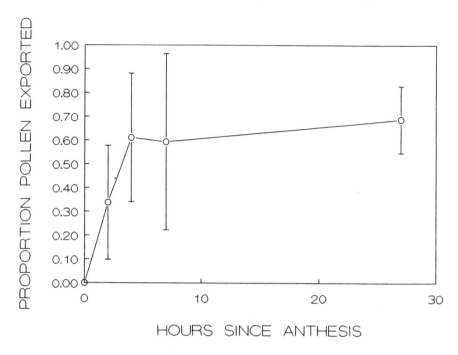

Figure 4.1. The proportion of pollen removed from *Raphanus sativus* flowers at successive time intervals. Means and standard deviations are shown. Sample sizes at each time interval varied from 22 to 29 flowers.

hour over the course of our observations. During the fourth trial, conducted 2 weeks later in the season, syrphid flies were unusually abundant, accounting for 55% of the visits observed. On most days, rates of flower visitation increased from early morning until late afternoon, corresponding with increasing honeybee activity late in the day. In contrast, syrphid fly activity tended to peak early in the morning.

Factors contributing to variation in pollen removal differed between freshly open flowers and flowers open for longer periods of time (Table 4.1). As expected, flower developmental stage made a statistically significant contribution to pollen export only for the first harvest interval. After 2 hr, developmentally older flowers had exported more pollen than their younger counterparts, presumably because their pollen had been available longer. The influence of initial flower developmental stage became statistically insignificant at later harvests. Pollen removal increased significantly with pollinator activity for the three time intervals over which pollinator visitation rates were monitored.

Flowers that produced more pollen exported more pollen. At all harvest intervals pollen production contributed significantly to pollen export, but

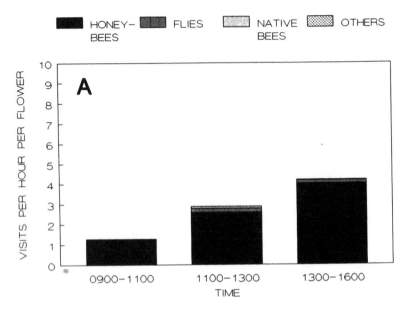

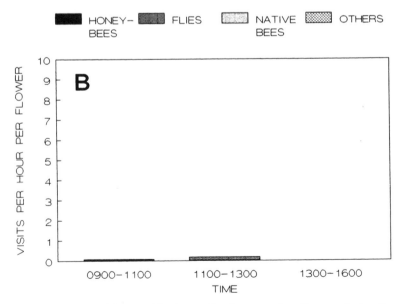

Figure 4.2. Rates of insect visitation to *Raphanus sativus* flowers during the pollen export experiment. Values shown are means for visit number per flower per hour. Observations were conducted hourly, during peak pollinator activity on the first day of each trial.

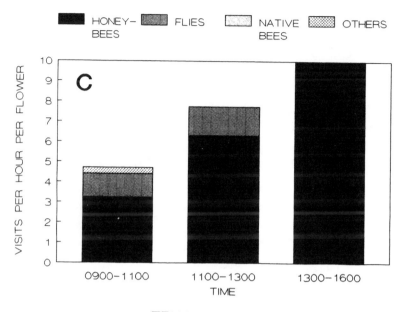

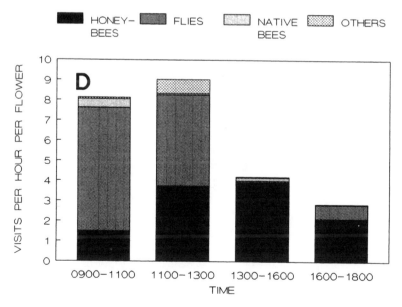

Figure 4.2. (*Continued*)

Table 4.1. Effects of Floral Characters and Pollinator Activity on Pollen
Removed from R. sativus flowers 2, 4, 7, and 27 hr after Opening.[a]

Time (hr)	Source of variation	df	F	r^2	p
2	Complete model	27	4.78	.454	.0059
	Pollen production		9.88	(.061)	.0046
	Petal size		1.70	(.040)	.2048
	Flower stage		5.53	(.099)	.0277
	Pollinator activity		12.09	(.253)	.0020
4	Complete model	28	6.35	.514	.0012
	Pollen production		17.81	(.236)	.0003
	Petal size		5.25	(.106)	.0311
	Flower stage		2.46	(.089)	.1302
	Pollinator activity		7.51	(.083)	.0114
7	Complete model	28	5.66	.485	.0024
	Pollen production		7.75	(.183)	.0103
	Petal size		1.24	(.025)	.2767
	Flower stage		0.15	(.003)	.6998
	Pollinator activity		18.60	(.274)	.0002
27	Complete model	21	7.99	.571	.0014
	Pollen production		20.52	(.544)	.0003
	Petal size		0.00	(.000)	.9866
	Flower stage		1.17	(.028)	.2944

[a]Because pollinator activity was not measured throughout the 27-hr interval, multiple regressions were conducted separately for each time period (PROC REG: SAS Institute, 1985). p shows the statistical significance of each predictor variable when added last into the full regression model. Partial r^2 values estimated for each predictor variable are shown in parentheses.

its predictive power tended to increase with flower age. In freshly open flowers harvested after 2 hr, pollen production explained only 6.1% of the variance in pollen removal. In contrast, pollen production explained over 50% of the variance in pollen removed over a flower's lifetime.

Larger petal size was associated with increased pollen export after 4 hr, but not at other time intervals. Its effects were probably diminished at the 2-hr harvest by variation in flower developmental stage, because apparent corolla size changes as a flower opens. Regression analyses for the 7- and 27-hr harvests suggested that total pollen removed from older flowers was determined primarily by pollen production and local pollinator activity, rather than by allocation to showy petals.

Discussion

In most studies of pollen removal, investigators have estimated either pollen exported for a known number of pollinator visits (Harder and Thomson, 1989; Harder, 1990; Young and Stanton, 1990; Wilson and Thomson, 1991;) or total pollen removed over a flower's lifetime (Queller, 1983; Bell, 1985; but see Cruzan et al., 1988). In this study, by harvesting flowers after different periods of time, we could differentiate between the *rate* of pollen removal achieved early in a flower's lifetime and the *total amount* of pollen removed. For flowers of *R. sativus*, total pollen removed is largely a function of pollen production. Because pollinator visitation rates are usually high, most flowers are stripped of the majority of their available pollen within 8 hr, regardless of their allocation to showy petals. In contrast, the rate of pollen removal early in a flower's lifetime is strongly influenced by developmental stage and corolla size.

Pollen removal will be a useful indicator of male reproductive success only so long as there are ovules available for fertilization. Distinguishing between total pollen export and rate of pollen export at different times of the day may be important for species like *R. sativus*, in which most effective pollination takes place within the first few morning hours (Ashman, Galloway, and Stanton, unpublished). Most flowers that open on any given day do so in the morning. At this site, 73% of the opening flowers have dehiscent anthers by 1000 hr and 87% make their pollen available before 1200 hr. Very few flowers open later in the day. Moreover, visitation rates are frequent enough that most flowers have visibly stripped anthers by late morning. Before 1000 hr, approximately 90% of the open flowers have obvious pollen available on anthers, indicating that they have been visited relatively few times. By 1300 hr, over 90% of the open flowers show signs of repeated or effective visitation. Because it takes fewer visits to fertilize all ovules than to export all available pollen in this species (Young and Stanton, 1990), our observations suggest that relatively few ovules are available for fertilization during the afternoon. Thus, although pollen export continues throughout the day, pollen removed in the morning should have greater opportunities for fertilization.

Which is likely to be a better predictor of male reproductive success— the rate of pollen removal from a young flower, or the number of pollen grains removed over that flower's lifetime? We have begun computer simulation studies to explore this question (Stanton et al., unpublished). Our simulation, in contrast to some models used by Harder and Thomson (1989) and Thomson and Thomson (1992), assumes that pollen export is a direct function of pollen availability. For simplicity, the model also assumes that pollen viability remains high through the pollination period (see Thomson and Thomson, 1992, for the consequences of varying pollen longevity schedules). Most importantly, unlike previous models that have assumed constant availability of ovules, we explicitly altered schedules of stigmatic presentation. The simulation results suggest that for species like *R. sativus*, in which floral presentation is relatively synchronous and pollinator visits are frequent, the rate of pollen export during the period of peak flower opening can be the most important determinant of male reproductive success. In contrast, in situations where flowers open continuously rather than synchronously through the day, lifetime pollen export is a much better predictor of male fitness. Our analysis suggests that, although lifetime pollen export is largely determined by pollen production in *R. sativus*, flower attractiveness and time of opening are likely to affect paternity by influencing rates of pollen removal early in the day.

The Fates of Pollen Grains: Radioactive Marker Studies

Background

Rates of pollen removal may be influenced by floral traits and pollinator activity, but studies of pollen export yield little information about the fates of pollen grains once they leave their anthers of origin. To explore pollen export as a component of male fitness, Young and Stanton (1990) measured pollen removal from *R. sativus* flowers after known numbers of visits by honeybees. Although variance in pollen removal and pollen deposition was very high, an average honey bee visit removed approximately 10,000 pollen grains and deposited approximately 60 grains. These data suggest that a pollen grain removed by a honeybee has less than a 1% chance of being delivered to a stigma in this system.

To study the fates of pollen grains in more detail, we conduced an experiment to trace the movements of grains after removal by honeybees. We labeled pollen grains from "donor" flowers using radioactive methionine, allowed honeybees to visit labeled flowers, then measured radioactivity on stigmas from subsequently visited flowers to estimate the amount of pollen deposited on them. We addressed several specific questions: (1) What proportion of pollen grains exported reach the stigmas of subsequently visited flowers? (2) What is the pattern of pollen carryover (i.e., how rapidly does pollen deposition decrease as a honeybee visits subsequent flowers)? (3) To what extent can pollen deposition be predicted from estimates of pollen removal from the donor flower?

Methods

Reinke and Bloom (1979) demonstrated that pollen grains could be labeled *in vivo* by injecting a ^{14}C-labeled amino acid cocktail into flower buds within 24–48 hr before anthesis. The labeled amino acids are incorporated into tapetal proteins deposited onto the grains shortly before their presentation, making grains detectable by microautoradiography. Their results and those from two previous studies (Colwell, 1951; Schlising and Turpin, 1971) suggested that radiolabeling could be used to measure pollen movement in natural populations. These studies, however, relied on microautoradiography or Geiger counter activity to trace labeled pollen, neither of which is appropriate for quantitative analyses of pollen movement. Geiger counters give only a rough estimate of radioactivity present. Microautoradiography of hundreds or thousands of stigmas presents great technical challenges, and film exposure cannot resolve thousands of pollen grains left on anthers or on the bodies of pollinators. Because of these problems, we used scintillation counting to estimate numbers of labeled pollen grains.

Use of scintillation counting required substantial modification of previous methods, because injection of labeled amino acids into buds or topical application of radioactive solutions to anthers resulted in very uneven labeling of pollen grains. After injecting an aqueous solution of [^{35}S]methionine into buds of *R. sativus*, we found up to a 300-fold range in labeling intensity between different anthers within the same flower. This would be inconvenient for autoradiographic counts of small numbers of labeled grains, but for scintillation counts to be used to estimate numbers of pollen grains, consistent labeling intensity is a necessity.

We developed the following technique to increase the consistency of pollen labeling within flowers. Early each morning, flowering stems were collected from wild plants growing near the University of California at Davis campus and placed immediately into florist's water pics. When treated in this way, stems of *R. sativus* continue to develop and open flowers normally for 2–3 days (also see Stanton, 1987; Young and Stanton, 1990). Two stems of comparable size and developmental stage were collected from each plant: one to be used as an unvisited control and the other as a pollen donor in experimental pollinations. Stems were taken directly to the lab, where all open flowers, opening buds, and very young buds were pinched off. Two opening buds were sampled to count pollen grains; their anthers were removed and placed into microcentrifuge tubes for pollen counts. Two buds, chosen so that they would open within the next 24–30 hr, were left on each stem. Cut stems were then placed individually into water pics containing 0.5 ml of deionized water and 20 μCi of [^{35}S]methionine. Stems were placed upright within a large clear acrylic box under lights in a growth chamber. As the stems transpired, the small volume of methionine solution was quickly taken up as a labeling pulse. After approximately 1 hr, all water pics were filled to capacity with deionized water to ensure that stem development could continue. Buds began to open the following morning.

Our first goal was to predict the amount of radioactivity contained in a donor flower's pollen by sampling a single anther and measuring the labeling intensity for that flower. Like other members of the Brassicaceae, flowers of R. *sativus* produce six anthers: four of these have long filaments, and two slightly later-dehiscing anthers have short filaments. To test for pollen labeling consistency, we placed one short anther from each of 20 freshly open flowers individually into scintillation vials. We then placed the remaining five anthers together into a second series of vials. Immediately thereafter, 0.5 ml of a protein-solubilizing agent (Solvable; NEN Research Products; DuPont) was pipetted into each vial. Vials were placed into a sonicating bath for 30 sec to disperse pollen grains in the solution, and then incubated at 40°C for 5 hr in a circulating water bath. Vials were then filled with 10 ml of scintillation cocktail (Atomlight; NEN Research Products; Dupont) and briefly vortex-mixed to homogenize the mixture. Mean cpm (counts per minute) were calculated from three readings on each sample using a Beckman LS9800 liquid scintillation counter set for optimal ^{35}S detection. Blank vials containing just solubilizing agent and scintillation cocktail were used to determine background counts.

Although pollen was labeled less intensively than with the bud injection method, using the stem's natural vascular system to deliver amino acids to developing pollen grains resulted in much more consistent pollen labeling within flowers. A regression of cpm for the five remaining anthers on cpm for the single short anther was highly significant, explaining 78.8% of the variance in remaining counts ($p < 0.0001$). Thus, although labeling intensity sometimes varied 2- to 3-fold between the two buds on a stem, we could estimate total label present within the pollen of any single flower quite accurately by measuring the radioactivity of one of its short anthers.

Once this protocol was established, we began conducting experiments with pollinators. Flowering stems were collected from field plants, processed, and labeled as before. We sampled a single lower anther from one labeled flower on each stem to estimate the amount of radioactivity that would be presented to an insect visitor. Mean pollen grain numbers measured on two other (unlabeled) buds from each stem were used to estimate the numbers of pollen grains present in the labeled flower (also see Young and Stanton, 1990). These two estimates together allowed us to calculate the numbers of pollen grains per cpm for each flower used in the experiment. In this way, we could use scintillation counts to estimate pollen grain numbers on bees, anthers, and stigmas for each pollination sequence.

Immediately after lower anthers were taken from labeled flowers, stems were placed back into acrylic boxes and transported to the field site at Davis. Field experiments were conducted in a field cage (2.5 × 2.5 × 1.8

m), into which a small honeybee colony had been placed several days previously. To habituate honeybees to foraging on *R. sativus* flowers within the confined setting, large bouquets of freshly cut stems were placed in the cage. On each morning of the experiment, the feeding bouquets were removed and replaced by an array of small stems in water pics. One location within the array was left vacant; a single radioactively labeled stem was placed there at the start of each experimental run. Each labeled stem had only a single open flower. Because we collected two labeled stems from any given plant in the field, one member of the pair was arbitrarily selected as the experimental flower to be visited by a bee, and the other stem was designated as its unvisited control.

Honeybees foraging on the array were watched until the labeled donor flower was visited, signaling the start of an experimental pollination run. A portable tape recorder was used to time the bee's movements as it visited up to 10 additional flowers. The donor flower was removed immediately from the array to prevent a second visit. The bee was captured after 10 additional visits or until it left the array, whichever came first, signaling the end of that pollination sequence. Scintillation vials containing 0.5 ml of solubilizing agent were used to collect all samples at that time. First, the bee was placed into a vial so that we could estimate the numbers of pollen grains remaining on its body. (Background counts for this estimate were determined for seven honeybees that had not visited labeled flowers.) Next, stigmas of all visited (recipient) flowers were placed individually into vials. For 13 runs, we sampled parts of recipient flowers other than the stigmas to determine how much pollen had been "lost" on nonreceptive floral surfaces. Unvisited flowers and flower parts were sampled to determine background counts. Finally, the anthers of the labeled donor flower were placed into one vial and the anthers from its matched control flower were placed into another vial. By handling experimental (visited) and control (unvisited) flowers similarly, we hoped to maximize our ability to estimate pollen removed by the honeybee visit. To avoid contact with or inhalation of labeled pollen, we wore disposable coveralls, pollen masks, and surgical gloves throughout these experiments. In all, we conducted 26 experimental runs.

Results

As seen in preliminary laboratory tests, the intensity of pollen labeling varied dramatically among flowers used in field experiments. Based on pollen counts and scintillation count data from lower anthers, a single cpm corresponded to an average of 1.26 pollen grains. This value ranged from 0.24 to 6.08 grains/cpm among flowers, so it was essential to have a within-flower measure of labeling intensity. Using unvisited control stems from the field experiment, we performed a new regression analysis to predict cpm in remaining anthers from the cpm in the lower anther of the labeled flower. In this sample, scintillation counts in the lower anther explained 93.0% of the variance in radioactivity within the remaining anthers (Fig. 4.3: $y = 3.591x + 19879$; $F_{1,22} = 264.8$; $p < 0.0001$). This least-squares linear regression was then used to estimate the radioactivity remaining in anthers on the experimental flowers before they were visited.

Despite the apparently good fit of the regression model, we were unable to estimate pollen removed from individual donor flowers very accurately using this method. On average, approximately 12% of the available pollen appeared to have been removed by a single honeybee visit. This corresponded to a mean of 12,900 pollen grains removed, but our estimates of pollen exported from individual donor flowers ranged from $-36,129$ to 76,645 grains! Negative removal rates were estimated for 5 out of 26 donor flowers, apparently because these flowers had unusually high levels of radioactivity in remaining anthers, compared with the lower anther that

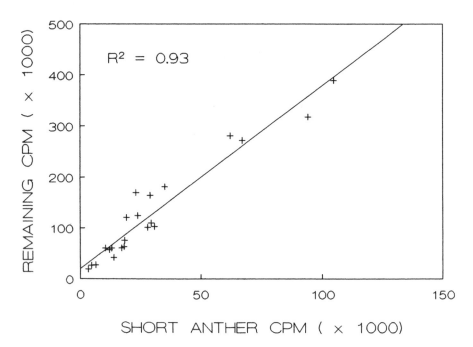

Figure 4.3. Predicting radioactivity of remaining anthers from scintillation counts on a single short anther from the same flower. Data are from anthers of unvisited control flowers of *Raphanus sativus* harvested at the same times as anthers from flowers used in experimental pollinations. Twenty-three control flowers were used for the analysis.

was sampled. Given the great uncertainty in our estimates of pollen removed, it is not surprising that we found no statistically significant relationships between estimated pollen removal and several other variables of
interest: estimated pollen deposited on recipient stigmas, estimated pollen
remaining on the honeybee after the visitation sequence, or time spent by
the bee foraging on the donor flower. We conclude that it may be difficult
to use radioactive labeling and scintillation counting to estimate rates of
pollen removal from flowers, at least in this system.

Although the inaccuracy of our individual pollen removal estimates was
disappointing, the study did yield useful information about the efficiency
of pollen transfer in *R. sativus*. Pooled over the 13 runs in which both
stigmas and other parts of recipient flowers were tested for the presence
of labeled pollen, we were able to account for only 6.2% of the pollen
removed from donor flowers (Fig. 4.4). That is, after 10 subsequent visits,
almost 95% of the pollen removed was lost from the pollination system.
Most of this "wasted" pollen probably dropped to the ground (also see
Harder and Thomson, 1989). Of the pollen that was detected later, 33%
was found on nonstigmatic surfaces of recipient flowers, 5% was deposited
on stigmas, and 63% remained on the pollinator's body. Overall, less than
0.5% of the pollen removed was deposited on stigmas of the next 10 flowers
visited. Fate is apparently not kind to the vast majority of pollen grains
removed from flowers of *R. sativus* by honeybees.

Although not conclusive, our results suggest that honeybee pollination
in this species could lead to long-distance gene flow via pollen carryover.
Even after visiting 10 unlabeled flowers, a bee's body retained an average

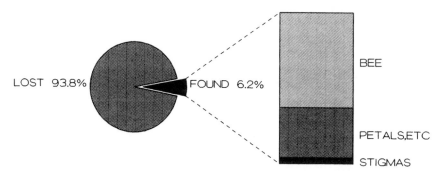

Figure 4.4. Schematic representation of the fates of pollen grains removed
from labeled donor flowers of *Raphanus sativus*. Over 93% of the exported
pollen was not recovered after 7–10 flower visits. Numbers of pollen grains on
the honeybee, on visited stigmas, and on nonstigmatic surfaces of visited flowers
("petals, etc.") were estimated in each trial using the calculated pollen grain/
cpm ratio for that donor flower.

of 533 donor pollen grains, more than 10 times that deposited on all 10 recipient stigmas combined. We do not know what proportion of that pollen still had access to stigmas, as grains may become lodged in recesses on the bee's surface. Still, we stress that honeybees typically visit *R. sativus* for nectar, rather than pollen. In this experiment, all honeybees appeared to be foraging solely for nectar: anthers were contacted only inadvertently as bees probed deeply into flowers. Obvious packing of pollen sacs was not observed, and only one bee groomed pollen from its body during our study. That single run was excluded from our analyses. It seems possible, then, that some of the pollen remaining on bees after 10 visits still had access to stigmas that might have been visited subsequently.

In addition to apparently high pollen carryover, we saw relatively little decline in pollen deposition over the first 10 recipients visited (Fig. 4.5). Although the first several stigmas visited tended to receive more donor pollen, a regression of estimated pollen grains deposited on order in the sequence was not statistically significant [$y(1, 204) = -0.71x + 7.9$; $r^2 = 0.02$; $p < 0.06$]. As for every other stage in the pollination process, variance in numbers of pollen grains deposited was very large. This vari-

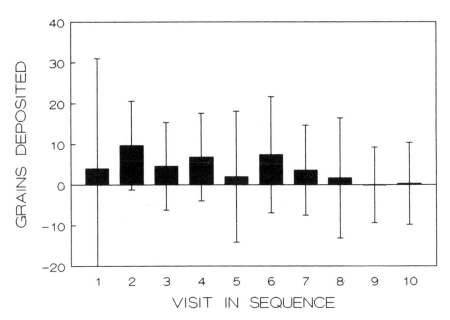

Figure 4.5. Estimated pollen grains deposited on stigmas of 10 *Raphanus sativus* flowers visited after a labeled donor flower. Means and standard deviations are shown for 26 visitation sequences.

ability reflects substantial variation in deposition patterns among individual bees (Fig. 4.6).

Conclusions

Our results for honeybee-pollinated *Raphanus sativus* converge in several respects with previous studies of bumblebee-pollinated *Erythronium grandiflorum* (Liliaceae). Although we generally think of insect pollination as a relatively efficient means of transferring pollen from one flower to another, the vast majority of exported pollen grains never reach a receptive stigma (also see Wyatt, 1976). Most pollen removed from a flower is rapidly lost from the system, being dropped to the ground or deposited on non-receptive floral parts. Flowers with pollen grains dispersed in pollinaria may provide an exception to this pattern of low pollen transfer efficiency. Broyles and Wyatt (1990) found a statistically significant association between pollinarium removal and male reproduction in a field population of *Asclepias exaltata*, possibly reflecting a more efficient pollination mecha-

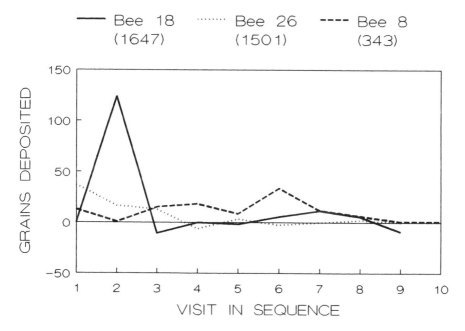

Figure 4.6. Pollen deposited on stigmas of recipient *Raphanus sativus* flowers during three visitation sequences. Negative pollen deposition estimates occurred when the cpm for a recipient stigma were less than mean cpm for background samples containing stigmas uncontaminated with labeled pollen. Numbers in parentheses are estimated numbers of pollen grains remaining on each honeybee at the end of her visitation sequence.

nism. For species with individually dispersed pollen grains, the rather stunning stochasticity of pollen transfer is likely to weaken the relationship between pollen removal and male success (but see Galen, 1991).

It has been almost dogma in plant reproductive biology that gene flow by pollen is highly restricted (Levin, 1981). When Ellstrand and colleagues began publishing evidence for long-distance gene flow by pollen in *R. sativus* (Ellstrand and Marshall, 1985), their results were therefore surprising to many. Our data suggest a mechanism that could lead to this kind of long-distance pollen movement. As honeybees visit flowers, relatively few pollen grains from any given flower are deposited on each stigma, and other grains may remain on the pollinator's body for relatively long periods. If those grains remain accessible to stigmas, long-distance pollen flow could result. Other genetic analyses using *R. sativus* have also suggested that pollen carryover in this system is pronounced. Most fruits contain seeds sired by more than a single pollen parent (Ellstrand, 1984), and controlled pollination studies suggest that multiple paternity is most likely to arise from deposition of single, mixed pollen loads, rather than from sequential visits (Marshall and Ellstrand, 1985). Mixed pollen loads will be generated when pollen grains remain accessible on a pollinator's body for extended periods of time, and our experiments with labeled pollen show that this may occur.

Our mixed results using radioactivity to label pollen grains imply that this technique may be of only limited use in quantitative studies of pollen movement. We were able to determine the fates of a small proportion of the pollen grains that were removed from donor flowers by honeybees, but we found it difficult to label pollen grains consistently enough to estimate both pollen deposition and pollen removal accurately. Because of environmental concerns, flowers bearing labeled pollen can be exposed to pollinators only under very controlled and restricted conditions. Thus, long-distance pollen movements should not be studied this way (but see Colwell, 1951, for a hair-raising account of long-distance dispersal in pollen labeled with ^{32}P). Even in controlled outdoor settings, working with radioactive pollen grains, which can be inhaled easily, presents a serious health hazard unless proper precautions are taken. The search for the perfect pollen marker continues.

The Bottom Line: Is Pollen Export Worth Measuring?

Pollen removal is one of the few components of male fitness that can be measured in almost any natural plant population, including those for which pollen marking techniques are impractical. For self-incompatible species, only pollen grains that leave the anthers have any chance of fertilizing ovules, but our results and those of Thomson and co-workers (e.g., Thomson and Thomson, 1992) suggest that the probability of an exported pollen grain reaching a receptive stigma is very small. The unpredictable nature of the pollen transport process inevitably leads to some uncoupling between pollen removal and paternity. Future studies should recognize that pollen removal will be, even under the best circumstances, only a rough indicator of relative male success. Nonetheless, we argue that it is better than having no indicator at all.

Pollen removal studies will be most useful and informative when they are conducted in the context of detailed analyses of pollination dynamics. On average, we expect to find some relationship between pollen removal and ovule fertilization, but that relationship can be dramatically altered in several ways. To minimize the chances that pollen removal measurements will give a misleading picture of male fitness, some biases and errors inherent in the method must be recognized.

Studies of pollen removal should be conducted concurrently with measurements of ovule availability and seed-set. Whenever mate availability is limited, one expects to see diminishing fitness gains with increases in pollen export (Lloyd, 1984; Nakamura et al., 1989). This saturation in the male fitness gain curve is most likely to occur when pollen movement is relatively restricted and/or when pollen deposition is great enough to saturate seed production. Similarly, pollen exported during times of low ovule availability or low seed-set is less likely to be successful (Thomson and Barrett, 1981; Devlin and Stephenson, 1987; Mazer et al., 1989; Thomson et al., 1989). This is an especially important issue in populations of *R. sativus*, because most seed production takes place early in the season and most effective pollination takes place during a few early hours each day. Plant traits that increase pollen export early in the season and early in the day should, on average, lead to increased male success.

Pollen export studies are most likely to be informative when conducted in populations dominated by a single pollinator or by a group of pollinators that pick up and deposit pollen in a similar way. Thomson and Thomson (1992) use simulation studies to illustrate this point. When pollen is removed by a less efficient pollinator, it is unavailable to a more efficient pollinator. Thus, in mixed pollination systems, higher rates of pollen removal can actually lead to lower rates of pollen deposition and seed paternity. In large populations of *R. sativus*, honeybees are the dominant flower visitors for most of the season, and their visitation patterns are very likely to determine mating patterns and reproductive success. In small populations, however, where honeybees account for only a minor fraction of flower visits, their visits can actually reduce paternal success (Stanton et al., 1991). Presumably, other visitors to these small populations are more efficient pollinators. Wilson and Thomson (1991) also have presented measurements of pollen deposition and removal, suggesting that honeybees can be more inefficient pollinators than other floral visitors.

There may well be opposing selective forces acting on schedules of pollen presentation and export. For species whose pollen loses viability very rapidly once anthers have dehisced and pollinator visits are relatively infrequent, selection will favor plants that disseminate pollen gradually, rather than simultaneously (Thomson and Thomson, 1992). Similarly, greater pollen wastage may occur when large amounts of pollen are available at the same time (Harder and Thomson, 1989). In contrast, our data for *R. sativus* suggest that selection for prudent pollen dispersal might be counterbalanced by selection for rapid pollen export during periods of simultaneous flower opening and ovule availability. Understanding the relative importance of these factors requires detailed information on the behaviors of pollinators, the dynamics of pollen transport and longevity, and the schedules of pollen and ovule availability.

Literature Cited

Adams, W.T., D.S. Birkes, and V.J. Erickson. 1992. Measuring gene flow and pollen dispersal in forest tree orchards with genetic markers. *In* R. Wyatt, ed., Ecology and Evolution of Plant Reproduction: New Approaches. Chapman & Hall, New York, pp. 37–60.

Barrett, S.C.H., J.R. Kohn, and M.B. Cruzan. 1992. Experimental approaches to the study of mating-system evolution. *In* R. Wyatt, ed., Ecology and Evolution of Plant Reproduction: New Approaches. Chapman & Hall, New York, pp. 192–230.

Bell, G. 1985. On the function of flowers. Proc. R. Soc. London Ser. B 224:223–265.

Bertin, R.I. 1982. Paternity and fruit production in the trumpet creeper (*Campsis radicans*). Am. Nat. 119:694–709.

Bertin, R.I. 1988. Paternity in plants. *In* J. Lovett Doust and L. Lovett Doust, eds., Plant Reproductive Ecology: Patterns and Strategies. Oxford University Press, New York, pp. 30–59.

Broyles, S.B., and R. Wyatt. 1990. Paternity analysis in a natural population of *Asclepias exaltata*: Multiple paternity, functional gender, and the "pollen-donation hypothesis." Evolution 44:1454–1468.

Burke, T. 1989. DNA fingerprinting and other methods for the study of mating success. Trends Ecol. Evol. 4:139–144.

Campbell, D.R. 1989. Measurements of selection in a hermaphroditic plant: Variation in male and female pollination success. Evolution 43:318–334.

Clegg, M.T., and B.K. Epperson. 1988. Natural selection of flower color polymorphisms in morning glory populations. *In* L.D. Gottlieb and S.K. Jain, eds., Plant Evolutionary Biology. Chapman & Hall, London, pp. 255–273.

Colwell, R.N. 1951. The use of radioactive isotopes in determining spore distribution patterns. Am. J. Bot. 38:511–523.

Cruzan, M.B., P.R. Neal, and M.F. Willson. 1988. Floral display in *Phyla incisa*: Consequences for male and female reproductive success. Evolution 42:505–515.

Devlin, B. and N.C. Ellstrand. 1990a. Male and female fertility variation in wild radish, a hermaphrodite. Am. Nat. 136:87–107.

Devlin, B. and N.C. Ellstrand. 1990b. The development and application of a refined method for estimating gene flow from angiosperm paternity analyses. Evolution 44:248–259.

Devlin, B. and A.G. Stephenson. 1987. Sexual variations among plants of a perfect-flowered species. Am. Nat. 130:199–218.

Devlin, B., K. Roeder, and N.C. Ellstrand. 1988. Fractional paternity assignment: Theoretical development and comparison to other methods. Theor. Appl. Genet. 76:369–380.

Ellstrand, N.C. 1984. Multiple paternity within the fruits of the wild radish, *Raphanus sativus*. Am. Nat. 123:819–828.

Ellstrand, N.C. and D.R. Marshall. 1985. Interpopulation gene flow by pollen in wild radish, *Raphanus sativus*. Am. Nat. 126:606–616.

Galen, C. 1992. Pollen dispersal dynamics in an alpine wildflower, *Polemonium viscosum*. Evolution, in press.

Galen, C. and M.L. Stanton. 1989. Bumble bee pollination and floral morphology: Factors influencing pollen dispersal in the alpine skypilot, *Polemonium viscosum* (Polemoniaceae). Am. J. Bot. 76:419–426.

Handel, S.N. 1976. Restricted pollen flow of two woodland herbs determined by neutron-activation analysis. Nature (London) 260:422–423.

Handel, S.N. 1983. Contrasting gene flow patterns and genetic subdivision in adjacent populations of *Cucumis sativus* (Cucurbitaceae). Evolution 37:760–771.

Handel, S.N., and J. Le Vie Mishkin. 1984. Temporal shifts in gene flow and seed set: Evidence from experimental populations of *Cucumis sativus*. Evolution 36:1350–1356.

Harder, L.D. 1990. Pollen removal by bumble bees and its implications for pollen dispersal. Ecology 71:1110–1125.

Harder, L.D., and J.D. Thomson. 1989. Evolutionary options for maximizing pollen dispersal of animal-pollinated plants. Am. Nat. 133:323–344.

Harper, J.L. 1977. Population Biology of Plants. Academic Press, New York.

Hayes, J.L., and C.L. Claussen. 1988. Marking Lepidoptera and their offspring: Trace element labelling of *Colias eurytheme* (Pieridae) with rubidium. J. Lepidopt. Soc. 42:196–203.

Hodges, S.A. 1990. The roles of nectar variation, hawkmoth behavior, and pollen movement on natural selection for nectar production in *Mirabilis multiflora*. Ph.D. Dissertation, University of California, Berkeley, California.

Horovitz, A. 1978. Is the hermaphrodite flowering plant equisexual? Am. J. Bot. 65:485–486.

Horovitz, A., and J. Harding. 1972. The concept of male outcrossing in hermaphroditic higher plants. Heredity 29:223–236.

Knox, R.B., C. Suphioglu, T. Hough, and M. Singh. 1992. Genetic and molecular dissection of male reproductive processes. In R. Wyatt, ed., Ecology and Evolution of Plant Reproduction: New Approaches. Chapman & Hall, New York, pp. 231–254.

Levin, D.A. 1981. Dispersal versus gene flow in plants. Ann. Missouri Bot. Gard. 68:233–253.

Lloyd, D.G. 1984. Gender allocations in outcrossing cosexual plants. In R. Dirzo and J. Sarukhan, eds., Principles of Plant Population Ecology. Sinauer, Sunderland, MA, pp. 277–300.

Marshall, D.R., and N.C. Ellstrand. 1985. Proximal causes of multiple paternity in wild radish, Raphanus sativus. Am. Nat. 126:596–605.

Marshall, D.R., and M.W. Folsom. 1992. Mechanisms of nonrandom mating in wild radish. In R. Wyatt, ed., Ecology and Evolution of Plant Reproduction: New Approaches. Chapman & Hall, New York, pp. 91–118.

Mazer, S.J., R.R. Nakamura, M.L. Stanton. 1989. Seasonal changes in components of male and female reproductive success in wild radish. Oecologia 81:345–353.

Meagher, T.R. 1986. Analysis of paternity within a natural population of Chamaelirium luteum. I. Identification of most likely male parents. Am. Nat. 128:199–215.

Meagher, T.R., and E. Thompson. 1987. Analysis of parentage for naturally established seedlings of Chamaelirium luteum (Liliaceae). Ecology 68:803–812.

Morgan, M.T., and S.C.H. Barrett. 1990. Outcrossing rates and correlated mating within a population of Eichhornia paniculata (Pontederiaceae). Hereditary 64:271–280.

Morris, W.F. 1992. The consequences of plant spacing and biased movement for pollen dispersal by honey bees. Ecology, in press.

Mulcahy, D.L., G. Mulcahy, and K.B. Searcy. 1992. Evolutionary genetics of pollen competition. In R. Wyatt, ed., Ecology and Evolution of Plant Reproduction: New Approaches. Chapman & Hall, New York, pp. 25–36.

Muller-Starck, G., and M. Ziehe. 1984. Reproductive systems in conifer seed orchards. 3. Female and male fitnesses of individual clones realized in Pinus sylvestris. L. Theor. Appl. Genet. 69:173–177.

Murawski, D.A. 1987. Floral resource variation, pollinator response, and potential pollen flow in Psiguria warscewiczii. Ecology 68:1273–1282.

Nakamura, R.R., M.L. Stanton, and S.J. Mazer. 1989. Mate size and paternal success in plants. Ecology 70:71–76.

Nybom, H., and B.A. Schaal. 1990. DNA "fingerprints" reveal genotypic distributions in natural populations of blackberries and raspberries (Rubus, Rosaceae). Am. J. Bot. 77:883–888.

Price, M.V., and N.M. Waser. 1982. Experimental studies of pollen carryover: Hummingbirds and Ipomopsis aggregata. Oecologia 54:353–358.

Queller, D.C. 1983. Sexual selection in a hermaphroditic plant. Nature (London) 284:450–451.

Reinke, D.C., and W.L. Bloom. 1979. Pollen dispersal in natural populations: A method for tracking individual pollen grains. Syst. Bot. 4:223–229.

Ritland, K. 1989. Correlated matings in the partial selfer, *Mimulus guttatus*. Evolution 43:848–860.

Roeder, K.M., B. Devlin, and B.G. Lindsay. 1989. Application of maximum likelihood methods to population genetic data for the estimation of individual fertilities. Biometrics 45:363–379.

SAS Institute. 1985. SAS User's Guide: Statistics, Version 5. SAS Institute. Cary, North Carolina.

Schaal, B.A. 1980. Measurement of gene flow in *Lupinus texensis*. Nature (London) 284:450–451.

Schemske, D.W. 1980. Evolution of floral display in the orchid *Brassavola nodosa*. Evolution 34:489–493.

Schlising, R.A., and R.A. Turpin. 1971. Hummingbird dispersal of *Delphinium cardinale* pollen treated with radioactive iodine. Am. J. Bot. 58:401–406.

Schoen, D.J., and M.T. Clegg. 1985. The influence of flower color on outcrossing rate and male reproductive success in *Ipomoea purpurea*. Evolution 39:1242–1249.

Schoen, D.J., and S.C. Stewart. 1986. Variation in male reproductive investment and male reproductive success in white spruce. Evolution 40:1109–1120.

Stanton, M.L. 1987. The reproductive biology of petal color variants in wild populations of *Raphanus sativus* L.: I. Pollinator response to color morphs. Am. J. Bot. 74:178–187.

Stanton, M.L., and R.E. Preston. 1988. Ecological consequences and phenotypic correlates of petal size variation in wild radish, *Raphanus sativus* (Brassicaceae). Am. J. Bot. 75:528–539.

Stanton, M.L., A.A. Snow, and S.N. Handel. 1986. Floral evolution: Attractiveness to pollinators influences male fitness. Science 232:1625–1627.

Stanton, M.L., A.A. Snow, S.N. Handel, and J. Bereczky. 1989. The impact of a flower-color polymorphism on mating patterns in experimental populations of wild radish (*Raphanus raphanistrum* L.). Evolution 43:335–346.

Stanton, M.L., H.J. Young, N.C. Ellstrand, and J. Clegg. 1991. Consequences of floral variation for male and female reproduction in experimental populations of wild radish, *Raphanus sativus* L. Evolution 45:268–280.

Stephenson, A.G., T.-C. Lau, M. Quesada, M. and J.A. Winsor. 1991. Factors that affect pollen performance. *In* R. Wyatt, ed., Ecology and Evolution of Plant Reproduction: New Approaches. Chapman & Hall, New York, pp. 119–136.

Thomson, J.D., and S.C.H. Barrett 1981. Temporal variation of gender in *Aralia hispida* Vent. (Araliaceae). Evolution 35:1094–1107.

Thomson, J.D., and R.C. Plowright. 1980. Pollen carryover, nectar rewards, and pollinator behavior with special reference to *Diervilla lonicera*. Oecologia 46:68–74.

Thomson, J.D., and B.A. Thomson. 1989. Dispersal of *Erythronium grandiflorum* pollen by bumblebees: Implications for gene flow and reproductive success. Evolution 43:657–661.

Thomson, J.D., and B.A. Thomson. 1992. Pollen presentation and viability schedules. *In* R. Wyatt, ed., Ecology and Evolution of Plant Reproduction: New Approaches. Chapman & Hall, New York, pp. 1–24.

Thomson, J.D., M.V. Price, N.M. Waser, and D.A. Stratton. 1986. Comparative studies of pollen and fluorescent dye transport by bumble bees visiting *Erythronium grandiflorum*. Oecologia 69:561–566.

Thomson, J.D., K.R. Shivanna, J. Kenrick, and R.B. Knox. 1989. Sex expression, breeding system, and pollen biology of *Ricinocarpos pinifolius:* A case of androdioecy? Am. J. Bot. 76:1048–1059.

Waser, N.M., and M.V. Price. 1982. A comparison of pollen and fluorescent dye carry-over by natural pollinators of *Ipomopsis aggregata* (Polemoniaceae). Ecology 63:1168–1172.

Williams, J.G.K., A.R. Kubelik, K.J. Livak, J.A. Rafalski, and S.V. Tingey. 1990. DNA polymorphisms amplified by arbitrary primers are useful as genetic markers. Nucleic Acids Res. 18:6531–6535.

Willson, M.F., and P.W. Price. 1977. The evolution of inflorescence size in *Asclepias* (Asclepiadaceae). Evolution 31:495–511.

Willson, M.F., and B.J. Rathcke. 1974. Adaptive design of the floral display in *Asclepias syriaca* L. Am. Midl. Nat. 92:47–57.

Willson, P., and J.D. Thomson. 1991. Heterogeneity among floral visitors leads to discordance between removal and deposition of pollen. Ecology 72:1503–1507.

Wyatt, R. 1976. Pollination and fruit-set in *Asclepias*: A reappraisal. Am. J. Bot. 63:845–851.

Young, H.J., and M.L. Stanton. 1990. Influences of floral variation on pollen removal and seed production in wild radish. Ecology 71:536–547.

5

Mechanisms of Nonrandom Mating in Wild Radish

Diane L. Marshall and Michael W. Folsom
University of New Mexico

Introduction

Traditionally, studies of plant mating systems have described the amount of inbreeding and outbreeding or the mechanisms that determine these processes (reviewed by Richards, 1986). However, the possibilities for mating in plants are far wider. During mating there may be differential success of self versus outcross pollen (e.g., de Nettancourt, 1977), closely related versus unrelated mates, one compatible donor versus another (e.g., Marshall, 1990), very distant versus nearby mates (Waser, 1993), and mates from the same versus different species. A comprehensive view of plant mating systems must include understanding of the degree of sorting at each of those levels and knowledge of the similarity or dissimilarity of the mechanisms that control mating at all levels. Here we cannot include all of the possible levels of sorting among mates, but we do go beyond the traditional consideration of inbreeding and outbreeding to consider sorting among compatible mates.

Mating is nonrandom among compatible mates whenever the paternity of seeds differs from that which would be produced by random success of the available pollen (Marshall, 1990). Nonrandom mating is important because it can affect the fitness of pollen parents, seed parents, and offspring. The fitness of pollen parents is affected because nonrandom success of pollen produces differences in the number of seeds sired by each pollen donor. The fitness of the seed parent and the seeds will be altered by nonrandom mating if the available pollen donors differ in quality.

Sorting among compatible mates may be effected by a variety of mechanisms that differ in time of operation, possibilities for precise control, and identity of the controlling agent (Stephenson and Bertin, 1983; Willson and Burley, 1983; Marshall and Folsom, 1991). Differential success of compatible mates may become manifest at any stage from pollen presentation through fruit and seed maturation, but it is more difficult to interpret sorting in the later stages. Prepollination mechanisms can affect only the amount and diversity of pollen available on stigmas, whereas postpollination mechanisms can involve precise signaling between maternal tissue and pollen tubes. Early-acting mechanisms can be influenced by the pollen donor and pollen tubes or the maternal tissue, whereas later-acting mechanisms can be influenced by the maternal tissue, the paternal genotype, the interaction between maternal and paternal tissues, or the developing embryo. Determining which of these mechanisms are most important is one of the major challenges of plant reproductive biology.

In this chapter we confine our consideration to mechanisms of nonrandom mating that occur after pollen has been deposited on stigmas. These include differential germination of pollen, differential growth of pollen tubes, nonrandom fertilization of ovules, nonrandom abortion of seeds and fruits, and differential provisioning of seeds and fruits based on the paternity of their seeds. There is ample evidence that these processes can occur (reviewed by Stephenson and Bertin, 1983; Willson and Burley, 1983; Bertin, 1988; Marshall and Folsom, 1991). To ascertain which of these mechanisms are important in any particular mating event, however, a number of specific issues must be considered.

Problems of Interpretation

First, to what degree do the maternal plant and the pollen donor control mating? Both have clear fitness interests in the process that may determine, respectively, the quality and number of seeds produced. Differences in pollen donor ability can be estimated *in vitro* by growing pollen tubes in culture (Mazer, 1987a; Elgersma et al., 1989). This allows measurement of potential differences in pollen tube growth in the absence of maternal tissue, but pollen tube growth *in vitro* may not reflect pollen tube growth and pollen donor success *in vivo* (Cruzan, 1990b; Mazer, 1987b). In contrast, maternal effects on mating cannot be observed in the absence of pollen and pollen tubes. Hence, tests of maternal ability to sort among mates are always confounded with the possibility of pollen donor effects on mating or of maternal plant by paternal plant interactions.

Second, once ovules are fertilized, a variety of genetically distinct entities (i.e., the maternal plant, the endosperm, and the embryo) have fitness interests in the mating process (Westoby and Rice, 1982; Queller, 1983; Mazer, 1987a; Haig and Westoby, 1988; Shaanker et al., 1988). Thus, determining which of these genetic entities control differential success of seeds and fruits is complex. At this point, even if the maternal plant controls most sorting, it is difficult to determine whether sorting is based on the paternal contribution to seeds or on the diploid genotype of the seed (i.e., is the maternal plant sorting among mates, offspring, or both?).

Third, postpollination mechanisms of nonrandom mating are difficult to observe because they occur inside stigmas, styles, and ovaries. Thus, flowers and fruits must be sacrificed to make observations. Moreover, because marking of pollen tubes from individual donors is difficult, mechanisms that sort among mates after mixed pollinations must be inferred from performance of pollen in single-donor pollinations.

Fourth, because deposition of mixed pollen loads (due to pollen carryover) appears to be common in nature (Schaal, 1980; Thomson and Plowright, 1980; Levin, 1981; Ennos and Clegg, 1982; Handel, 1982; Price and Waser, 1982; Waser and Price, 1984; Geber, 1985; Thomson et al., 1986), it is important to know whether mixed pollen loads result in nonrandom mating. Genetic markers for seed paternity after mixed pollination are an essential tool in such investigations.

Finally, the mechanisms that produce sorting among compatible mates may be related to the mechanisms that regulate self-pollination or to the mechanisms that operate at other levels of nonrandom mating. If this is true, then it is possible that sorting among compatible mates is confounded by subtle effects of S-alleles or by subtle effects of the degree of relatedness among pollen donors and maternal plants. Thus, in order to understand the mechanisms of sorting in compatible matings, it may also be important to gain further understanding of the operation of incompatible matings, inbred matings, and extremely outbred matings (crosses involving physically distant mates).

Potential Approaches to the Study of Nonrandom Mating in Plants

Various approaches have been suggested to solve the problems in interpreting mating patterns in plants. These range from statistical aspects of experimental design (Lyons et al., 1989) to more careful observation of the intermediate stages of mating. Because each approach constrains the experiment, it is unlikely that all can be applied simultaneously.

The first problem involves distinguishing between maternal and paternal effects on nonrandom mating. Maternal, paternal, and interaction effects can be distinguished by appropriate design and statistical treatment (Lyons et al., 1989). Powerful designs will include several maternal and several paternal plants. Often, however, the number of maternal and paternal plants, and hence the ability to resolve these effects, is quite small (Snow and Mazer, 1988) or unequal (e.g., Marshall, 1988, 1990). Suitable levels of replication of crosses are difficult to implement. It seems important to repeat crosses several times on each maternal plant so that a chance error in pollen application will not bias the results. Moreover, sorting due to maternal mechanisms may occur only when the maternal plant has limited resources to mature fruits (Marshall, 1988). Accordingly, completing only a few crosses on plants that produce many flowers may be inappropriate. Repeated crosses on the same plant, however, will not be entirely independent. How independent they are depends on the level of physiological integration of the plant (e.g., Watson and Casper, 1984; Marshall and Oliveras, 1990).

Statistical interpretation of the designs of experiments is also complex. For example, in cross-factor analyses of variance, maternal effects on mating can be due either to genetic differences among maternal plants in mating characteristics or to environmental differences among maternal plants. A statistically significant paternal effect may indicate differences in pollen donor ability due to either the genotype or the condition (e.g., Young and Stanton, 1990) of the pollen donor. Alternatively, it could be caused by maternal plants all sorting among pollen donors in the same way. Likewise, significant maternal by paternal interaction effects may indicate that some combinations of maternal and paternal plants have complementary genes (Waser et al., 1987) or that maternal plants sort among pollen donors in different ways (Stephenson and Bertin, 1983).

It may be possible to dissect out the effects of the seed parent by manipulation of the maternal plants. If changing maternal condition changes mating patterns, then maternal effects must have been involved (Marshall and Ellstrand, 1988; Marshall, 1988). Maternal condition might be changed in many ways. Lyons et al. (1989) suggest turning off the incompatibility system and observing mating patterns. Unfortunately, such a design assumes that all mating patterns are related to incompatibility or that the mechanisms of sorting self from nonself pollen are the same as those that sort among compatible pollen. Alternatively, maternal condition can be altered by changing resource availability (e.g., reducing water or nutrients). When maternal plants have limited resources for reproduction, mating may be more selective (Marshall, 1988). In the study reported here, we have chosen to use low water as a treatment to elucidate maternal effects on mating.

The postpollination mechanisms by which nonrandom mating occurs are amenable to microscopic observation. It is relatively straightforward to collect appropriate tissue and stain for pollen germination, pollen tube growth, and ovule fertilization (Fig. 5.1). Unfortunately, it is usually necessary to sacrifice tissue to make these observations so that pollen germination and pattern of fertilization, for example, cannot be observed in the same tissue. This is avoidable in some species in which removal of the stigma and style for staining is possible while leaving the ovary in place to

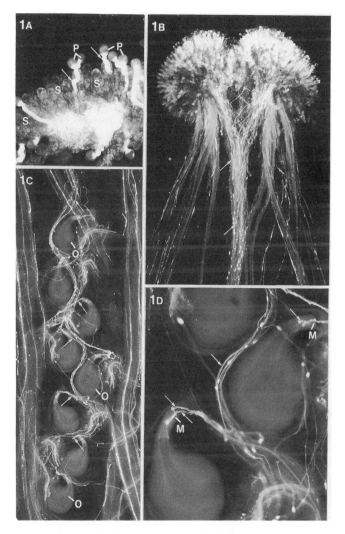

Figure 5.1. Pollen tube (arrows) growth and fertilization in wild radish. (A) Pollen (P) germination and early phases of pollen tube growth. Note that pollen tubes enter the tissue of the stigma between papillose cells (S). 180×. (B) Path of pollen tube growth from the bilobed stigma to the base of the ovary. The structure of the stigma is such that pollen tubes from both lobes are directed toward the transmitting tissue of the style. 60×. (C) Basal portion of the radish ovary showing rows of ovules (O) on both sides of the septum. It is through the septum that pollen tubes grow. The base of the ovary is toward the bottom of the micrograph. 30× (D) Micrograph showing the third and fourth ovules from the base of the ovary seen in C. In the case of the third ovule, as in numerous others, more than one pollen tube can be seen in the micropyle (M). 90×.

mature so that patterns of pollen germination and ovule fertilization in single crosses can be related (Waser and Price, 1991).

Observation of these intermediate stages after mixed pollination is particularly problematical because it is generally not possible to distinguish which pollen tubes came from particular pollen donors. For some plants, it is possible to pollinate different parts of the stigma so that pollen tubes from different donors grow, distinguishably, within the same style (Cruzan, 1990a,b). However, interactions among mixtures of pollen may occur over short time spans (see below) so that growing pollen tubes in different sections of the style may not be completely informative. Finally, the time constraints of detailed observation of all of the stages of mating force reductions in sample size over simpler experimental protocols.

Our designs offer an intermediate sample size, timed collection of crosses, and correlation of observations of mechanisms with number of seeds sired in mixed crosses. Although we could not distinguish the identity of pollen tubes in styles, we studied both mixed and single pollinations to determine whether there were any fundamental differences in the two kind of crosses.

Determining whether differential seed provisioning and seed abortion are due to maternal, paternal, or offspring control is also difficult. Our approach to this problem is mainly structural. Examination of the process of seed abortion can reveal, for example, whether the embryo begins to wither while the structures that supply nutrients remain intact (Cooper et al., 1937; Brink and Cooper, 1947; Cave and Brown, 1954, 1957). This pattern suggests that the embryo is unable to mature because of genetic or phenotypic defects. Alternatively, callose may form barriers in the maternal tissues that supply nutrients to developing seeds prior to any evidence of deterioration of the embryo (Bradbury, 1929; Tukey, 1933; Pimienta and Polito, 1982). This pattern suggests that the maternal tissue controls seed abortion.

Consideration of the timing and nature of seed and fruit abortion may also eliminate some possibilities for control. For example, seed abortion is position dependent in numerous species (Horovitz et al., 1976; Bawa and Webb, 1984; Nakamura, 1988). In *Raphanus raphanistrum*, fertilization of ovules has been shown to be nonrandom, with central ovules being fertilized first (Hill and Lord 1987; Mazer et al., 1986). In *Raphanus sativus*, pollen donors tend to fertilize seeds nonrandomly after mixed pollination; some donors tend to sire seeds in the stylar parts of fruits, whereas others tend to sire seeds in the basal parts of fruits. This pattern, coupled with a strong tendency to abort stylar ovules before basal ovules, results in seed abortion that is selective among pollen donors (Marshall and Ellstrand, 1988). This seed abortion is likely to be based on the characteristics of the pollen donors rather than the offspring because it is due to patterns of fertilization, not to patterns of offspring growth. Likewise,when seed and fruit filling depend on the number of seeds per fruit (Marshall and Ellstrand, 1986; Marshall, 1988), these mechanisms cannot be a response to particular offspring genotypes.

Finally, the interaction between self-incompatibility and other forms of nonrandom mating is problematical. At present, we do not know enough about the underlying processes that control the various forms of mating to determine whether they are functionally related. The approach we have taken is to ensure that all maternal and paternal lineages involved in our experiments have different S-allele genotypes. Thus, subtle effects of these alleles are unlikely to confound our results. Ultimately, direct comparisons of the processes that operate in incompatible and compatible matings will be needed to solve the problem. In sporophytically incompatible species it is possible to perform bud-pollinations that allow fertilization before the incompatibility system in the maternal tissue is active. Similarly, it may be possible to remove the recognition molecules from pollen by washing the pollen with a solvent prior to pollination (Roggen, 1974). This would allow us to turn off parts of the incompatibility system and ask how elimination of those functions alters mating among compatible mates.

In this paper we attempt to answer some of the questions about nonrandom mating using wild radish, *Raphanus sativus* L., as a model. Specifically, we concentrate on isolation of male and female effects on mating, on observation of the mechanisms of nonrandom mating, and on use of genetic markers to identify the paternity of seeds. The data that we will discuss do not address the issue of maternal versus offspring control of seed filling, but we will describe how we intend to answer that question in the future. Finally, in the study described below, we attempt to factor out any confounding effects of self-incompatibility; we will also discuss experiments designed to address the relationship of incompatibility to the sorting processes we observed.

Wild Radish as a Model System

Wild radish, *Raphanus sativus*, is a weedy annual found in abandoned fields and along roadsides in California. Plants are visited by an array of pollinators including honeybees, syrphid flies, and lepidopterans (Kay, 1976; Stanton, 1987a,b; Stanton and Preston, 1988; D. Marshall, personal observation). In California, plants germinate with the onset of winter rains and flower from January until late spring or summer (in wet seasons). Given adequate water, plants may live for more than one year in either the field or the greenhouse.

Most plants in southern California, the source of seeds for our studies, have sporophytic self-incompatibility (Hinata and Nishio, 1980). Self-incompatibility facilitates experimental pollinations because emasculation of flowers is not necessary. Fruits are multiseeded, making multiple paternity of fruits possible, yet there are not so many seeds per fruit that it is impossible to determine the paternity of all of the seeds. There are hundreds of flowers per plant, allowing many kinds and replications of crosses. Seed weights, about 10 mg, are large enough that individual seeds can be weighed precisely. Seeds germinate rapidly, making studies of offspring success feasible. Wild radish has high levels of isozyme polymorphism, making paternity analysis of seeds very tractable (Ellstrand, 1984; Ellstrand and Marshall, 1985a,b, 1986).

In spite of its many useful features, wild radish is not a perfect model system. Its pollen tends to stick together in clumps, making studies that vary pollen grain number per stigma difficult. Wild radish is not easily cloned, so replicate genotypes are difficult to obtain. Furthermore, this species in California is the result of escape from cultivation and introgression with *R. raphanistrum* (Panestos and Baker, 1967). Past introgression with another species may be the reason that *R. sativus* expresses an unusually high degree of genetic variation. In any event, wild populations have existed for many years and the high degree of polymorphism provides many markers for paternity analysis. Thus, wild radish is an excellent model for asking whether variation in mating characters, such as pollen tube growth and fertilization ability, and paternal success exists.

Previous studies of the reproductive ecology of wild radish indicate that opportunities for nonrandom mating exist in the field and suggest that, given the opportunity, greenhouse plants do not mate at random. Most open-pollinated fruits contain seeds sired by more than one donor, as many as four, and mate number per plant ranges from 4 to 14 (Ellstrand, 1984; Ellstrand and Marshall, 1986). Multiple paternity of fruits probably results from deposition of mixed pollen loads through pollen carryover, rather than from sequential deposition of pollen from different donors (Marshall and Ellstrand, 1985). Given the high frequency of multiple paternity and pollen carryover as the likely cause, most stigmas probably receive mixed pollen loads. Thus, field plants will frequently have opportunities for postpollination sorting among mates.

Data from paternity analysis of field populations indicate that plants often receive pollen from other populations, as well as from multiple donors (Ellstrand and Marshall, 1986; Ellstrand et al., 1989). Within populations, pollen donors sire unequal numbers of seeds (Devlin and Ellstrand, 1990), but the mechanisms by which that occurs are not known.

Greenhouse studies reveal that postpollination sorting among pollen donors or tubes is a possible mechanism of nonrandom mating. When maternal plants are pollinated with mixed pollen loads, pollen donors sire unequal numbers of seeds (Marshall and Ellstrand, 1986, 1988; Marshall,

1990). In most of these experiments one pollen donor sires the most seeds across all maternal plants. However, when some maternal plants were given limited water, the degree of unevenness of seed paternity increased relative to seed paternity on other, related maternal plants that were given adequate water (Marshall and Ellstrand, 1988; Marshall, 1988). Thus, the maternal tissue appears to be involved in nonrandom mating.

In addition to the differences in number of seeds sired, studies with both *R. sativus* and *R. raphanistrum* indicate that paternal identity can have a small, but significant effect on seed weight (Marshall and Ellstrand, 1986; Mazer et al., 1986). Fruits in which seeds are sired by multiple pollen donors have a greater total seed weight than fruits sired singly by any of the donors found in multiply-sired fruits. (Marshall and Ellstrand, 1986; Marshall, 1988). This result is difficult to explain as a result of pollen competition, but could be due to selective allocation of resources by maternal plants to multiply-sired over singly-sired fruits (selective filling). The result could also be due to changes in competitive regimes among seeds developing in multiply-sired fruits as compared to singly-sired fruits.

Selective filling of multiply-sired fruits acts to increase mate number per fruit, but some other mechanism appears to keep mate number at an intermediate level. When the number of pollen donors per cross increases, actual mate number per fruit levels off with a strong mode of 2 (Marshall and Ellstrand, 1989).

Previous studies only allow us to make inferences about the mechanisms by which mate choice can occur. Taken together, these studies suggest that most of the postpollination mechanisms of nonrandom mating have the potential to act in this system. Mature *Raphanus* pollen is trinucleate and germinates rapidly. In compatible crosses over 70% pollen germination occurs within 15–30 min (Dickinson and Lewis, 1973). Compatible pollen tubes grow rapidly through the solid transmitting tissue of the style and septum, which acts as a mechanical and cytochemical guide for pollen tube growth in *R. raphanistrum* (Hill and Lord, 1987). The tubes exit the septum within the ovary, grow along the septum surface, and are guided to the micropyle of the ovules by the obturator, a small piece of tissue that grows out from the placenta.

There is some evidence for variation in pollen tube growth rates in *Raphanus*. Pollen from two donors of *R. sativus* grew much further through the stylar tissue in 4 hr than pollen from a third donor (Marshall and Ellstrand, 1986). This third donor sired few seeds overall, and many of these seeds were located in the stylar portion of the ovary. In contrast, Mazer et al., (1986) found no variation among four donors of *R. raphanistrum* in the extent of ovule fertilization after 10 hr. The degree to which any of the variation in pollen tube growth is affected by maternal tissue is unknown.

In cultivated *R. sativus* fertilization may occur within 10 hr (Pundir et al., 1983). In the closely related *R. raphanistrum* the earliest fertilizations may occur within 4 hr, and most ovules are fertilized within 24 hr (Hill and Lord, 1986; Mazer et al., 1986). Thus, there may be as much as a 20-hr lapse between fertilization of the first and last ovules. Ovules closest to the style are not always fertilized first; the first ovules to be fertilized may be those in the central region of the ovary (Hill and Lord, 1987; Mazer et al., 1986). The order of fertilization depended on both paternal and maternal identity, but the mechanisms by which the order of fertilization was determined could not be elucidated. Hill and Lord (1986) suggested that there was a negative correlation between pollen tube growth rate and the probability of a pollen tube exiting the transmitting tissue of the septum. This would result in the most vigorous pollen tubes fertilizing primarily central and basal ovules and would produce a strong paternal component to fertilization. Alternatively, variation in order of fertilization could be due to differences in maturation of ovules within ovaries (Palser et al., 1989), but variation in ovule development has not yet been examined.

Several experiments indicate that pollen donors vary in the location of seeds sired within multiply sired fruits, with donors tending to sire seeds in either the basal or stylar portion of the ovary preferentially (Marshall and Ellstrand, 1986, 1988; Marshall, 1991). Unfortunately, these experiments could not detect whether this was due to inherent properties of pollen tubes, to interactions among competing pollen tubes, or to interactions with the maternal tissue.

Location of seeds fertilized can be important from the perspective of the pollen donors. Seed abortion in stressed plants is more frequent in the developing stylar seeds (Marshall and Ellstrand, 1988), and seed weight is often lower in the stylar portions of the fruit (Stanton, 1984).

Abortion of developing seeds can also sort among mates in wild radish. In greenhouse plants, flowers typically contain 6–10 ovules at anthesis, whereas at maturity flowers usually contain 5–6 seeds, indicating that up to half of the ovules per ovary might be aborted. Most ovule abortion is reported to occur within 10 days after pollination and can be difficult to detect in mature fruits (Nakamura and Stanton, 1987). Stress treatments applied after 10 days can increase ovule abortion significantly and ovules in stylar positions are most likely to abort (Marshall and Ellstrand, 1988). Embryo development in *Raphanus* has not been studied in detail, so the developmental processes leading to abortion of ovules and, hence, whether maternal control or offspring failure cause abortion are not known.

Experimental Design

We selected plants for experimental crosses in two steps. First, in the spring before the experiment, a large number of plants were screened for compatibility (Karron et al., 1990). A total of 200 plants, two from each of 50 lineages from each of two field populations, were grown to maturity. Over 10,000 crosses were performed on these plants in order to find a set of 16 plants that was fully intercompatible. That is, each plant was reciprocally compatible with all 15 of the other plants. Given that genetic self-incompatibility in wild radish is sporophytic, this means that at least 32 S-alleles were present in the two populations screened. These 16 plants were the parents of the plants used in the experimental crosses (Fig. 5.2).

The 16 plants were used to form eight unique pairs that served as parents of the four maternal and four paternal lineages used in the experiment. Thus, all maternal and paternal lineages had different S-alleles. The pairs were also selected on the basis of electrophoretic phenotype. From these lineages, 16 maternal plants (1–16) (four from each of the four full-sibships) and four pollen donors (A–D) (one from each pollen donor lineage) were selected. Plants were grown in the greenhouse in 3.8-liter pots. Two maternal plants from each family were randomly assigned to full-water (control) and half-water (low-water) groups (Fig. 5.2). Control pots were watered to capacity three times a day, whereas low-water plants were given half the control amount of water.

The four pollen donors were used to make four kinds of single pollinations (A, B, C, D) and six kinds of mixed pollinations (A+B, A+C, A+D, B+C, B+D, C+D). Each single pollination was repeated 31 times on each maternal plant, four times for each of three ovary collection intervals (16, 20, and 24 hr), four times for measurement of pollen germination, and 15 times to generate mature fruits. Mixed pollinations were repeated four times per ovary collection interval, four times for measurement of pollen germination, and 25 times to generate fruits for analysis of seed paternity and allocation to seeds. A total of 5920 pollinations were performed (Fig. 5.2).

Single pollinations were performed by collecting pollen in a small petri dish and applying the pollen with tissue-wrapped forceps until stigmas were thoroughly coated with pollen. Pollen mixes for mixed pollinations were made by collecting pollen from equal numbers of anthers from each donor in a small petri dish and then mixing thoroughly with tissue-covered forceps. Pollen was then applied as in single pollinations. This technique results in the application of large numbers of pollen grains and approximately equal numbers of pollen grains are applied in all crosses.

To measure pollen germination, pollen was applied evenly to stigmas of emasculated, fully open flowers that were free of self-pollen. After pollination the style was tapped lightly to remove excess pollen from the stigmatic surface. At 0.5 hr the stigma was removed from the flower and placed in a drop of stain containing 1 mg/ml of both water-soluble aniline blue and sodium azide in 0.033 M K_3PO_4. Sodium azide was added to

Experimental Design

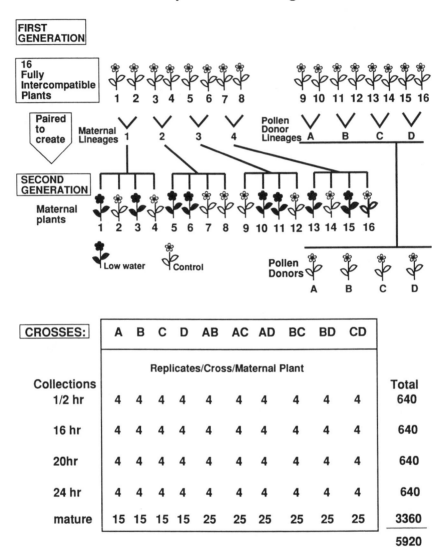

Figure 5.2. Crossing design for experiments with wild radish. First generation plants were screened for complete intercompatibility, so all 16 must have had different S-alleles. First generation plants were paired to produce the second generation. The pairings indicated are shown for convenience; pairs were actually selected based on their electrophoretic phenotypes. Procedures for pollinating the maternal plants and collecting tissue are described in the text.

inhibit further pollen germination. Pollen grains were scored as germinated if a pollen tube could be seen at $100 \times$ magnification (Fig. 5.1A).

To score fertilized ovules, tissue collected 16, 20, and 24 hr after pollination was placed in Farmer's fixative (3:1, 70% ethanol:glacial acetic acid), held for several days, transferred to 70% ethanol, and stored. Prior to staining, the tissue was dehydrated to H_2O and ovaries were dissected and treated with 0.8 M NaOH for 45 min at 60°C. Subsequently the tissue was washed in H_2O, placed in the staining solution (1 mg/ml water-soluble aniline blue and 0.1 mg/ml ethidium bromide in 0.1 M K_3PO_4) and held at 4°C until examined. Ovules were scored as fertilized if a pollen tube was observed in the micropyle at $100 \times$ magnification (Fig. 5.1C, D).

Paternity of seeds resulting from mixed pollination was analyzed by starch-gel electrophoresis. Paternity was assigned on the basis of progeny alleles at a single locus of phosphoglucose isomerase (PGI) using the methods of Marshall and Ellstrand (1986, 1988).

To compare fertilizations and seed paternity among regions of the fruits, fruits were divided into the stylar, middle, and basal thirds. When seeds or ovules per fruit was not evenly divisible by three, extra seeds or ovules were added first to the stylar third and then to the middle third of the fruit. Hence, there are slightly more stylar and middle than basal ovules overall.

Variables that could be treated as continuous (proportion of pollen germinated and mean seed weight per fruit) were analyzed by analysis of variance (ANOVA). Independent variables were transformed as appropriate to improve normality. Dependent variables are specified for each design in the data tables. Variables that could not be treated as continuous (number of ovules fertilized and number of seeds sired) were analyzed by contingency tables. These could not be treated as continuous because of the low number of possible values within each fruit.

Results

1. *Do pollen germination and order of fertilization vary due to pollen donor identity, maternal identity, or the interaction between the two?* Pollen germination differed significantly among both pollen donors (Table 5.1) and maternal plants (Table 5.2); however, the effect of the maternal plant ($r^2 = 0.24$) was considerably larger than that of the pollen donor ($r^2 = 0.024$). (Coefficients of determination, r^2, were calculated as the type III sums of squares for the independent variable divided by the total sums of squares.) Numbers of ovules fertilized after 24 hr varied strongly among maternal plants and weakly among pollen donors. The pollen donor effect was significant in a categorical model that included maternal plant, pollen

Table 5.1. Variation among Pollen Donors of Wild Radish in Pollen Germination, Ovule Fertilization, Number of Mature Seeds (in Mixed Pollinations), and Mean Seed Weight per Fruit.[a]

		Pollen donor			
Variable	*n*	A	B	C	D
Mean proportion of pollen germinated	256	0.11[ab]	0.10[b]	0.12[a]	0.09[b]
(standard deviation)		(0.12)	(0.13)	(0.12)	(0.12)
Proportion of ovules fertilized after 16 hr	1731	0.26	0.25	0.27	0.31
Proportion of ovules fertilized after 20 hr	1727	0.34	0.34	0.40	0.39
Proportion of ovules fertilized after 24 hr	1742	0.52	0.54	0.58	0.50
Proportion of seeds sired in mixed pollinations (means of pairwise crosses)	5892	0.14	0.25	0.34	0.28
Mean individual seed mass (mg)	1298	8.99[a]	7.60[c]	8.45[b]	7.90[c]
(standard deviation)		(3.34)	(2.85)	(3.16)	(2.91)

[a]The arcsine square root transformation of proportion of pollen germinated and mean seed weight per fruit were dependent variables in ANOVAs, in which maternal plant, pollen donor, replicate, and their two-way interactions were independent variables. Means with different superscripts are significantly different using Tukey's studentized range tests. Data on ovule fertilization and seed paternity are presented as proportions to facilitate comparison with pollen germination. However, analyses of these variables were not based on proportions. Number of ovules fertilized and number of seeds sired were compared by contingency table analyses. For proportion of ovules fertilized across pollen donors, $X^2 = 10.3$ ($p < 0.02$) from a multiway χ^2 analysis that included maternal plant, pollen donor, and collection interval. There are no significant differences in ovule fertilization by the pollen donors at any of the individual collection intervals. For seed paternity, $X^2 = 587$ ($p < 0.0001$) across all mixed pollinations.

donor, replicate, and their interactions, but not in a simple contingency table analysis of differences in number of seeds fertilized by pollen donors. There were no significant maternal plant by pollen plant donor interaction effects.

Patterns of fertilization of ovules can vary by location within fruits, as well as in overall frequency. There was a strong tendency for ovules in the central third of the fruit to be fertilized before basal and stylar ovules and for stylar ovules to be fertilized last (Fig. 5.3). This trend was consistent across maternal plants and the direction of the trend was consistent across pollen donors as well (Table 5.3). Pollen donor D showed a slightly greater pattern of early fertilization of central ovules.

Table 5.2. Variation among Maternal Plants of Wild Radish in Pollen Germination and Ovule Fertilization.[a]

Maternal plant	Proportion of pollen germinated (standard deviation)	Proportion of ovules fertilized after		
		16 hr	20 hr	24 hr
1	0.031[fg] (0.046)	0.084	0.169	0.392
2	0.121[bc] (0.066)	0.186	0.436	0.600
3	0.155[ab] (0.072)	0.221	0.235	0.432
4	0.065[defg] (0.064)	0.244	0.394	0.570
5	0.204[a] (0.321)	0.269	0.238	0.352
6	0.122[bc] (0.092)	0.133	0.136	0.206
7	0.090[cde] (0.062)	0.129	0.324	0.477
8	0.120[bc] (0.062)	0.452	0.456	0.510
9	0.057[efg] (0.037)	0.745	0.857	0.942
10	0.114[bcd] (0.047)	0.426	0.556	0.768
11	0.117[bcd] (0.067)	0.395	0.390	0.752
12	0.081[cdef] (0.061)	0.505	0.613	0.883
13	0.056[efg] (0.073)	0.105	0.080	0.385
14	0.028[g] (0.028)	0.083	0.172	0.427
15	0.060[efg] (0.080)	0.182	0.229	0.317
16	0.097[cde] (0.059)	0.087	0.271	0.343
n	640	4350	4367	4359

[a]The arcsine square root transformation of proportion of pollen germinated was the dependent variable in an ANOVA, in which maternal plant, pollen donor, replicate, and their two-way interactions were the independent variables. For pollen germination, $n = 640$, 4 flowers/pollination/plant. Means with different superscripts are significantly different using Tukey's studentized range tests. Whereas proportion of ovules fertilized is presented to facilitate comparisons with proportion of pollen germinated, statistical tests were performed by comparing the numbers of ovules fertilized and not fertilized using contingency table analyses ($X^2 = 752$, $p < 0.0001$; $X^2 = 714$, $p < 0.0001$, for ovule fertilization at 16, 20, and 24 hr, respectively).

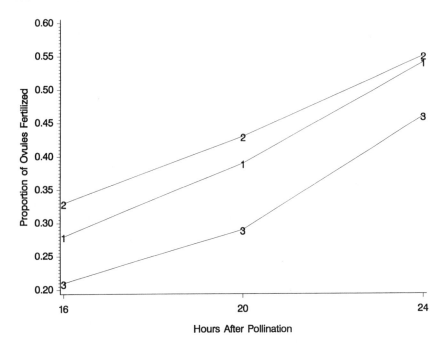

Figure 5.3. Proportion of wild radish ovules fertilized 16, 20, and 24 hr after pollination. Ovaries were divided into basal (1), central (2), and stylar (3) regions. Whereas proportion of ovules is presented for each comparison, statistical tests were based on the numbers of ovules fertilized. At all three time intervals, the numbers of ovules fertilized were significantly different among regions (16 hr, $X^2 = 38.5$, $p < 0.0001$; 20 hr, $X^2 = 34.2$, $p < 0.0001$; 24 hr, $X^2 = 6.7$, $p < 0.035$).

Table 5.3. Proportion of Ovules Fertilized in the Basal, Central, and Stylar Regions of Fruits of Wild Radish 16 hr after Single Pollinations by Pollen Donors A, B, C, and D.

| Pollen donor | n | Region of ovary | | | X^2 | p |
		Basal	Central	Stylar		
A	433	0.20	0.34	0.21	9.25	0.010
B	424	0.25	0.32	0.18	8.64	0.013
C	439	0.23	0.36	0.21	10.51	0.005
D	435	0.30	0.41	0.23	12.07	0.002

2. *Are patterns of pollen germination and pollen tube growth the same after single versus mixed pollinations?* The proportion of germinated pollen after 30 min is significantly less in mixed than in single pollinations (Table 5.4). Whereas the absolute difference is small (1.9%), the relative difference is large: pollen in single loads had 22% higher germination than pollen in mixed loads.

Differences in pollen germination are mirrored by differences in number of ovules fertilized 24 hr after pollination. However, the single pollen loads resulted in only 6% greater fertilization, and there was no significant difference in fertilization 16 and 20 hr after single and mixed pollinations. Thus, although pollen germination is slower after mixed pollination, pollen tubes appear to have "caught up" during pollen tube growth.

3. *Does the rate of pollen germination determine the rate of ovule fertilization?* If rapid pollen germination sets the stage for rapid ovule fertilization, then the proportion of ovules fertilized should correlate strongly with the proportion of pollen grains germinated. However, correlations between mean pollen germination and mean ovule fertilization on maternal plants are not significant ($r = -0.21$, $p < 0.44$; $r = -0.01$, $p < 0.97$; and $r = -0.11$, $p < 0.68$, respectively, for correlation of pollen germination

Table 5.4. Proportion of Pollen Grains Germinated and Proportion of Ovules Fertilized in Single and Mixed Pollinations and Total Seed Weight per Fruit in Singly- and Multiply-Sired Fruits of Wild Radish.[a]

		Pollination type or paternity	
Variable	*n*	Single	Mixed
Proportion of			
pollen germinated	640	0.106[a]	0.087[b]
(standard deviation)		(0.123)	(0.099)
Proportion of ovules			
fertilized after 24 hr	4359	0.53	0.50
Total seed weight			
per fruit (mg)	2022	32.8[b]	41.3[a]
(standard deviation)		(20.3)	(20.3)

[a]The arcsine square root transformation of proportion of pollen germinated was the dependent variable in an ANOVA, in which maternal plant, pollination type, replicate, and their two-way interactions were the independent variables. Total seed weight per fruit was used as the dependent variable in an ANOVA, in which maternal plant, replicate, paternity, and their two-way interactions were the independent variables. Means with different superscripts are significantly different using Tukey's studentized range tests. Although proportion of ovules fertilized is presented for ease of comparison with pollen germination, statistical analyses were done by comparing numbers of ovules fertilized and not fertilized in each region of the ovary using contingency tables ($X^2 = 4.75$, $p < 0.03$).

with fertilization at 16, 20, and 24 hr; $n = 16$ for each correlation). Likewise, mean pollen germination for each of the 10 kinds of pollinations is not significantly correlated with ovule fertilization ($r = 0.14$, $p < 0.70$; $r = -0.15$, $p < 0.68$; and $r = 0.56$, $p < 0.09$, respectively, for correlation of pollen germination with fertilization at 16, 20, and 24 hr; $n = 10$ for each correlation).

4. *Do pollen germination and order of fertilization after single pollination predict the number and location of ovules sired in multiply-sired fruits?* Differences in the abilities of pollen donors to sire seeds might be a result of differences among donors in the rate of pollen germination and in the number and location of ovules fertilized as measured in single-pollen-donor crosses. Therefore, we used the performance of four pollen donors in single pollinations to predict the numbers and locations of seeds sired after mixed pollination (Tables 5.5, 5.6). The pattern of pollen germination correctly predicted the relative numbers of seeds sired in mixed pollinations in three of the six crosses. The number of ovules fertilized after 16 hr in single pollinations correctly predicted the pattern of seed paternity in four of the six cases. The relative numbers of central ovules fertilized 16 hr after single pollinations did not correctly predict the relative proportions of central seeds sired in any of the six crosses. Thus, early measures of performance, especially those obtained from single pollinations, were poor predictors of the performance of pollen donors in mixed pollinations.

5. *Does reduction of water to maternal plants alter patterns of pollen germination, ovule fertilization, and seed paternity?* Reduction of water alters the condition of maternal plants without a similar alteration of pollen parents. Thus, if reduction of water for maternal plants alters the performance of pollen donors, this must be a maternal effect. Mean percentage of pollen germinated was higher (10.7%) in plants that received less water than in control plants (8.2%) ($p < 0.001$ in an ANOVA with maternal

Table 5.5. Predicted and Observed Patterns of Seed Paternity in Pairwise Mixtures of Pollen from Donors A, B, C, and D of Wild Radish.[a]

Prediction from pollen germination	Prediction from ovule fertilization (16 hr)	Actual paternity	X^2
A > B	A > B	A:B = 32:68	123.0
A < C	A < C	A:C = 19:81	370.0
A > D	A < D	A:D = 25:75	261.0
B < C	B < C	B:C = 37:63	64.8
B > D	B < D	B:D = 44:56	15.4
C > D	C < D	C:D = 61:39	22.9

[a]The χ^2 analysis tested for departures from an expected ratio of 50:50. All values are significant ($p < 0.0001$).

Table 5.6. Predicted and Observed Patterns of Central Seeds Sired in Wild Radish, Based on the Pattern of Ovules Fertilized in Single-Pollen-Donor Crosses.[a]

Prediction from ovules fertilized at 16 hr	Observed proportion of central seeds	X^2	p
A > B	A:B = 36 < 43	13.6	0.001
A < C	A:C = 38 = 41	4.76	0.093
A < D	A:D = 38 = 41	1.09	0.58
B < C	B:C = 40 = 40	3.34	0.89
B < D	B:D = 39 > 38	18.1	0.0012
C < D	C:D = 41 > 36	6.49	0.039

[a]The χ^2 analysis compared the relative proportions of seeds sired in basal, middle, and stylar regions of the fruit. Only paternity of the central region is shown.

Table 5.7. Proportion of Ovules Fertilized 24 hr after Single Pollinations of Control and Low-Water Wild Radish Plants.[a]

Treatment	Pollen donor			
	A	B	C	D
Control	0.57	0.59	0.64	0.59
Low-water	0.46	0.49	0.51	0.40

[a]$N = 128$ flowers per treatment (4 replicates × 4 crosses × 8 plants). Although the proportion of ovules fertilized is presented for ease of comparison, statistical tests are based on χ^2 analyses of the number of ovules fertilized by the four pollen donors.

family, cross, replicate, and treatment as the main effects). However, there was no significant treatment by pollination interaction effect, indicating that relative performance of pollen donors was not altered. Fertilization of ovules was reduced in treated maternal plants (Table 5.7, Fig. 5.4), and there was a significant pollination by treatment interaction effect (in a categorical model). The most noticeable difference in number of ovules fertilized is the change in rank of pollen donor D from control to low-water plants.

Of the six kinds of pairwise mixed pollinations, frequencies of seeds sired by the pollen donors changed from control to low-water plants in only one case, but this cross did not involve donor D (as the fertilization data might have suggested). Rather, in mixed pollinations with donors A and B, donor A sired 29% of the seeds on control plants and 36% of the seeds on low-water plants ($X^2 = 5.32$, $p < 0.025$). Donor A also sired slightly more seeds on low-water plants when its pollen was mixed with that of donor C (17 vs. 21%), but the trend was not statistically significant ($p < 0.093$).

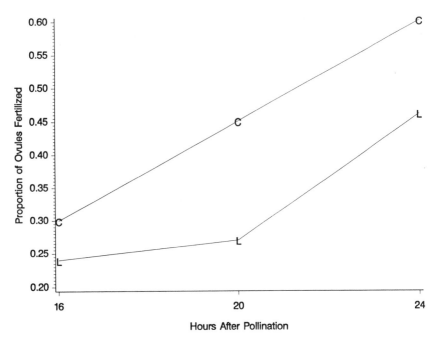

Figure 5.4. Proportion of wild radish ovules fertilized 16, 20, 24 hr after pollination of control (C) and low-water (L) plants, that received half the water of controls. Whereas proportion of ovules fertilized is shown to illustrate trends, statistical analyses are based on contingency table analyses of number of ovules fertilized and not fertilized. At all time intervals the number of ovules fertilized and unfertilized were not equal across control and low-water plants (16 hr X^2 = 9.78, $p < 0.002$; 20 hr, X^2 = 57.2, $p < 0.0001$; 24 hr, X^2 = 31.8, $p < 0.0001$).

Discussion

Determining which of the potential mechanisms of nonrandom mating operate and whether maternal plants, pollen donors, or embryos influence the outcome of mating is a complex problem. Our observations of the process of mating in wild radish provide only a partial answer for this system. Both pollen donors and maternal plants show statistically significant effects on pollen germination and ovule fertilization. There were, however, no pollen donor by maternal plant interaction effects, indicating that pollen donor performance varied little across maternal plants. This could be interpreted as an overwhelming influence of pollen donor abilities, but maternal effects were far stronger than paternal effects and alteration of maternal condition affected patterns of fertilization by pollen donors and seed paternity in one kind of pairwise mixed pollination. Other experiments with wild radish (Marshall and Ellstrand, 1988) have confirmed that maternal condition can affect seed paternity and that it is therefore possible for the maternal plant to regulate mating.

Perhaps our most striking result is that mixed and single pollinations resulted in different mating patterns from the earliest stages of the interaction. Pollen germinated more slowly and ovules were fertilized somewhat later after mixed pollinations. This suggests that direct interference among pollen grains may alter mating patterns. Although pollen allelopathy among species has been reported (Kanchan and Jayachandra, 1980; Thomson et al., 1981; Jimenez et al., 1983; Murphy and Aarssen, 1989), these appear to be the first data to suggest intraspecific interference.

Despite this initial delay in the mating process, at maturity multiply-sired fruits have a greater total seed weight than singly-sired fruits. Perhaps initial interference competition among pollen grains in mixed pollinations results in selection for more vigorous pollen phenotypes. If so, this may explain why multiply-sired fruits are typically filled to greater weights than singly-sired fruits of wild radish (Marshall and Ellstrand, 1986; 1988, 1990, 1991).

Early measures of pollen performance in single pollinations were very poor predictors of the outcome of mixed pollinations. In face, the earliest measure of pollen performance, pollen germination, was not correlated with ovule fertilization. This may be because we had an insufficient number of pollen donors or because interactions among pollen tubes and among pollen tubes and the maternal tissue after mixed pollinations play an important role in mating. Other studies have found that early measures of pollen performance are more closely related to fertilization (e.g., Waser and Price, 1991).

Another explanation for the lack of correspondence among early measures of performance and estimates of number of seeds sired is that the early measures are largely indicators of speed of pollen germination and pollen tube growth, whereas other characters are important in determining final seed paternity. In all crosses, there was a strong pattern of fertilization of central ovules first, followed by basal and then stylar ovules. This tendency for central ovules to be fertilized first is consistent with previous studies of *R. raphanistrum* (Hill and Lord, 1986; Mazer et al., 1986). Because stylar ovules produce smaller seeds (Stanton, 1984) and are more susceptible to abortion (Marshall and Ellstrand, 1988), where pollen tubes grow may be as important as how fast they grow.

Alternatively, our measures of pollen performance may not have been good predictors because the techniques themselves are limited. Assignment of pollen grains or tubes to a donor was not possible after mixed pollination. Thus, differences in performance due to interaction among the pollen tubes could not be measured directly. To address the questions posed by plant reproductive ecology, techniques are needed that permit visualization of structures and processes that normally occur deep inside living tissues of higher plants. The relationships between pollen tubes, maternal tissues, and young sporophytes are complex, and the complexity of this situation may increase due to physiological stress and mixed pollen loads.

To explore these relationships better and especially to understand events occurring after mixed pollinations, techniques are needed that allow visualization of pollen tubes deep within maternal tissue and establishment of their genetic identity. By use of vital dyes this need can, in part, be addressed. Hough et al. (1985) were able to visualize pollen tube nuclei within maternal tissue. Pollen grains were labeled with the vital fluorescent dye Hoechst 33258 and then used to pollinate a flower. The stained nuclei were visualized *in situ* with a conventional fluorescent microscope after pollen tubes had grown into the style. By staining nuclei of one of a pair of pollen donors, it may be possible to differentiate the pollen tubes that are the products of mixed pollinations.

The ability to establish the paternity of a pollen tube provides a powerful new technique for the study of plant reproductive biology; but our ability to make observations of pollen tubes deep within maternal tissue is still limited. A new type of microscope, the laser scanning confocal microscope (LSCM), may solve this problem. The LSCM can produce "optical sections" of tissue that possess a high degree of clarity. It has been used to study the embryology of an orchid species (Fredrikson, 1990), and we have employed it in combination with staining and clearing techniques to visualize the relationships between ovules, their micropyles, and the septum in intact radish ovaries. Our observations suggest that the LSCM provides an opportunity to examine events deep within intact tissues. This capability, plus that provided by the vital dyes, allows questions of male/female interactions to be explored at levels that were previously thought impossible.

In this experiment we attempted to exclude confounding effects of the self-incompatibility system by ensuring that all maternal and paternal lineages had different S-alleles. Thus, it is very unlikely that we were inadvertently measuring subtle effects of the incompatibility system. However, a complete understanding of nonrandom mating will require analysis of processes at all of the levels of nonrandom mating. For example, two levels of nonrandom mating might interact such that sorting among compatible mates is achieved by mechanisms that are physiologically similar to those that sort compatible from incompatible mates. To address that issue, we are performing experiments with both compatible and incompatible donors. This can be done in wild radish because it is possible to effect pollination with incompatible donors if stigmas are pollinated while the flower is still in bud. This results in completion of the initial stages of pollen tube growth before the incompatibility system is activated in the stigma. By this means we can perform crosses in a factorial design where both shared S-alleles and the possibility for inbreeding vary among donors. Ultimately, it may be possible to use other techniques, such as washing the recognition molecules from pollen (Roggen, 1974), to disable other parts of the incompatibility system. The critical questions will be whether disabling all or part of the incompatibility system affects the degree of sorting among compatible, as well as among compatible and incompatible, donors.

There is also considerable debate about the degree to which the kinds of mechanisms we study can produce sexual selection in plants (Stephenson and Bertin, 1983; Willson and Burley, 1983; Queller, 1987; Charlesworth et al., 1987; Lyons et al., 1989). In wild radish, there is nonrandom seed paternity and, therefore, nonrandom mating success of pollen donors. Hence, sexual selection may result. However, we have not yet collected the data to resolve whether the sorting is based on pollen donor or embryo qualities, and we have not yet tested whether the sorting is based on heritable characteristics of the pollen donors. We do know that pollen donor identity typically affects offspring growth in *R. sativus* (Marshall and Whittaker, 1989; Marshall, 1990; Karron and Marshall, 1990). Although these data suggest that mating pattern can alter maternal fitness, they are neither necessary nor sufficient to demonstrate the operation of sexual selection. We must measure the mating characteristics of the offspring.

Finally, all of the studies of nonrandom mating in wild radish reported on and referred to here have been done using greenhouse-grown plants. They demonstrate only that nonrandom mating is possible, not that it occurs in the field. Data from several field populations (Ellstrand, 1984; Ellstrand and Marshall, 1986) show that multiple paternity per fruit is common in the field. This probably results from mixed pollination and suggests that opportunities for nonrandom mating within flowers are common in wild radish. Additionally, we now have data showing that experimental mixed pollinations of field plants result in nonrandom seed paternity, just as do similar pollinations of greenhouse plants (Marshall and Fuller, unpublished data). Nevertheless, studies of wild radish and of other plants that document opportunities for nonrandom mating provided by typical patterns of pollinator visits to field plants are essential.

Acknowledgments

The research reported here was supported by National Science Foundation Grants BSR-8818552 and BSR-8958233 to D.L. Marshall. We thank Diana Oliveras and Oller Fuller for organizing and performing greenhouse crosses. We also thank Lynette Michelson, Susan Wolterstorff, Kerry Parker, Chris Mazer, and Joy Avritt for assistance in performing experimental pollinations and in counting and weighing seeds. We thank Beth Dennis for drafting Figure 5.2 and Jeff Hill for providing valuable comments on an earlier draft of the manuscript.

Literature Cited

Bawa, K.S., and C.J. Webb. 1984. Flower, fruit, and seed abortion in tropical forest trees: Implications for the evolution of paternal and maternal reproductive patterns. Am. J. Bot. 71:736–751.

Bertin, R.I. 1988. Paternity in plants. *In* J. Lovett Doust and L. Lovett Doust, eds., Plant Reproductive Ecology. Oxford University Press, New York, pp. 30–59.

Bradbury, D. 1929. A comparative study of the developing and aborting fruits of *Prunus cerasus*. Am. J. Bot. 16:525–542.

Brink, R.A., and D.C. Cooper. 1947. The endosperm in seed development. Bot. Rev. 13:423–541.

Cave, M.S., and S.W. Brown. 1954. The detection and nature of dominant lethals in *Lilium* II. Cytological abnormalities in ovules after pollen irradiation. Am. J. Bot. 41:469–483.

Cave, M.S., and S.W. Brown. 1957. The detection and nature of dominant lethals in *Lilium* III. Rates of early embryogeny in normal and lethal ovules. Am. J. Bot. 44:1–8.

Charlesworth, D., D.W. Schemske, and V.L. Sork. 1987. The evolution of plant reproductive characters: Sexual versus natural selection. *In* S.C. Stearns, ed., Evolution of Sex. Birkhauser, Basel, pp. 317–335.

Cooper, D.C., R.A. Brink, and H.R. Albrecht. 1937. Embryo mortality in relation to seed formation in alfalfa (*Medicago sativa*). Am. J. Bot. 24:203–213.

Cruzan, M.B. 1990a. Pollen-pollen and pollen-style interactions during pollen tube growth in *Erythronium grandiflorum* (Liliaceae). Am. J. Bot. 77:116–122.

Cruzan, M.B. 1990b. Variation in pollen size, fertilization ability, and postfertilization siring ability in *Erythronium grandiflorum*. Evolution 44:843–856.

Devlin, B. and N.C. Ellstrand. 1990. Male and female fertility variation in wild radish, a hermaphrodite. Am. Nat. 136:87–107.

Dickinson, H.G., and D. Lewis. 1973. Cytochemical and ultrastructural differences between intraspecific compatible and incompatible pollinations in *Raphanus*. Proc. R. Soc. London Ser B 183:21–38.

Elgersma, A., A.G. Stephenson, and A.P.M. den Nijs. 1989. Effects of genotype and temperature on pollen tube growth in perennial ryegrass (*Lolium perenne* L.). Sex. Plant Reprod. 2:225–230.

Ellstrand, N.C. 1984. Multiple paternity within the fruits of the wild radish, *Raphanus sativus*. Am. Nat. 123:819–828.

Ellstrand, N.C., B. Devlin, and D.L. Marshall. 1989. Gene flow by pollen into small populations: Data from experimental and natural stands of wild radish. Proc. Natl. Acad. Sci. U.S.A. 86:9044–9047.

Ellstrand, N.C. and D.R. Marshall. 1985a. Interpopulation gene flow by pollen in wild radish, *Raphanus sativus*. Am. Nat. 126:606–616.

Ellstrand, N.C. and D.R. Marshall. 1985b. The impact of domestication on distribution of allozyme variation within and among cultivars of radish *Raphanus sativus* L. Theor. Appl. Genet. 69:393–398.

Ellstrand, N.C. and D.R. Marshall. 1986. Patterns of multiple paternity in populations of *Raphanus sativus*. Evolution 40:837–842.

Ennos, R.A., and M.T. Clegg. 1982. Effect of population substructuring on estimates of outcrossing rate in plant populations. Heredity 48:283–292.

Fredrikson, M. 1990. Embryological study of *Herminium monorchis* (Orchidaceae) using confocal scanning laser microscopy. Am. J. Bot. 77:123–127.

Geber, M.A. 1985. The relationship of plant size to self-pollination in *Mertensia ciliata*. Ecology 66:762–772.

Haig, D., and M. Westoby. 1988. Inclusive fitness, seed resources, and maternal care. *In* J. Lovett Doust and L. Lovett Doust, eds., Plant Reproductive Ecology. Oxford University Press, New York, pp. 60–79.

Handel, S.N. 1982. Dynamics of gene flow in an experimental population of *Cucumis melo* (Cucurbitaceae). Am. J. Bot. 69:1538–1546.

Hill, J.P., and E.M. Lord. 1986. Dynamics of pollen tube growth in the wild radish, *Raphanus raphanistrum* (Brassicaceae). I. Order of fertilization. Evolution 40:1328–1333.

Hill, J.P., and E.M. Lord. 1987. Dynamics of pollen tube growth in the wild radish, *Raphanus raphanistrum* (Brassicaceae). II. Morphology, cytochemistry, and ultrastructure of transmitting tissues, and rate of pollen tube growth. Am. J. Bot. 74:988–997.

Hinata, K., and T. Nishio. 1980. Self-incompatibility in crucifers. *In* S. Tsunoda, K. Hinata, and C. Gomez-Campo, eds., Brassica Crops and Wild Allies: Biology and Breeding. Japanese Scientific Societies, Tokyo, pp. 223–234.

Horovitz, A., L. Meiri, and A. Beiles. 1976. Effects of ovule positions in fabaceous flowers on seed set and outcrossing rates. Bot. Gaz. 137:250–254.

Hough, T., P. Bernhardt, R.B. Knox, and E.G. Williams. 1985. Applications of fluorochromes to pollen biology. II. The DNA probes ethidium bromide and Hoechst 33258 in conjunction with the callose-specific aniline blue fluorochrome. Stain Tech. 60:155–162.

Jimenez, J.J., K. Schultz, A.L. Anaya, J. Hernandez, and O. Espejo. 1983. Allelopathic potential of corn pollen. J. Chem. Ecol. 9:1011–1025.

Kanchan, S., and Jayachandra. 1980. Pollen allelopathy—a new phenomenon. New Phytol. 84:739–746.

Karron, J.D., and D.L. Marshall. 1990. Fitness consequences of multiple paternity in wild radish, *Raphanus sativus*. Evolution 44:260–268.

Kay, Q.O.N. 1976. Preferential pollination of yellow-flowered morphs of *Raphanus raphanistrum* by *Pieris* and *Eristalis* sp. Nature (London) 261:230–232.

Levin, D.A. 1981. Dispersal versus gene flow in plants. Ann. Missouri Bot. Gard. 68:233–253.

Lyons, E.L., N.M. Waser, M.V. Price, J. Antonovics, and A.F. Motten. 1989. Sources of variation in plant reproductive success and implications for concepts of sexual selection. Am. Nat. 134:409–433.

Marshall, D.L. 1988. Post pollination effects on seed paternity: Mechanisms other than microgametophyte competition operate in wild radish. Evolution 42:1256–1266.

Marshall, D.L. 1990. Nonrandom mating in a wild radish, *Raphanus sativus*. Plant Species Biol. 5:143–156.

Marshall, D.L. 1991. Nonrandom mating in wild radish: Variation in pollen donor success and effects of multiple paternity among one- to six-donor pollinations. Am. J. Bot. 78:1404–1418.

Marshall, D.L., and N.C. Ellstrand. 1985. Proximal causes of multiple paternity in wild radish, *Raphanus sativus*. Am. Nat. 126:596–605.

Marshall, D.L., and N.C. Ellstrand. 1986. Sexual selection in *Raphanus sativus*: Experimental data on nonrandom fertilization, maternal choice, and consequences of multiple paternity. Am. Nat. 127:446–461.

Marshall, D.L., and N.C. Ellstrand. 1988. Effective mate choice in wild radish: Evidence for selective seed abortion and its mechanism. Am. Nat. 131:736–759.

Marshall, D.L., and N.C. Ellstrand. 1989. Regulation of mate number in fruits of wild radish. Am. Nat. 133:751–765.

Marshall, D.L., and M.W. Folsom. 1991. Mate choice in plants: an anatomical to population perspective. Annu. Rev. Ecol. Syst. 22:37–63.

Marshall, D.L., and D.M. Oliveras. 1990. Is regulation of mating within branches possible in wild radish (*Raphanus sativus*. L.)? Func. Ecol. 4:619–627.

Marshall, D.L., and K.L. Whittaker. 1989. Effects of pollen donor identity on offspring quality in wild radish, *Raphanus sativus*. Am. J. Bot. 76:1081–1088.

Mazer, S.J. 1987a. Maternal investment and male reproductive success in angiosperms: Parent-offspring conflict or sexual selection? Biol. J. Linn. Soc. 30:115–133.

Mazer, S.J. 1987b. Parental effects on seed development and seed yield in *Raphanus raphanistrum*: Implications for natural and sexual selection. Evolution 41:355–371.

Mazer, S.J., A.A. Snow, and M.L. Stanton. 1986. Fertilization dynamics and parental effects upon fruit development in *Raphanus raphanistrum*: Consequences for seed size variation. Am. J. Bot. 73:500–511.

Murphy, S.D., and L.W. Aarssen. 1989. Pollen allelopathy among sympatric grassland species: *In vitro* evidence in *Phleum pratense* L. New Phytol. 112:295–305.

Nakamura, R.R. 1988. Seed abortion and seed size variation within fruits of *Phaseolus vulgaris*: Pollen donor and resource limitation effects. Am. J. Bot. 75:1003–1010.

Nakamura, R.R., and M.L. Stanton. 1987. Cryptic seed abortion and the estimation of ovule fertilization. Can. J. Bot. 65:2463–2465.

Nettancourt, D. de. 1977. Incompatibility in Angiosperms. Springer-Verlag, Berlin.

Palser, B.F., J.L. Rouse, and E.G. Williams. 1989. Coordinated timetables for megagametophyte development and pollen tube growth in *Rhododendron nuttallii* from anthesis to early postfertilization. Am. J. Bot. 76:1167–1202.

Panetsos, C.P., and H.G. Baker. 1967. The origin of variation in "wild" *Raphanus sativus* (Cruciferae) in California. Genetica 38:243–274.

Pimienta, E., and V.S. Polito. 1982. Ovule abortion in 'nonpareil' almond (*Prunus dulcis* [Mill.] D.A. Webb). Am. J. Bot. 69:913–920.

Price, M.V., and N.M. Waser. 1982. Experimental studies of pollen carryover: Hummingbirds and *Ipomopsis aggregata*. Oecologia 54:353–358.

Pundir, N.S., R.F. Abbas, and A.A.A. Al-Attar. 1983. Pollen germination and pollen tube growth in *Raphanus sativus* L. following self- and cross-pollinations. Phyton 43:127–130.

Queller, D.C. 1983. Kin selection and conflict in seed maturation. J. Theor. Biol. 100:153–172.

Queller, D.C. 1987. Sexual selection in flowering plants. *In* J.W. Bradbury and M.B. Andersson, eds., Sexual Selection: Testing the Alternatives. John Wiley, New York, pp. 165–181.

Richards, A.J. 1986. Plant Breeding Systems. George Allen & Unwin, London.

Roggen, H. 1974. Pollen washing influences (in)compatibility in *Brassica oleracea* varieties. *In* H.F. Linskens, ed., Fertilization in Higher Plants. North Holland, Amsterdam, pp. 273–278.

Schaal, B.A. 1980. Measurement of gene flow in *Lupinus texensis*. Nature (London) 284:450–451.

Snow, A.A., and S.J. Mazer. 1988. Gametophytic selection in *Raphanus raphanistrum*: A test for heritable variation in pollen competitive ability. Evolution 42:1065–1075.

Stanton, M.L. 1984. Seed variation in wild radish: Effect of seed size on components of seedling and adult fitness. Ecology 65:1105–1112.

Stanton, M.L. 1987a. The reproductive biology of petal color variants in wild populations of *Raphanus sativus* L. I. Pollinator response to color morphs. Am. J. Bot. 74:178–187.

Stanton, M.L. 1987b. The reproductive biology of petal color variants in wild populations of *Raphanus sativus* L. II. Factors limiting seed production. Am. J. Bot. 74:188–196.

Stanton, M.L., and R.E. Preston. 1988. Ecological correlates of petal size variation in wild radish, *Raphanus sativus* (Brassicaceae). Am. J. Bot. 75:528–539.

Stephenson, A.G., and R.I. Bertin. 1983. Male competition, female choice, and sexual selection in plants. *In* L. Real, ed., Pollination Biology. Academic Press, Orlando, pp. 109–149.

Thomson, J.D. 1986. Pollen transport and deposition by bumble bees in *Erythronium*: Influences of floral nectar and bee grooming. J. Ecol. 74:329–341.

Thomson, J.D., and R.C. Plowright. 1980. Pollen carryover, nectar rewards, and pollinator behavior with special reference to *Diervilla lonicera*. Oecologia 46:68–74.

Thomson, J.D., B.J. Andrews, and R.C. Plowright. 1981. The effect of a foreign pollen on ovule development in *Diervilla lonicera* (Caprifoliaceae). New Phytol. 90:777–783.

Thomson, J.D., M.V. Price, N.M. Waser, and D.A. Stratton. 1986. Comparative studies of pollen and fluorescent dye transport by bumble bees visiting *Erythronium americanum*. Oecologia 55:251–257.

Tukey, H.B. 1933. Embryo abortions in the early ripening varieties of *Prunus avium*. Bot. Gaz. 94:433–468.

Uma Shaanker, R., K.N. Ganeshaiah, and K.S. Bawa. 1988. Parent-offspring conflict, sibling rivalry, and brood size patterns in plants. Annu. Rev. Ecol. Syst. 19:177–205.

Waser, N.M. 1993. Population structure, optimal outbreeding, and assortative mating in angiosperms. *In* N.W. Thornhill, ed., The Natural History of Inbreeding and Outbreeding: Theoretical and Empirical Perspectives. University of Chicago Press, Chicago (in press).

Waser, N.M. and M.V. Price. 1984. Experimental studies of pollen carryover: Effects of floral variability in *Ipomopsis aggregata*. Oecologia 62:262–268.

Waser, N.M. and M.V. Price. 1991. Outcrossing distance effects in *Delphinium nelsonii*: Pollen loads, pollen tubes, and seed set. Ecology 72:171–179.

Waser, N.M., M.V. Price, A.M. Montalvo, and R.N. Gray. 1987. Female mate choice in a perennial herbaceous wildflower, *Delphinium nelsonii*. Evol. Trends Plants 1:29–33.

Watson, M.A., and B.B. Casper. 1984. Morphogenetic constraints on patterns of carbon distribution in plants. Annu. Rev. Ecol. Syst. 15:233–258.

Westoby, M., and B. Rice. 1982. Evolution of the seed plants and inclusive fitness of plant tissues. Evolution 36:713–724.

Willson, M.F., and N. Burley. 1983. Mate Choice in Plants: Tactics, Mechanisms and Consequences. Princeton University Press, Princeton, NJ.

Young, H.J., and M.L. Stanton. 1990. Influence of environmental quality on pollen competitive ability in wild radish. Science 248:1631–1633.

6

Factors That Affect Pollen Performance

Andrew G. Stephenson, Tak-Cheung Lau,
Mauricio Quesada, and James A. Winsor
The Pennsylvania State University

The growth of pollen tubes in the pistils of higher plants is one of the most remarkable feats in the plant kingdom. For example, in the common zucchini (*Cucurbita pepo* L.) a pollen grain with a diameter of 140 μm will germinate and grow to a length of 10 cm (from the stigma to the lowermost ovules) in 24 hr at 30°C. If the same exponential growth were to persist for only 4 days, a pollen grain that landed on a stigma in central Pennsylvania would fertilize an ovule in San Francisco—and pass through Rome and Tokyo on its trek! Although such a statement is obviously ridiculous, it does serve to illustrate the phenomenal growth rates that are attained by microgametophytes.

Whereas we know little of the details that generate such phenomenal growth, we do know that mature pollen grains are packed (primed) with proteins and mRNAs, that these gene products are derived primarily from the genome of the microgametophyte, that the gene products in the microgametophyte range from the common "cell-housekeeping" genes found during both stages of the life cycle to unique products (5–15% of the total) that are expressed only by the microgametophyte (e.g., Tanksley et al., 1981; Hoekstra, 1983; Mascarenhas, 1989; Willing and Mascarenhas, 1984; Sari-Gorla et al., 1986; Pederson et al., 1987; Willing et al., 1988; Ottaviano and Mulcahy, 1989). We also know that mature pollen contains a variety of "storage compounds" (i.e., starches, lipids, and phytic acid) and that many of these compounds are rapidly metabolized on germination (Stanley and Linskens, 1974; Bertin, 1988). Finally, we know that there are complex biochemical interactions between the pollen and pistil that lead to recognition or rejection of the microgametophyte, involve the uptake of stylar resources necessary for pollen tube growth, and give direction to the growth of the pollen tube (Ottaviano and Mulcahy, 1989). In short, the sprint from the stigma to the ovules requires the microgametophyte to be adequately provisioned by the parent plant. It also requires expression of a large portion of the genome (20,000 to 23,000 genes) of the microgame-tophyte, proper functioning of myriad metabolic pathways within the mi-crogametophyte, and facilitation of pollen tube growth by the pistil.

It is therefore reasonable to expect that the performance of microga-metophytes (speed of germination, pollen tube growth rate, and ability to penetrate the ovule and achieve fertilization) can potentially be affected by their genetic composition, by their interactions with the pistil through which they grow, by environmental conditions, and by the resources pro-vided to them by their pollen-producing parent. Moreover, any genetic or environmental factor that affects pollen performance could result in non-random fertilization, particularly when more pollen is deposited onto a stigma than is necessary to fertilize all of the ovules. It should also be noted that, to the extent that pollen performance is influenced by genes expressed during both stages of the life cycle, nonrandom fertilization can affect progeny performance/vigor. Finally, nonrandom fertilization may have profound evolutionary implications because it directly affects the transmission of genes from one generation to the next and may indirectly affect the vigor of the resulting progeny.

Nevertheless, there are few direct experimental demonstrations that ge-netic and environmental factors can lead to differential success of micro-gametophytes in growth and fertilization. Such studies have been hampered by the inherent difficulties of studying small, morphologically indistin-guishable organisms that grow entirely within the tissues of another or-ganism. To circumvent these difficulties, researchers have attempted to determine differences in pollen tube growth and ability to effect fertilization by using pollen mixtures in which the paternity of the resulting mature seeds is determined from genetic markers. Interpretation of these studies, however, is often fraught with ambiguity because differences in the number of seeds sired could also be due to differences in pollen viability, postzygotic differences in seed viability, or inconsistent (unequal) mixtures of the two types of pollen.

In this Chapter, we briefly review the literature on genetic and environ-mental factors that affect pollen performance. In addition, we describe a series of experiments designed to determine which ovules are fertilized first within the ovary of zucchini. We then use the results from these experiments to develop an assay to quantify differences in pollen perform-ance that is free of the ambiguities inherent in previous studies. Finally, we employ the assay in a series of projects to determine if pollen genotype, soil fertility levels, and temperature affect pollen performance in C. pepo.

Effects of Pollen Genotype on Pollen Performance

There is evidence that speed of germination and pollen tube growth depend, in part, on the genotype of the microgametophyte. Various major mutations (chromosomal abnormalities and so-called "gametophyte factors") are known to affect pollen performance and to result in distorted segregation ratios in the offspring of maize and other crops (Pfahler, 1967; Grant, 1975; Ottaviano and Mulcahy, 1989). For several species, direct measurements of pollen growth have shown that the tubes of various lines/ genotypes differ in their growth either *in vitro* (e.g., Barnes and Cleveland, 1963a; Pfahler, 1970; Pfahler and Linskens, 1972; Dane and Melton, 1973; Sari-Gorla et al., 1976; Schemske and Fenster, 1983; Bookman, 1984; Mazer, 1987) or *in vivo* (e.g., Barnes and Cleveland, 1963b; Sari-Gorla et al., 1976; Gawel and Robacker, 1986; Elgersma et al., 1989). These studies, however, have not actually shown that fast pollen tubes are more likely to achieve fertilization. Other studies have revealed that microgametophytes growing *in vitro* express sensitivities and tolerances to selective agents similar to the sporophytes that produced them, such as tolerance to fungal toxins (Bino and Stephenson, 1988; Bino et al., 1988), copper and zinc (Searcy and Mulcahy, 1985a), salinity (Eisikowitch and Woodell, 1975; Sacher et al., 1983), herbicides (Smith and Moser, 1986), antibiotics (Bino et al., 1987), ozone (Feder, 1986), and acidity and trace elements (Cox, 1986). Finally, a few studies have altered the proportion of progeny resistant to an environmental challenge by applying the challenge to a mixed load of sensitive and resistant pollen (Zamir et al., 1981, 1982; Searcy and Mulcahy, 1985b; Zamir and Gadish, 1987; Sari-Gorla et al., 1989). For example, Sari-Gorla et al. (1989) crossed an herbicide-resistant line to an herbicide-sensitive line and deposited F_1 pollen onto the silks (stigmas) of maize that had been misted with the herbicide prior to pollination. Compared to controls, a larger proportion of resistant progeny were produced by experimental ears.

There is also evidence that pollen tube growth rates are influenced by the interaction of the pollen genotype with the pistil genotype. Gametophytic self-incompatibility is probably the most thoroughly studied pollen–pistil interaction (de Nettancourt, 1977), but more subtle genetically based pollen–pistil interactions lead to cryptic incompatibility in several species (e.g., Sayers and Murphy, 1966; Murakami et al., 1972; Levin, 1975; Weller and Ornduff, 1977, 1989; Ockendon and Currah, 1978; Eeninck, 1982; Bowman, 1987; Casper et al., 1988; Cruzan, 1989; Aizen et al., 1990). In these studies, self-pollen is capable of producing mature seeds, but cross-pollen outperforms self-pollen in mixtures. Many studies of cryptic incompatibility, however, have failed to control for differential survival of cross- and self-fertilized zygotes (Charlesworth, 1988; but see Levin, 1975; Bowman, 1987; Casper et al., 1988; Aizen et al., 1990). Even more subtle and less well-documented pollen–pistil interactions may influence the relative performance of pollen produced by different nonself donors. Direct measurements of pollen tube growth *in vivo* have revealed that both the identity (genotype) of the pollen donor and pollen–pistil interactions affect tube growth in cotton (Gawel and Robacker, 1986) and perennial rye grass (Elgersma et al., 1989). Pollen mixture studies have also shown that both the identity of the donor and pollen–pistil interactions influence the number of seeds sired in *Raphanus sativus* (Marshall and Ellstrand, 1986) and maize (Pfahler, 1967; Ottaviano et al., 1975; Sari-Gorla et al., 1976). None of these studies, however, has simultaneously excluded both differential pollen and zygote viability as possible explanations for the results.

Differences in pollen performance due solely to the genotype of the microgametophyte result in directional change. However, under conditions of pollen competition (more pollen tubes than ovules), heritable variation affecting microgametophyte performance should erode rapidly (Charlesworth et al., 1987; Snow and Mazer, 1988), with new variation arising by mutation or gene flow. In contrast, pollen–pistil interactions that cause variation in pollen performance could be maintained in the population because the performance of pollen would vary with the recipient plant (Stephenson and Bertin, 1983; Waser et al., 1987; Snow and Mazer, 1988).

In summary, direct experimental demonstrations that genetic differences among microgametophytes lead to differences in their ability to achieve fertilization have been limited, with only a few exceptions, to cases where the differences in performance are dramatic (e.g., major mutations, self-incompatibility, and a few cases of cryptic self-incompatibility). Consequently, we wanted to develop an assay to quantify more subtle effects of pollen genotype on fertilizaton. We also wanted to separate the effects of pollen performance from the effects of pollen and seed viability. It is possible to develop such an assay, we reasoned, if fast- and slow-growing pollen tubes fertilize ovules in different positions within an ovary.

Ovule Position, Pollen Genotype, and Nonrandom Fertilization

We used *C. pepo* (zucchini cultivars) for these studies because it offers many advantages for investigating microgametophyte performance: (1) It is monoecious. Stigmas are large and pollen production is high (ca. 30,000 grains/flower). Consequently, flowers are easy to manipulate and no emasculations are necessary. (2) Pollen is easy to manipulate because grains are massive (140 μm) and dry (not embedded in mucilage). (3) Because the stigma is up to 12 cm from the basal ovules and ovules are arranged linearly within the ovary, differences in pollen performance can be both manifested and detected experimentally. (4) Fruits are large (ca. 3 liters in volume), cylindrical, and contain ca. 300 seeds that store and germinate easily. Consequently, we can obtain large sample sizes from each cross. (5) We have developed a technique to transfer known quantities of pollen to the stigma rapidly and accurately (Winsor et al., 1987). Consequently, we can control the intensity of pollen competition, and pollen mixture experiments are more precise than with any other plant. (6) A variety of genetic markers are available, as are several genetically well-characterized cultivars.

To determine the order of fertilization within the ovules of zucchini, we gathered three types of data (Stephenson et al., 1988). First, we examined fruits produced under field conditions that had received either very large (>10,000 grains) or very small (240 ± 8.1 grains, mean ± standard error, $N = 20$) pollen loads (Winsor et al., 1987). The mature fruits were divided into three regions, each containing about one-third of the ovules (A = stylar end and C = peduncular end of the ovary). Seeds from large-pollen-load fruits were located in all three regions of the ovary, whereas the 30 or so seeds from the small-pollen-load fruits were located exclusively in the lower half of region A.

In the second experiment, we placed large pollen loads onto stigmas of plants growing in chambers, greenhouses, and the field. The gynoecia were harvested at 6, 12, 24, and 48 hr after pollination, softened in 8 N NaOH, and stained with aniline blue. Pollen tube growth was examined under a microscope with UV-fluorescent light (Martin, 1959). An ovule was considered to be fertilized if a pollen tube had entered the micropyle.

We found that the first ovules to be fertilized were located in the lower half of region A of the ovary. Ovules in the upper half of region A and those in region B were fertilized next, and the ovules in region C were fertilized last (Stephenson et al., 1988). Not surprisingly, we also found that pollen tube growth is influenced by temperature. For example, the first pollen tubes were found in the ovary 12 hr after pollination at 30°C but were not found in the ovary until 24 hr after pollination at 20°C. Together, these two experiments reveal that the order of fertilization differs by ovule position within the ovary of *C. pepo*.

To determine if the first ovules to be fertilized are indeed fertilized by the fastest growing pollen tubes, we performed a third experiment in which pistillate flowers from plants growing in the greenhouse at 20°C were given large pollen loads. After 24 hr (when, presumably, only the fastest growing tubes had entered the ovary), the style was excised. Later, the mature fruits were harvested and divided into regions A–C. A similar experiment was conducted in the field during the summer except that the timing of style excision was varied with ambient temperature.

Considering only those fruits that produced 40 or fewer seeds (i.e., those fruits in which the styles were excised after only a few pollen tubes had entered the ovary), seeds are located almost exclusively in the lower half of region A. When additional seeds are produced, they are located in the upper portion of region A and in region B of the ovary. These three experiments indicate that prezygotic differences in pollen performance sort pollen tubes into different ovule positions within the ovary of zucchini. Other species with linearly arranged ovules are also known to have a pattern of fertilization similar to that of zucchini. For example, in both *Raphanus raphanistrum* and *Phaseolus coccineus*, the first ovules to be fertilized are those that are close to, but not at, the stylar end of the ovary (Hill and Lord, 1986; Rocha and Stephenson, 1991).

Because pollen tubes with different growth rates fertilize ovules in different positions within the ovary of zucchini, it is possible to separate the effects of pollen performance from the effects of pollen viability, seed viability, and unequal mixtures on the ability to sire seeds in pollen mixture studies. Differences in pollen or seed viability or in the number of deposited pollen grains will result in different proportions of ovules sired by the two types of donors in the mixture. Nevertheless, differences in pollen tube growth rates will not only result in differences in the total proportion of seeds sired but also in the proportion of seeds sired within each region of the ovary. For example, differences in pollen viability will lead to differences in the number of seeds sired by the two donors in a mixture experiment, but the percentage of seeds sired in each region of the ovary will be similar unless the pollen from the two donors also differ in growth rate.

To determine directly if the genotype of the microgametophyte influences pollen performance and results in nonrandom fertilization, we performed a pollen mixture experiment using the Black Beauty (green ovaries) and Golden Burpee (yellow ovaries) cultivars of zucchini. Ovary color is controlled by a single locus with two alleles: yellow is dominant to green (Robinson et al., 1976). Both are inbred cultivars, and both are true-breeding for ovary color. We placed equal amounts of pollen from each cultivar (three pollen loads of 462 ± 48 grains from each of the two cultivars) onto the stigmas of Black Beauty recipient plants. The amount of pollen from each cultivar was sufficient to give a nearly complete seed-set, whereas the combined pollen load was more than sufficient to produce a fruit with a full complement of seeds. Because both cultivars are inbred and they differ genetically, genetic variation among pollen grains from each cultivar should be less than the variation between cultivars. Mature fruits were harvested, and seeds from regions A–C of the ovary were removed. We planted a sample of 40 seeds (where possible) from each region of each of 33 fruits and scored them for ovary color (Quesada et al., 1991).

We found that the cultivars differ in their ability to sire seeds: Black Beauty sired 53% of all seeds that were scored [$G(1) = 5.53$; $p < 0.05$]. A G-test for heterogeneity revealed that the progeny from regions A and C of the ovary deviate significantly from the overall proportion (53:47) of the scored progeny. In region A 57% of the progeny were sired by Golden

Burpee, whereas 61% of the progeny from region C were sired by Black Beauty.

Thus, the overall surplus of seeds sired by Black Beauty is not due to superior pollen performance. In fact, pollen from the Golden Burpee cultivar apparently achieved fertilization, germinated, and/or grew faster than that of the Black Beauty. The Golden Burpee pollen was overrepresented among progeny sired in the region of that ovary where seeds are sired by the fastest growing pollen tubes, whereas the Black Beauty pollen was overrepresented in that region of the ovary where seeds are sired by slower growing tubes. In short, genetic differences among the microgametophytes in this study appear to influence pollen performance and result in nonrandom fertilization by ovule position.

The experimental design does not allow us to determine if differences in pollen performance are due solely to differences in the genotype(s) of the pollen from each cultivar, or if they are due to pollen–pistil interactions. It is possible that the green cultivar would have sired the most seeds in region A if the Golden Burpee cultivar or a third cultivar had been used as the recipient parent.

We currently have a series of experiments, in various stages of completion, that use the "ovary region assay" (1) to differentiate between the effects of pollen genotype and pollen–pistil interactions on pollen performance, (2) to determine how pollen performance varies with heterozygosity in the pollen-producing parent, and (3) to demonstrate that pollen genotype and pollen–pistil interactions affect pollen performance in natural populations of *C*. texana.

Effects of the Environment on Pollen Performance

Environmental conditions during pollen development (e.g., temperature and soil fertility) are known to affect the chemical composition of pollen (Stanley and Linskens, 1974; Baker and Baker, 1979; Herpen and Linskens, 1981; Herpen, 1986; Bertin, 1988). For example, phytic acid, which is a reserve for the phosphorus and myoinositol that are needed in large quantities for cell wall and membrane synthesis, has been identified in pollen from many plant species (Jackson et al., 1982). In *Petunia hybrida* the concentration of phytic acid is known to vary with the temperature at which the pollen parent is grown (Herpen, 1986). Because of the rapid growth of pollen tubes, there is a high demand for cell wall and membrane precursors (Helsper et al., 1984), and phytate is known to be rapidly metabolized into precursors during pollen tube growth *in vitro* (Jackson and Linskens, 1982; Dickenson and Lin, 1986). Although there has been no direct demonstration of a relationship between phytate concentration and pollen performance, it is known that significant quantities of phytate are found in pollen from species with long styles. Conversely, little or no phytate is found in species with very short styles, such as some grasses and composites (Helsper et al., 1984).

Pollen grain size is also known to vary within and among plants in a population (e.g., Willson and Burley, 1983; Stanton and Preston, 1986; McKone and Webb, 1988), and some of this variation appears to be related to environmental conditions affecting the pollen-producing parent (Stanley and Linskens, 1974). Moreover, studies of distylous species (Ganders, 1979) and species producing both cleistogamous and chasmogamous flowers (Lord and Eckard, 1984) reveal a correlation between pollen grain diameter and style length. Barnes and Cleveland (1963a) found that the pollen of the three lines of alfalfa exhibiting the fastest pollen tube growth *in vitro* was larger than the pollen from the three slowest growing lines. It is conceivable that conditions affecting a pollen-producing plant could have an effect on pollen composition and pollen grain size, which, in turn, could affect pollen performance and fertilization.

To investigate the relationship between soil nutrient conditions and pollen performance, we initiated a pilot project in 1985. In this study, we found that the pollen from Black Beauty plants grown under low nutrient soil conditions and plants grown under higher nutrient conditions sired significantly different numbers of seeds when low pollen loads were applied to alternate pistillate flowers on recipient plants. Moreover, the differences in the number of seeds sired were not due to differences in pollen viability, as determined by viability stains. Recently, Young and Stanton (1990) showed that pollen from *Raphanus raphanistrum* plants grown under low nutrient conditions in a greenhouse sired fewer seeds than pollen from plants grown under high nutrient conditions when mixtures were applied to stigmas.

In 1989, we initiated a set of experiments using the Black Beauty and Golden Burpee cultivars. Twelve seeds from each of two families of each cultivar were sown into a field that soil analyses had shown to be very deficient in nitrogen but that had been fertilized to commercial production levels of all other nutrients. By adding slow release nitrogen fertilizer (40–0–0) to the soil around each plant, we created two nitrogen treatments, each consisting of six plants from each family of each cultivar. Flower and fruit production were recorded daily. Every Monday all staminate flowers (flowers last only one morning) were harvested, and pollen production per flower and pollen grain size were determined using an electronic particle counter.

We found that the N treatment had a significant effect on all measures of reproductive output: flowers, fruits, seeds, and pollen grains per flower. We also found that pollen from plants in the high-N treatment was significantly ($p < 0.05$) larger than that produced by plants in the low-N treatment, perhaps indicating differences in the chemical composition of the pollen (mixed effect repeated measure ANOVA; other sources of variance include block, cultivar, family nested within cultivar, and interactions: Fig. 6.1).

To examine the performance of pollen produced under the two N treatments, pollen mixture experiments were performed on Black Beauty recipient families growing in a nearby field. An equal amount of pollen from a high-N Golden Burpee plant and low-N Black Beauty plant was placed on the stigma of a Black Beauty recipient plant. On the same day, an equal amount of pollen from a low-N Golden Burpee plant (of the same family used in the first cross) and a high-N Black Beauty plant (same family as the first cross) was placed on the stigma of a second recipient plant (same family as the first cross). The mature fruits were harvested, and the seeds from regions A–C of the ovary were removed and stored. In the summer of 1990, 25 seeds from each ovary region from each pair of fruits (yellow high–green low-N and green high–yellow low-N) were planted and ovary

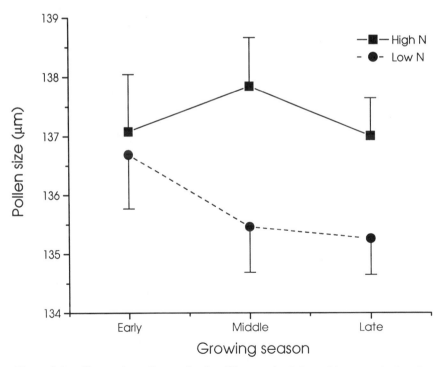

Figure 6.1. Change in pollen grain size (diameter) of *Cucurbita pepo* during the growing season on plants growing in high and low soil nitrogen conditions.

color was scored. Log-linear analyses revealed that the high-N treatment sired more seeds (54%) than expected by chance alone [$X^2(1) = 10.2$, $p < 0.001$]. Moreover, analyses of the proportion of progeny sired in the three regions of the ovary indicate that the N treatment affects pollen performance [$G(2) = 10.06$; $p < 0.01$] (Fig. 6.2). The pollen from the high-N treatment sired 59% of the seeds in region A (where the ovules are fertilized by the fastest growing tubes) and only 50% of the seeds in region C. This study clearly demonstrates a relationship between the availability of a single soil nutrient, mean pollen grain size, and pollen performance. We do not know, however, if the pollen grains from the two treatments differed in their chemical composition.

Consequently, in the summer of 1990 we began a series of experiments in which we varied phosphorus availability to Black Beauty and Golden Burpee plants. This study was similar in design to the N study and, at this

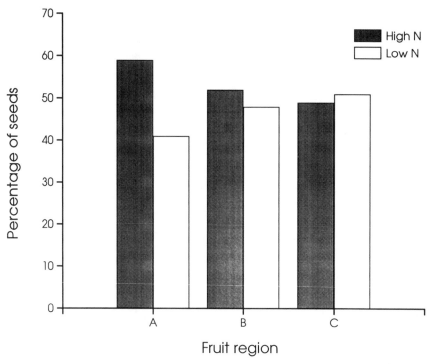

Figure 6.2. The percentage of seeds of *Cucurbita pepo* sired by pollen from plants growing in high and low soil nitrogen in each of three regions of the fruit (A = stylar end and C = peduncular end of the fruit) when equal mixtures of pollen were applied to sigmas.

time, we have not yet scored the progeny from the pollen mixture experiment. Preliminary analyses indicate, however, that total P and concentration of phytate in pollen vary with phosphorus availability.

In natural habitats soil fertility is known to vary from one microsite to another. Herbivory also could affect a plant's ability to provision pollen with resources. Therefore, it is possible that variation in pollen performance, due to variations in soil fertility and other factors that affect a plant's nutrient availability, are common in natural populations of plants. These variations, in turn, could affect male reproductive success and gene flow in natural populations.

Environmental conditions at the time of pollen germination are also known to affect the speed of germination and pollen tube growth rate both *in vitro* and *in vivo* (Gawel and Robacker, 1986; Herpen, 1986; Schlichting, 1986; Bertin, 1988; Elgersma et al., 1989). For example, pollen tube growth in cotton increased with temperature but was fastest at intermediate levels of relative humidity (Gawel and Robacker, 1986). Environmental conditions can also affect the strength of pollen–pistil interactions (Willson and Burley, 1983; Bertin, 1988). For example, in *Lolium perenne* the strength and timing of the self-incompatibility reaction vary with temperature (Elgersma et al., 1989).

Although many studies have demonstrated that an environmental condition, such as temperature, can affect various aspects of pollen performance, few studies have separated the effects of the environment during pollen development from those during pollen germination and growth. Consequently, in 1989 we initiated a pilot study designed to examine the effects of temperature, during both pollen development and pollen germination, on pollen performance. F_1 Black Beauty × Golden Burpee plants were grown in either warm (22°C—10 hr, 30°C—14 hr) or cool (14°C—10 hr, 22°C—14 hr) growth chambers. To examine the response of pollen to different germination and development temperatures, pollen samples from both growth chambers were germinated *in vitro* at either 30 or 22°C, producing four treatments: cool developed/cool germinated, cool/warm, warm/cool, and warm/warm. Pollen viability was estimated by staining samples of pollen with tetrazolium.

Analyses of variance revealed that the temperature during development had no effect on pollen viability, but both temperature during development ($p < 0.001$) and during germination ($p < 0.0001$) had significant effects on our measure of performance, percentage germination. Means were as follows: cool/cool, 11.1 ± 1.4; cool/warm, 18.9 ± 1.5; warm/cool, 17.4 ± 1.4; and warm/warm, 45.2 ± 1.4. Moreover, we detected a highly significant developmental by germination temperature interaction ($p < 0.001$), suggesting that the conditions under which pollen develops affects its response to conditions at the time of germination. We have initiated pollen mixture experiments to determine the effects of development and germination temperatures on pollen performance *in vivo*.

Random Factors That Influence Pollen Performance

Finally, pollen performance can be influenced by a variety of factors associated with (1) precise location of deposition on the stigma, (2) location of the pollen grain within a clump on a stigma, (3) composition of the pollen load (e.g., proportion of self- and cross-pollen), (4) age of pollen at the time of transfer, (5) density of the deposited pollen grains, and (6) differences in time of arrival of pollen loads to a stigma (e.g., Ockendon and Currah, 1978; Mulcahy, 1979; Cruzan, 1986; Epperson and Clegg, 1987; Knox et al., 1987; Ottaviano and Mulcahy, 1989; Thomson, 1989; Shivanna et al., 1991). The effects of these factors on pollen performance are unrelated to the intrinsic properties of a pollen grain, such as its genotype or its chemical composition, but instead are due to chance effects associated with pollen removal and deposition. For example, in *Erythronium grandiflorum*, pollen that lands in the fluid-filled crevices on the stigma germinates faster than pollen that lands among the papillae. Moreover, when pollen grains are deposited in clumps, those grains in direct contact with the stigma germinate faster than those that are not in direct contact (Thomson, 1989).

These random factors can obscure differences in pollen performance associated with resource provisioning by the paternal parent, pollen genotype, and genetically based pollen–pistil interactions. Such factors may, in fact, be at least partially responsible for fertilizations by pollen from low-N plants in region A of the ovary of zucchini (e.g., some pollen from low-N plants may germinate and begin to grow faster than pollen from high-N plants by virtue of being deposited in a favorable location on the stigma). Because of these random factors, studies of the genetic and environmental factors that influence pollen performance require a high degree of control over the system that is experimentally manipulated in order to detect differences in pollen performance.

A little more than a decade ago, Heslop-Harrison (1979) waggishly referred to the gametophyte as the "forgotten generation." Despite technical differences in assessing pollen performance and despite difficulties in separating differences in pollen performance from differences in pollen and seed viability, this decade has been marked by vigorous research, clarification of the terminology and issues, and healthy debate. Nevertheless, there remain serious gaps in our understanding of the factors that affect pollen performance, their potential to influence the transmission of genes in natural populations, and their effects on progeny performance. Although the obstacles are many, theoreticians, empiricists, and critics agree that the issues under investigation are genuinely important in both evolutionary biology and applied plant sciences.

Acknowledgments

We thank R. Oberheim, his staff, and the Department of Horticulture for use of the Agricultural Experiment Station; J. Dippery, L. Hartzel, R. Naugle, T. O'Malley, D. Phayre, and M. Schlank for help in the field and lab; and M. Johannsson, J. Madsen, M. Nordborg, C. Schlichting, and M. Uyenoyama for comments on an earlier draft. This work was supported by an H. D. Hill and J. B. Hill Fellowship to T-C Lau, a Fulbright-Hays Fellowship to M. Queseda, and NSF Grants BSR-8315612 A01 and BSR-8818184 to A. G. Stephenson and J. A. Winsor.

Literature Cited

Aizen, M.A., K.B. Searcy, and D.L. Mulcahy. 1990. Among- and within-flower comparisons of pollen tube growth following self- and cross-pollinations in *Dianthus chinensis* (Caryophyllaceae). Am. J. Bot. 77:671–676.

Baker, H.G., and I. Baker. 1979. Starch in angiosperm pollen grains and its evolutionary significance. Am. J. Bot. 66:591–600.

Barnes, D.K., and R.W. Cleveland. 1963a. Pollen tube growth of diploid alfalfa *in vitro*. Crop Sci. 3:291–295.

Barnes, D.K., and R.W. Cleveland. 1963b. Genetic evidence for non-random fertilization in alfalfa as influenced by differential pollen tube growth. Crop Sci. 3:295–297.

Bertin, R.I. 1988. Paternity in plants. *In* J. Lovett Doust and L. Lovett Doust, eds., Plant Reproductive Ecology. Oxford University Press, New York, pp. 30–59.

Bino, R.J., and A.G. Stephenson. 1988. Selection and manipulation of pollen and sperm cells. *In* H.J. Wilms and C.J. Keijzer, eds. Plant Sperm Cells as Tools for Biotechnology. Pudoc Press, Wageningen, Netherlands, pp. 125–136.

Bino, R.J., J. Hille, and J. Franken. 1987. Kanamycin resistance during *in vitro* development of pollen from transgenic tomato plants. Plant Cell Rep. 6:333–336.

Bino, R.J., J. Franken, H.M.A. Witsenboer, J. Hille, and J.J.M. Dons. 1988. Effects of *Alternaria alternaria* f. sp. *lycopersici* toxins on pollen. Theor. Appl. Genet. 76:204–208.

Bookman, S.S. 1984. Evidence for selective fruit production in *Asclepias*. Evolution 38:72–86.

Bowman, R.N. 1987. Cryptic self-incompatibility and the breeding system of *Clarkia unguiculata* (Onagraceae). Am. J. Bot. 74:471–476.

Casper, B.B., L.S. Sayigh, and S.S. Lee. 1988. Demonstration of cryptic incompatibility in distylous *Amsinckia douglasiana*. Evolution 42:248–253.

Charlesworth, D. 1988. A comment on the evidence for pollen competition in plants and its relationship to progeny fitness. Am. Nat. 132:298–302.

Charlesworth, D., D.W. Schemske, and V.L. Sork. 1987. The evolution of plant reproductive characters. Sexual vs. natural selection. *In* S.C. Stearns, ed., The Evolution of Sex and Its Consequences. Birkhauser Verlag, Basel, Switzerland, pp. 317–335.

Cox, R.M. 1986. *In vitro* and *in vivo* effects of acidity and trace elements on pollen function. *In* D.L. Mulcahy, G.B. Mulcahy, and E.M. Ottaviano, eds., Biotechnology and Ecology of Pollen. Springer-Verlag, New York, pp. 95–100.

Cruzan, M.B. 1986. Pollen tube distributions in *Nicotiana glauca*: Evidence for density dependent growth. Am. J. Bot. 73:902–907.

Cruzan, M.B. 1989. Pollen tube attrition in *Erythronium grandiflorum*. Am. J. Bot. 76:562–570.

Dane, F., and B. Melton. 1973. Effect of temperature on self- and cross-compatibility and *in-vitro* pollen growth characteristics in alfalfa. Crop. Sci. 13:587–591.

de Nettancourt, D. 1977. Incompatibility in Angiosperms. Springer-Verlag, Berlin.

Dickenson, D.B., and J.J. Lin. 1986. Phytases of germinating Lily pollen. *In* D.L. Mulcahy, G.B. Mulcahy, and E. Ottaviano, eds., Biotechnology and Ecology of Pollen. Springer-Verlag, New York, pp. 357–362.

Eeninck, A.H. 1982. Compatibility and incompatibility in witloof-chicory (*Cichorium intybus* L.). 3. Gametic competition after mixed pollinations and double pollinations. Euphytica 31:773–786.

Eisikowitch, D., and S.R.J. Woodell. 1975. Some aspects of pollination ecology of *Armeria maritima* (Mill.) Willd. in Britain. New Phytol. 74:307–22.

Elgersma, A., A.G. Stephenson, and A.P.M. den Nijs. 1989. Effects of genotype and temperature on pollen tube growth in perennial ryegrass (*Lolium perenne* L.) Sex. Plant Reprod. 2:225–230.

Epperson, B.K., and M.T. Clegg. 1987. First-pollination primacy and pollen selection in the morning glory *Ipomoea purpurea*. Heredity 58:5–14.

Feder, W.A. 1986. Predicting species responses to ozone using a pollen screen. *In* D.L. Mulcahy, G.B. Mulcahy, and E.M. Ottaviano, eds., Biotechnology and Ecology of Protein. Springer-Verlag, New York, pp. 89–94.

Ganders, F.R. 1979. The biology of heterostyly. N.Z. J. Bot. 17:607–635.

Gawel, N.J., and C.D. Robacker. 1986. Effects of pollen-style interaction on the pollen tube growth of *Gossypium hirsutum*. Theor. Appl. Genet. 72:84–87.

Grant, V. 1975. Genetics of Flowering Plants. Columbia University Press, New York.

Helsper, J.F.G., H.F. Linskens, and J.F. Jackson. 1984. Phytate metabolism in petunia pollen. Phytochemistry 23: 1841–1845.

Herpen, M.M.A. van. 1986. Biochemical alterations in the sexual partners resulting from environmental conditions before pollination regulate processes after pollination. *In* D.L. Mulcahy, G.B. Mulcahy, and E.M. Ottaviano, eds., Biotechnology and Ecology of Pollen. Springer-Verlag, New York, pp. 131–133.

Herpen, M.M.A. van, and H.F. Linskens. 1981. Effect of season, plant age, and temperature during plant growth on compatible and incompatible pollen tube growth in *Petunia hybrida*. Acta Bot. Neerl. 30:209–218.

Heslop-Harrison, J. 1979. The forgotten generation: Some thoughts on the genetics and physiology of angiosperm gametophytes. *In* R. Davies and D.A. Hopwood, eds., Proceedings of the Fourth John Innes Symposium. John Innes Inst., Norwich, England, pp. 1–14.

Hill, J., and E.M. Lord. 1986. Dynamics of pollen tube growth in the wild radish *Raphanus raphanistrum* (Brassicaceae). I. Order of fertilization. Evolution 40:1328–1334.

Hoekstra, F.A. 1983. Physiological evolution in angiosperm pollen: Possible role of pollen vigor. *In* D.L. Mulcahy and E. Ottaviano, eds., Pollen: Biology and Implications for Plant Breeding. Elsevier, New York, pp. 35–42.

Jackson J.F., and H.F. Linskens. 1982. Phytic acid in *Petunia hybrida* pollen is hydrolyzed during germination by a phytase. Acta Bot. Neerl. 31:441–447.

Jackson, J.F., G. Jones, and H.F. Linskens. 1982. Phytic acid in pollen. Phytochemistry 21:1255–1258.

Knox, R.B., M. Gaget, and C. Dumas. 1987. Mentor pollen techniques. Int. Rev. Cytol. 107:315–332.

Levin, D.A. 1975. Gametophytic selection in *Phlox*. *In* D.L. Mulcahy, ed. Gamete Competition in Plants and Animals. North-Holland, Amsterdam, pp. 207–217.

Lord, E.M., and K.J. Eckard. 1984. Incompatibility between the dimorphic flowers of *Collomia grandiflora*, a cleistogamous species. Science 223:695–696.

Marshall, D.L., and N.C. Ellstrand. 1986. Sexual selection in *Raphanus sativus*: Experimental data on nonrandom fertilization, maternal choice, and consequences of multiple paternity. Am. Nat. 127:446–461.

Martin, F.W. 1959. Staining and observing pollen tubes in the style by means of fluorescence. Stain Technol. 34:125–128.

Mascarenhas, J.P. 1989. The male gametophyte of flowering plants. Plant Cell 1:657–664.

Mazer, S.J. 1987. Parental effects on seed development and seed yield in *Raphanus raphanistrum*: Implications for natural and sexual selection. Evolution 41:355–371.

McKone, M.J., and C.J. Webb. 1988. A difference in pollen size between the male and hermaphrodite flowers of two species of Apiaceae. Aust. J. Bot. 36:331–337.

Mulcahy, D.L. 1979. The rise of the angiosperms: A genecological factor. Science 206:20–23.

Murakami, K., M. Yamada, and K. Takayanagi. 1972. Selective fertilization in maize *Zea mays* L. I. Advantage of pollen from F_1 plants in selective fertilization. Jpn. J. Breeding 22:203–208.

Ockendon, E.M., and L. Currah. 1978. Time of cross- and self-pollination affects the amount of self-seed set by partially self-incompatible plants of *Brassica oleracea*. Theor. Appl. Genet. 52:223–237.

Ottaviano, E., and D.L. Mulcahy. 1989. Genetics of angiosperm pollen. Adv. Genet. 26:1–64.

Ottaviano, E.M., M. Sari-Gorla, and D.L. Mulcahy. 1975. Genetic and intergametophytic influences on pollen tube growth. *In* D.L. Mulcahy, ed., Gamete Competition in Plants and Animals. North-Holland, Amsterdam, pp. 125–134.

Pederson, S., V. Simonsen, and V. Loeschere. 1987. Overlap of gametaphytic and sporophytic gene expression in barley. Theor. Appl. Genet. 75:200–206.

Pfahler, P.L. 1967. Fertilization ability of maize pollen grains. II. Pollen genotype, female sporophyte and pollen storage interactions. Genetics 57:513–521.

Pfahler, P.L. 1970. In vitro germination and pollen tube growth of maize (*Zea mays*) pollen. III. The effect of pollen genotype and pollen source vigor. Can. J. Bot. 48: 111–115.

Pfahler, P.L., and H.F. Linskens. 1972. In vitro germination and pollen tube growth of maize (*Zea mays*) pollen. VI. Combined effects of storage and the alleles at the waxy (wx), sugary (su_1), and shrunken (sh_2) loci. Theor. Appl. Genet. 42:136–140.

Quesada, M., C.D. Schlichting, J.A. Winsor, and A.G. Stephenson. 1991. Effects of genotype on pollen performance in *Cucurbita pepo*. Sex. Plant Reprod. 4:208–214.

Robinson, R.W., H.M. Munger, T.W. Whitaker, and G.W. Bohn. 1976. Genes of the Cucurbitaceae. Hort. Sci. 11:554–568.

Rocha, O.J., and A.G. Stephenson. 1991. Order of fertilization within the ovary of *Phaseolus coccineus* L. (Leguminosae). Sex. Plant Reprod. 4:126–131.

Sacher, R.F., D.L. Mulcahy, and R.C. Staples. 1983. Developmental selection during self-pollination of *Lycopersicon* × *Solanum* F_1 for salt tolerance of F_2. *In* D.L. Mulcahy and E.M. Ottaviano, eds., Pollen: Biology and Implications for Plant Breeding. Elsevier, New York, pp. 329–334.

Sari-Gorla, M., E.M. Ottaviano, and D. Faini. 1976. Genetic variability of gametophytic growth rate in maize. Theor. Appl. Genet. 46:289–294.

Sari-Gorla, M., C. Frova, G. Binella, and E. Ottaviano. 1986. The extent of gametophytic-sporophytic gene expression in maize. Theor. Appl. Genet. 72:42–47.

Sari-Gorla, M., E.M. Ottaviano, E. Frascaroli, and P. Landi. 1989. Herbicide-tolerant corn by pollen selection. Sex. Plant Reprod. 2:65–69.

Sayers, E.R., and R.P. Murphy. 1966. Seed set in alfalfa as related to pollen tube growth, fertilization frequency, and post fertilization ovule abortion. Crop Sci. 6:365–368.

Schemske, D.W., and C. Fenster. 1983. Pollen-grain interactions in a neotropical *Costus*: Effects of clump size and competitors. *In* D.L. Mulcahy and E. Ottaviano, eds., Pollen: Biology and Implications for Plant Breeding. Elsevier, New York, pp. 405–410.

Schlichting, C.D. 1986. Environmental stress reduces pollen quality in *Phlox*: Compounding the fitness deficit. *In*: D.L. Mulcahy, G.B. Mulcahy, and E.M. Ottaviano, eds., Biotechnology and Ecology of Pollen. Springer-Verlag, New York, pp. 483–488.

Searcy, K.B., and D.L. Mulcahy. 1985a. The parallel expression of metal tolerance in pollen and sporophytes in *Silene dioica* (L.) Clairv., *S. alba* (Mill.) Krause and *Mimulus guttatus*. Theor. Appl. Genet. 69:597–602.

Searcy, K.B., and D.L. Mulcahy. 1985b. Selection for heavy metal tolerance during pollen development among pollen grains from a single individual. Am. J. Bot. 72:1700–1706.

Shivanna, K.R., H.F. Linskens, and M. Cresti. 1991. Pollen viability and pollen vigor. Theor. Appl. Genet. 81:38–42.

Smith, G.A., and H.S. Moser. 1986. Sporophytic-gametophytic herbicide tolerance in sugarbeet. Theor. Appl. Genet. 71:231–37.

Snow, A.A., and S.J. Mazer. 1988. Gametophytic selection in *Raphanus raphanistrum*: A test for heritable variation in pollen competitive ability. Evolution 42:1065–1075.

Stanley, R.G., and H.F. Linskens. 1974. Pollen: Biology, Biochemistry, Management. Springer-Verlag, New York.

Stanton, M.L., and R.E. Preston. 1986. Pollen allocation in wild radish: Variation in pollen grain size and number. *In* D.L. Mulcahy, G.B. Mulcahy, and E.M. Ottaviano, eds., Biotechnology and Ecology of Pollen. Springer-Verlag, New York, pp. 461–466.

Stephenson, A.G., and R.I. Bertin. 1983. Sexual selection in plants. *In* L. Real, ed., Pollination Biology. Academic Press, New York, pp. 109–150.

Stephenson, A.G., J.A. Winsor, and C.D. Schlichting. 1988. Evidence for nonrandom fertilization in the common zucchini, *Cucurbita pepo*. *In* M. Cresti, P. Gori, and E. Pacini, eds., Sexual Reproduction in Higher Plants. Springer-Verlag, Berlin, pp. 333–338.

Tanksley, S., D. Zamir, and C.M. Rick. 1981. Evidence for extensive overlap of sporophytic and gametophytic gene expression in *Lycopersicon esculentum*. Science 213: 453–455.

Thomson, J.D. 1989. Germination schedules of pollen grains: Implications for pollen selection. Evolution 43:220–223.

Waser, N.M., M.V. Price, A.M. Montalvo, and R.N. Gray. 1987. Female mate choice in a perennial herbaceous wildflower, *Delphinium nelsonii*. Evol. Trends Plants 1:29–33.

Weller, S.G., and R. Ornduff. 1977. Cryptic self-incompatibility in *Amsinckia grandiflora*. Evolution 31:47–51.

Weller, S.G., and R. Ornduff. 1989. Incompatibility in *Amsinckia grandiflora* (Boraginaceae): Distribution of callose plugs and pollen tubes following inter- and intramorph crosses. Am. J. Bot. 76:277–282.

Willing, R.P., and J.P. Mascarenhas. 1984. Analysis of the complexity and diversity of mRNAs from pollen and shoots of *Tradescantia*. Plant Physiol. 75:865–868.

Willing, R.P., D. Bashe, and J.P. Mascarenhas. 1988. An analysis of the quantity and diversity of messenger RNAs from pollen and shoots of *Zea mays*. Theor. Appl. Genet. 75:751–753.

Willson, M.F., and N. Burley. 1983. Mate Choice in Plants: Tactics, Mechanisms and Consequences. Princeton University Press, Princeton.

Winsor, J.A., L.E. Davis, and A.G. Stephenson. 1987. The relationship between pollen load and fruit maturation and the effect of pollen load on offspring vigor in *Cucurbita pepo*. Am. Nat. 129:643–656.

Young, H.J., and M.L. Stanton. 1990. Influence of environmental quality on pollen competitive ability in wild radish. Science 248:1631–1633.

Zamir, D., and I. Gadish. 1987. Pollen selection for low temperature adaptation in tomato. Theor. Appl. Genet. 74:545–548.

Zamir, D., S.D. Tanksley, and R.A. Jones. 1981. Low temperature effect on selective fertilization by pollen mixtures of wild and cultivated tomato species. Theor. Appl. Genet. 59:235–238.

Zamin, D., S.D. Tanksley, and R.A. Jones. 1982. Haploid selection for low temperature tolerance of tomato pollen. Genetics 101:129–132.

7

Evolutionarily Stable Strategies of Reproduction in Plants: Who Benefits and How?

David G. Lloyd
University of Canterbury

Introduction

We are in the midst of a revolution in studies of plant reproduction. The rise of evolutionary ecology has brought an emphasis on evolutionarily stable strategies (ESSs) for the deployment of pollination and dispersal mechanisms. In both floral biology and seed biology, new concepts such as Bateman's Principle, kin selection, and sib competition are now regularly employed to explain the selection of adaptive strategies. They are being applied to novel or revitalized topics, including strategies for paternity, phenology, gender, reproductive allocations, seed number, and many other aspects of reproduction.

It is imperative that the selection hypotheses that are put forward to explain adaptive strategies rest on a sound theoretical basis. Hence on occasion we must examine the theoretical justification for the practices involved in modeling the process of natural selection. During the past 30 years, theorists have confronted a number of major problems concerning the operation of natural selection. Here, in the context of the reproductive ecology of plants, I examine two of these concerns: the subject of selection (whose fitness is maximized?) and the way fitness is attributed to individuals (whether by lineal transmission to direct descendants as in conventional selection models or by inclusive fitness as in kin selection models).

The first topic, choice of the appropriate subject, or subjects, of selection is particularly troublesome for characters involved in the reproduction of plants. Parent and offspring generations, male and female sexes, and haploid and diploid stages of the life cycle are all involved. Genes may be expressed in either alternative of each dichotomy, so that any of the possible combinations may function as the subject of a selection process. Potential conflicts between the sexes (Trivers, 1972), generations (Trivers, 1974), and life history stages (Beach and Kress, 1980) are all considered below. Note that we are dealing here with alternative individuals as *subjects* (units) of selection, not with the separate issue of whether selection occurs basically at the *level* of genes or individuals.

The second aspect of selection models that is considered here, lineal versus inclusive fitness, concerns selection of the "social" acts of plants or animals (i.e., those behaviors that directly affect the fitnesses of conspecific individuals other than the "actors" performing a behavior). The two phenomena that are discussed below, parental investment in offspring and reproductively disabled pollen and seeds (those with intrinsically reduced or zero fitness), are both social acts by this definition. Until recently these phenomena were considered, if they were considered at all, by models of the lineal fitness of individuals or genes, measured as the number of offspring matured. Following the introduction of the concept of kin selection (Hamilton, 1963, 1964), an alternative assignment of the fitness effects of social acts is now frequently used. Hamilton's rule, the usual formulation of kin selection, weights effects on the fitness of participants in social acts by the relatednesses, r_{ai}, of the actor, a, to the various participants, i ($r_{ai} = 1$ in the case of effects on the actor itself). Thus the effects of a social act are redistributed according to the genetic relatednesses of the actor to the individuals involved. The kin selection formulation often allows the analysis of social acts to be greatly simplified, and it has provided exciting new interpretations of interactions between relatives in both the animal and plant kingdoms. Nevertheless, there have been sporadic concerns about the validity and importance of various applications of the concept of kin selection (e.g., Alexander, 1974; Evans, 1977; Crozier, 1982; Stubblefield and Charnov, 1986; Guilford, 1988).

In this chapter, I examine the validity and merit of kin selection formulations of some social acts of plants. In his exposition of kin selection, Hamilton (1964) pointed out that the results of inclusive fitness calculations are equivalent to those of conventional calculations based on lineal fitness. To show how the selection of social acts can be described very simply in terms of the lineal transmission of a gene to relatives from a common ancestor, the Appendix introduces a "collective fitness rule." The rule is derived from the definition of the relative fitness of alleles at a single locus. It states that an allele that causes a social act is selected if the sum of its effects (e_i, both positive benefits and negative costs) on the various individuals involved in the social act, weighted by the numbers of copies of the allele in each individual, g_i, is positive; that is,

$$\sum e_i g_i > 0 \tag{1}$$

The collective fitness rule is really only an elaboration of the well-known axiom that an allele will spread if its average fitness is greater than that of competing alleles. The collective fitness rule is used below to show that Hamilton's rule provides a valid formulation of selection only when genetic relatedness (usually calculated as identity by descent, IBD) accurately weights the effects of selection on the various individuals affected by a social act. This is not the case with IBD formulations of relatedness of some social acts, such as parental investment in equally competing offspring (considered below) and the expression of endosperm genes (Lloyd, 1992). The collective fitness rule provides analyses of social acts that can cope with situations where IBD formulations of relatedness fail. In this respect it is similar to kin selection models using covariance formulations of relatedness (Queller, 1989), but it is simpler and more intuitively obvious.

The social acts of plants are particularly suitable for comparisons of collective and inclusive fitness because they are confined to behaviors within immediate families (i.e., parents and their progeny). The pathways by which collective or inclusive fitness are calculated are therefore short. As a result, comparisons of the two methods of calculating fitness are much simpler than when more distant relationships are involved, and the way in which kin selection operates is more evident.

Alternative subjects of selection and methods of assigning fitness are examined in the following analyses of parental investment in offspring and self-incompatible pollen and other kinds of reproductively disabled individuals and organs. The process of selection is described in genetic models of evolutionary stability and in collective and inclusive fitness formulations of the same events. The models show that parent–offspring conflicts can be explained just as simply by a collective fitness formulation as by kin selection. The importance of parent–offspring conflict in causing deviations from parental strategies in these family matters is questioned. Instead, the role of collaboration between parents and their progeny is emphasized. The collective fitness models demonstrate why for some social acts kin selection formulations using IBD measures of relatedness agree with the genetic models, and are then valid, and why in other cases the kin selection models disagree and are not valid. The collective fitness rule also provides a short cut of full genetic models that is an alternative to kin selection formulations. The rule is applicable to more circumstances than kin selection formulations using IBD measures of relatedness, but it must be applied to specific models of inheritance.

Parental Investment Strategies

The amount of resources that a parent provides to each offspring is important both to the parent, which obtains its fitness from the progeny it matures, and to the offspring, which are dependent on parental invest-ment for their survival. In a landmark paper, Trivers (1974) examined the amounts of resources that parents are selected to invest in each offspring and that offspring are selected to seek from their parents. On the basis of kin selection, he argued that because parents are equally related to all their offspring, they should optimally continue to nourish an offspring as long as the fitness benefit it received b, exceeded the fitness cost to another offspring, c; that is, while $b > c$. Offspring are selected to be more selfish in seeking care because they are less closely related to their sibs ($r = \frac{1}{2}$ for full sibs) than to themselves ($r = 1$). Hence, by Hamilton's rule they should seek further care as long as $b > rc$, or $b > \frac{1}{2}c$ (i.e., beyond the point to which parents are selected to provide resources). Trivers (1974) concluded that there is a conflict between parents and their offspring over the amount of care that is optimally provided. The idea has been applied to many aspects of parent–offspring interactions in animals and plants.

A seed parent provides maternal resources to developing seeds; paternal investment is absent in plants. Here I compare maternal and progeny strategies in terms of the interactions that take place while a family of seeds is being raised. The progeny solicit resources by presenting "begging" signals; developing plant embryos signal by producing hormones. A parent responds to the signals by feeding or ignoring the young; in the case of seed plants, a parent treats a hormone signal as either sufficient or insuf-ficient for the offspring to be fed as an assimilate sink. The models examine the threshold levels for offspring to cease begging and for parents to stop nurturing.

Parker and Macnair (1978) and others have recognized that there are two major situations under which care is provided. They differ in the way offspring compete with each other and in how contrasting phenotypes and their controlling genes bear the cost of progeny selfishness. Offspring in a family are described here as "equally competing" if they beg on equal terms; there are no differences in their positions or the times at which they were initiated that give one an intrinsic advantage over others. This would apply, for instance, to two equivalently placed seeds that were initiated in a fruit at the same time or to two fruits that were produced at the same time at one node. In contrast, two offspring compete "unequally" when one has an intrinsic advantage over the other in gaining parental resources. This can occur when a fruit is initiated earlier in a season than others or when a seed or fruit is in a physiologically favored position (Stephenson, 1981). In reality, the seed crop of a plant is likely to be a complex mixture of individual offspring with varying degrees of inequality.

Unequally Competing Offspring

Assume for simplicity that half the offspring in a family have an intrinsic advantage and have become "larger" than their disadvantaged "smaller" sibs (cf. Maynard Smith, 1989). Consider the parental strategy for feeding the unequally competing offspring over an interval when a fixed amount of parental resources is allocated to the N progeny in a family. At a locus controlling feeding rates, a recessive allele f in the homozygous state causes parents to feed both sets identically so that at the end of the period the larger and smaller individuals have fitnesses 1 and k, respectively. When the identical-feeding allele is fixed as the resident allele, the fitness of the allele is the total number of progeny a parent produces. Then

$$w_f = w_{ff} = N[\tfrac{1}{2} + \tfrac{1}{2}k] = \tfrac{1}{2}N(1 + k)$$

A dominant allele, F, is introduced into the population as a single mutant. The mutant allele causes the parent to feed the larger offspring more so that they receive a disproportionate share of the parental resources. The fitnesses of the larger and smaller progeny then become $1 + b$ and $k - c$, respectively, where b and c measure the benefit and cost of differential feeding. The fitness of the mutant allele.

$$w_F = w_{Ff} = N[\tfrac{1}{2}(1 + b) + \tfrac{1}{2}(k - c)] = \tfrac{1}{2}N(1 + k + b - c)$$

The mutant can invade the population (increase when rare) when $w_F > w_f$, i.e.,

$$b > c \tag{2}$$

As F spreads, FF and Ff parents continue to behave differently from ff parents, regardless of their frequencies. Thus, Eq. (2) also specifies the condition for the continued spread and fixation of an unevenly feeding parent.

The collective fitness rule specifies the parental ESS even more simply. A parental gene has an equal chance of appearing in any offspring. Hence a greater investment in larger offspring is favored if the benefits to the larger offspring exceed the costs to the smaller ones. Thus, F invades when $\Sigma e_i g_i > 0$; that is, when $\tfrac{1}{2} Nb - \tfrac{1}{2} Nc > 0$ or $b > c$, as in the genetic model.

Parental selection could also be examined by a kin selection model, but there is no effect on the actor (the parent) and the actor is related equally to all recipients (the offspring). Thus the relatednesses of the actor to the recipients ($r = 1/2$) is identical to the probability of a parental gene being passed to an offspring. Consequently, inclusive fitness formulations of parental selection using IBD measures of relatedness give the same answer as the collective fitness calculations presented above. For parental strategies, kin selection is valid but superfluous.

Now consider how the distribution of care affects the fitnesses of two offspring when they have reached various sizes. In a fitness curve (Fig. 7.1), the fitness of a single offspring is graphed against its size, which is assumed to be proportional to the care received. Provided that the total parental fitness is proportional to the number of progeny produced, the parental ESS is at the point where the fitness curve touches a line from the origin (Smith and Fretwell, 1974; Lloyd, 1987). Hence the curve must have a region of nil or low rate of increase in fitness when the progeny are small, followed by an interval of higher fitness gains, which then decline again beyond a certain size. Assume the curve is smoothly S-shaped (Fig. 7.1A). In such situations, increments of parental expenditure will bring higher marginal returns over the later stages of seed growth than over the initial stages. Consider the case where a shortfall of parental resources occurs when two progeny that are very unequal in size, say of sizes 1 and 2, are being fed simultaneously. If z_1 is at the point where the marginal fitness (the slope of the fitness curve) is the same as at the ESS size at independence (z), we expect that expenditure on any developing seed that is larger than z_1 (but does not exceed the ESS size) would bring a higher marginal fitness than that from any seed smaller than z_1. When a shortfall

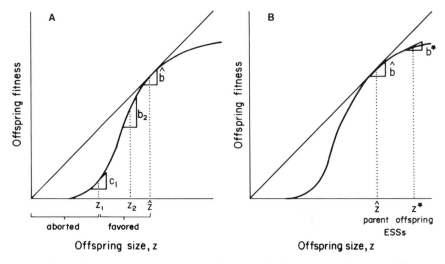

Figure 7.1. Fitness curves relating the fitness of single seeds to their size. The optimal seed size is where the curve touches a line from the origin. (A) The benefit, b, and cost, c, of switching an increment of expenditure from a smaller offspring at size z_1 to a larger one at size z_2. For a parent to abort the younger offspring, $b_2 > c_1$. (B) The narrow difference in size between the parental strategy, z, where an investment of expenditure brings a benefit, b, and the offspring strategy, z^* where the same increment brings a benefit $b = b/2$.

of resources occurs, a parent should then concentrate its resources on the larger seed and abort the smaller seed. In Figure 7.1A, parents should favor larger offspring at the expense of smaller ones whenever $b_2 > c_1$ for a given increment of expenditure. As seeds grow and increase in fitness only slowly at the beginning of their development, there should be frequent occasions when the abortion of smaller seeds is favored by the parent. Such patterns of brood reduction are common (Stephenson, 1981).

The relative advantages of feeding unequally competing offspring are different when both the larger and smaller offspring are approaching their mature size. Because the fitness curve is decelerating around the ESS size, the marginal benefits of increments of parental resources decrease with increasing size. Thus if there are two sizable but unequal offspring competing for parental resources, a parent obtains greater benefit by preferentially feeding the slightly smaller one so that they both reach the same size, even if it is not the optimal size.

Now consider the strategy of the *unequally competing offspring*. It is clear from Figure 7.1 that by growing to a size greater than z_2, individual offspring would benefit from receiving more resources than the parent is prepared to give. But does this necessarily mean that larger progeny in a family will be selected to take parental resources from their smaller sibs if they are able to do so? For simplicity I consider that there are equal numbers of larger and smaller offspring (the results can easily be extended to unequal subfamilies). The average fitness of the resident allele, h, is

$$\overline{w}_h = \tfrac{1}{2}N(1 + k)$$

Consider the invasion of a dominant allele, H, that causes *larger* offspring carrying it to signal more strongly and thereby obtain more care at the expense of smaller sibs. A single dominant mutant allele occurs in a segregating family of the type $Hh \times hh$. The dominant allele appears in half the larger offspring and takes resources from all the smaller offspring. If the benefit and cost of the action of one larger offspring are b and c, respectively, each smaller offspring receives a cost of $c/2$ since there are twice as many recipients as actors. The average fitness of the dominant allele is

$$\overline{w}_H = \tfrac{1}{2}N[(1 + b) + (k - c/2)]$$

The allele invades the population $(\overline{w}_H > \overline{w}_h)$ when

$$b > \tfrac{1}{2}c \qquad\qquad (3)$$

By considering the invasion of the recessive allele, h, into a resident population of stronger signaling HH individuals, it can readily be shown that the condition for H to become fixed is the same as that for its invasion and is also specified by Eq. (3).

The signaling ESS for larger offspring can also be obtained by applying the collective fitness rule, which considers only the effects of a stronger signal by the larger offspring. Hence $\Sigma e_i g_i = b(N/2) - (c/2)(N/2) > 0$ when $b > \frac{1}{2}c$, as in the genetic model.

It can be shown in a similar way that a *recessive* stronger signaling allele will invade and become fixed if the same condition [Eq. (3)] is met. Moreover, additional offspring-expressed genes that cause even higher signal levels will be selected under the same conditions. Modifiers that reduce the signals controlled by stronger signaling genes are subject to the same disadvantage as hh weaker signallers and are powerless to prevent the spread of stronger signals even though they are disadvantaged by them.

Thus larger offspring will be selected to beg for a disproportionate amount of care even if this is against the parent's interest [when $c > b > \frac{1}{2}c$, as Trivers (1974) proposed]. In Figure 7.1A, larger offspring should continue to beg whenever $b_2 > \frac{1}{2}c_1$. These are also the conditions under which an older offspring is selected to try to abort a younger offspring. Siblicide is favored under more lenient conditions than for the parental strategy of infanticide ($b_2 > c_1$). The siblicide may be implemented by producing a stronger signal (exploitation competion) or by the older progeny producing a hormone that directly induces younger seeds to abort (reviewed by Lee, 1988), a manifestation of interference competition.

Using Hamilton's rule, larger offspring are selected to beg selfishly when $b > rc$ or $b > \frac{1}{2}c$ as in (3), where $r = \frac{1}{2}$ for full sibs. A kin selection model using an IBD measure of relatedness provides a valid formulation of events. Hamilton's rule is accurate because the IBD relatednesses accurately weight the benefits and costs of selfishness on the fitness of the action allele. The inclusive fitness formulation is very simple in that it does not depend on a particular mode of inheritance and examines only the differences in fitness between the phenotypes and ignores the unchanged elements. On the other hand, its logic is more complex because in calculating relatedness by identity by descent it starts with the actor and traces back and forth between generations (Figure 7.2). In contrast, collective fitness traces the transmission of genes as they occur, by lineal descent from parent to offspring.

The offspring strategies described here do not maximize the fitness of a family or the average fitness of the genes they contain. Selection favors selfish offspring even though the selfishness decreases the average fitness of the alleles responsible, because the selfishness decreases the fitness of the competing allele even more. Thus selfish offspring behavior outcompetes egalitarian behavior within a family. On the basis of my family se-

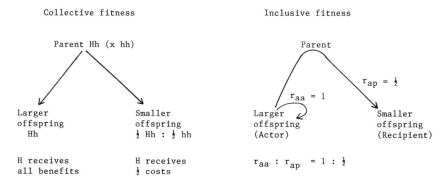

Figure 7.2. A diagrammatic comparison of collective and inclusive fitness formulations of selection of an allele, *H*, for a stronger begging signal for parental investment by the larger of two classes of unequally competing offspring. r_{aa}, relatedness of the actor to itself; r_{ap}, relatedness of the actor to the recipient.

lection models of parental care, I had previously concluded that offspring strategies conform to the parental strategy (Lloyd, 1979). The models considered the collective interests of all offspring in a family. Hence they relied on among-family comparisons and did not include the within-family component of selection. The earlier models were therefore incomplete and wrong.

Equally Competing Offspring

The parental strategy of feeding maturing progeny so that they all reach the same size and same fitness also operates when the progeny competing for parental investment have no intrinsic advantage over each other.

For the offspring ESS in this situation, consider the invasion of a dominant stronger signaling allele *H* into a resident population of weaker signaling *hh* individuals. In each family or part-family, two offspring compete on equal terms for a fixed quantity of parental resources. When the two offspring signal with equal intensity, they are fed identically and their resulting fitnesses are defined as one. Then the average fitness of the resident allele *h* is

$$\overline{w}_h = N[\tfrac{1}{2} + \tfrac{1}{2}] = N$$

When offspring signal unequally, a stronger signaler receives more resources and has fitness $1 + b$, whereas the weaker signaler receives fewer resources and has fitness $1 - c$. A single mutant allele *H* occurs as an *Hh* heterozygote, which mates with *hh* homozygotes. The average fitness of the mutant allele is

$$\overline{w}_H = \tfrac{1}{2}N[1 + (1 + b)]$$

Therefore $\overline{w}_H > \overline{w}_h$ when

$$b > 0 \tag{4}$$

Offspring are selected to solicit parental care as long as they continue to benefit individually from more resources. A stronger signaling strategy invades a population because the stronger signaling allele experiences benefits but never experiences the cost of its actions, which are borne entirely by the other allele. As *H* spreads and *HH* individuals appear, the *hh* offspring continue to bear the full cost of selfishness (in *Hh* × *hh* or *Hh* × *Hh* families). Thus, an allele that causes an offspring to seek parental resources whenever it benefits from them will become fixed. Even more extreme stronger signaling alleles that are subsequently introduced will also be selected. Hence we conclude that parent–offspring conflict should appear in more circumstances when progeny compete equally than when they compete unequally.

Again the collective fitness rule gives the simplest derivation of the offspring strategy. Here *H* invades when $\Sigma e_i g_i > 0$; that is, when $\tfrac{1}{2}Nb > 0$ or $b > 0$, as in the genetic model.

By Hamilton's rule, *H* invades under the same condition as for equally competing progeny, signaling for parental resources whenever $b > \tfrac{1}{2}c$. The rule is invalid because an identity by descent measure of relatedness does not correctly weight the benefits and costs of the social act on the action allele.

The Resolution of Parent–Offspring Conflict

Both unequally and equally competing offspring are selected to seek more parental resources than the parents optimally give. Several factors prevent offspring from achieving a marked departure from the parental strategy of caring evenly for all progeny, however.

1. The optimal investments sought by parents and offspring differ widely only if the fitness curve for single offspring decelerates slowly beyond the parental ESS. This may be true for animal young, which are typically active and mobile both before and after parental care ceases. But in plants, the mature seed performs a variety of unique tasks. A successful seed must be dispersed, land in a safe site, be buried (usually), remain dormant for varying

periods (frequently), germinate, and establish an independent life—
all while warding off potential predators, parasites, and competitors. These functions may have contradictory requirements for
optimal size and other properties (Harper et al., 1970; Stebbins,
1971). In particular, efficient dispersal often favors a smaller size
than that favored by optimal establishment. Hence the size interval
of the fitness curve over which offspring fitness increases rapidly
(when $\partial w/\partial z$ is high) may be very narrow and followed rapidly by
an interval of shallow slope or even decline (Fig. 7.1B). In this
case, there may be only a small size difference between the point
where $b = c$, the parental ESS ($\hat{z}$), and that where $b = c/2$, the
offspring ESS (z^*). There is then little scope for conflict between
the generations.

2. In the case of equally competing offspring, when a stronger signaling allele becomes fixed, all competing offspring have the same
    signaling strength, as was so before the allele invaded the population. Thus the parental resources will again be dispensed evenly
    in accordance with the parental strategy. Parent–offspring conflict
    is therefore self-correcting, having only a transitory effect on the
    distribution of resources.

3. With either type of offspring competition, parents may continue
    indefinitely to respond to escalations in offspring signals by simply
    raising the threshold required for a positive response. For plants,
    this means raising the hormone level required to treat a seed or
    fruit as a resource sink. Such escalation–retaliation cycles are
    likely to peter out because there are unequal costs for parents and
    offspring. Parental retaliation is virtually costless and can continue
    indefinitely, whereas offspring escalation carries considerable
    physiological costs (and for animals, predator costs) and is therefore limited. The greater staying power of parental retaliation is
    likely to ensure that the parental strategy is closely approximated.

Thus there are three factors (similar selection targets for the two generations, self-correction of uneven feeding, and parental retaliation) that
work toward the parent's even-handed strategy being followed. In many
species, the only prominent and permanent result of the introduction of
strong signaling alleles is likely to be that both the offspring signal and the
parental threshold for response will escalate. On this scenario, the residue
of the conflict will be the evolution of strong, seemingly exaggerated begging signals. This frequently appears to occur in both animals and plants.
Zoologists have often remarked on the extraordinary intensity of the begging signals of many animal young (e.g., Trivers, 1985). The hormone
levels produced by growing seeds are equally remarkable; the seed signals

are so strong that seeds have been used as a major source of plant hormones (Nitsch, 1952; Goodwin, 1978). The introduction and subsequent nullification of stronger signals that are postulated above can explain the strong signals.

Following the innovative leads of Hamilton (1964) and Trivers (1974), much of the theoretical work over the last decade on the selection of seed strategies has used the concept of conflict, not only between parents and their offspring but also among the other genetic entities involved in fertilization and seed development, the male and female gametophytes and the endosperm (Charnov, 1979; Westoby and Rice, 1982; Law and Cannings, 1984; Queller, 1983, 1984; Haig and Westoby, 1987; Uma Shaanker et al., 1988; Haig, 1990). The theories have been constructive and imaginative, but not very realistic because they argue as though the genetic entities are all present at one time and compete against each other in all combinations. The identity by descent measures of relatedness that have been used are also invalid for the endosperm because the two maternal genomes are identical and are not separate random samples of the parental genes. Moreover, the three arguments advanced above against offspring causing marked departures from the parental strategy apply also to the other nonparental entities. It will be argued elsewhere that the properties of the endosperm, like those of seeds generally, are better explained by theories of cooperation than of conflict (Lloyd, 1992).

Rearing a family of seeds or other young is a complex operation that requires continual interactions between the participants. Parents must dispense their resources to a number of young, often at widely different stages of development at one time. Parents cannot automatically know how much each offspring would benefit from being fed at a given time. On their part, the offspring must signal to a parent whose level of resources is unknown to them. In both plants and animals, the signal–response system enables an input of information from both participants. The pooling of information generally benefits the relative fitnesses of both the parents and the offspring.

The concept of parent–offspring conflict has dominated recent theories of parental care. But if this conflict is relatively minor and has minimal effects even though it is virtually ubiquitous, as is argued here, a postulate of cooperation between the generations has more heuristic value than one of conflict. Many aspects of seed biology, such as the regulation of seed numbers, the distribution of seed sizes, and the roles of various tissues, could be enlightened by examining the nurturing of seeds as a cooperative enterprise for mutual benefit rather than a conflict between adversaries with different interests.

An illustration of cooperation between parents and their progeny is provided by structural adaptations of seeds. An angiosperm seed is not just an offspring of its parents; the embryo (the true offspring) is enveloped by parental tissues, the seed coat, and sometimes the nucellus. In indehiscent fruits (fleshy fruits and some dry fruits) the composite seed is wrapped in another layer of parental material, the fruit wall. Some fruits are wrapped in still further parental layers, being furnished with extrafloral bracts such as the glumes of grass fruits. All these parental layers routinely (if not invariably) furnish the means of protection, dispersal, burial, and dormancy of the embryo. Structures such as the flesh and stones of stone fruits, the hooks of externally animal-dispersed fruits, and the plumes and wings of wind-dispersed fruits are all composed of parental materials. The selection of these structures must be attributed to parental strategies even though they accompany and assist the young embryos after their release from their maternal parents. The influence of parental manipulation on the selection of seed characters is pervasive. When they are dispersed, the offspring are separated from their parent and from each other. Hence in contrast to predispersal investment, postdispersal benefits to one offspring do not directly harm others. The parental and offspring interests coincide, rather than conflict, and the parental strategy provides for both.

Reproductively Disabled Individuals and Organs

Kin selection has attained its greatest acclaim by providing an elegant explanation of the occurrence of sterile castes in eusocial animals. In addition to sterile castes, there are diverse types of plant and animal individuals and structures that are reproductively "disabled." Disabled individuals and organs have reduced or zero reproductive fitness because they are partially or totally sterile, or are potentially functional but are sacrificed or diverted to nonreproductive functions that assist the reproduction of other family members or are inviable although considerable resources are invested in them. In certain insects, for example, inviable eggs are laid beside the viable ones, apparently providing chemical protection. In other species, trophic eggs provide food for their sibs (Wilson, 1971; West-Eberhard, 1975). In a considerable number of animal species, sterile "helper" sperm accompany the fertile sperm and may outnumber them. The sterile sperm assist the fertile sperm in reaching the eggs, and they may help in other ways. Reproductively disabled animals, including sterile castes, were reviewed by Trivers (1985).

Plants differ from animals in two respects that increase the kinds of disabilities that can occur. First, the reproduction of plants requires external agents for both pollination and dispersal. Consequently, their reproductive organs carry out a number of accessory functions that facilitate these processes, in addition to bringing gametes together and nurturing offspring. There are usually many sex organs on a plant, and some flowers or fruits may be sterilized and thereby enhance the pollination or dispersal of others. Second, most plants perform both male and female functions; the partial or total disabling of one function may aid the other. Because of these two features, there is a remarkable variety of disabled plant individuals and organs. The major types are discussed below, with the exception of endosperm, which will be discussed elsewhere.

Self-Incompatibility, a Conditional Pollen Disability

The selection of outcrossing versus selfing is the central topic of floral biology. Floral characters are primarily adaptations that restrict self-pollination and encourage outcrossing, thereby giving progeny the advantage of hybrid vigor (Knight, 1799; Darwin, 1876). Yet many plants are regularly self-pollinated. Where self-fertilization has evolved, the advantage of outcrossing must be countered by contrary factors favoring self-fertilization. The traditional explanation of self-fertilization (e.g., Müller, 1883) is that it provides reproductive assurance when pollinators are scarce or unreliable. More recently, it has been recognized that self-fertilization has another advantage because selfed progeny contribute two genomes to each progeny, whereas outcrossed progeny contribute only one (Fisher, 1941; Kimura, 1959). The 2-fold disadvantage for outcrossing is a form of the cost of meiosis (Lively and Lloyd, 1990).

Multiallelic self-incompatibility is one of the most common mechanisms that prevent self-fertilization. Like other outcrossing mechanisms, self-incompatibility is usually interpreted as a parental strategy. Yet the operation of gametophytic self-incompatibility, the most widespread type, requires the matching of specificities in a parental tissue, the style, and an "offspring," the male gametophyte. Moreover, self-incompatibility may benefit the female function by providing fitter seeds, but it is disastrous to the fitness of the inhibited pollen tubes. Their reproductive capacity is disabled as effectively as that of sterile castes, as Beach and Kress (1980) and Hamilton (1987) recognized. Self-incompatibility is tantamount to a suicide–infanticide pact. A question arises that is parallel to that for sterile castes—which Darwin (1859) considered the most severe problem facing the theory of natural selection. If natural selection operates by rewarding reproductive success, why should a pollen grain connive in its own death in an incompatible pistil?

Beach and Kress (1980) proposed that self-incompatibility is the parent's response to a conflict between the gametophytic and sporophytic generations; the male gametophyte growing into a style is faced with a "no option" situation (Charnov, 1979), whereas the sporophyte benefits by the replacement of selfed progeny by outcrossed ones. In a slightly different vein,

Willson (1983) described self-incompatibility as a male–female conflict. Hamilton (1987) postulated that self-incompatibility is an act of altruism by pollen, whereas Uyenoyama (1988) considered it to be a parental mechanism for regulating outcrossing distance.

Here I examine selection of the strength of a gametophytic self-incompatibility system by deriving the conditions for a decrease in selfing that is achieved through a strengthening of the incompatibility reaction by modifying genes acting in the style or in pollen. Inbreeding depression, the cost of meiosis, and reproductive assurance are incorporated into the following models.

Consider first genes modifying the *diploid* stylar reaction of a *maternal parent*. The resident genotype in a population, a recessive homozygote *tt*, causes a fraction *s* of ovules to be selfed before outcrossing is possible. The seeds from outcrossing and selfing have fitnesses 1 and $1 - \delta$, respectively, where δ is the degree of inbreeding depression. Ovules that are not selfed have a probability *p* of being subsequently outcrossed. The fitness of residents is the sum of two gamete contributions to selfed ovules and one contribution each to outcrossed ovules and outcrossed pollen that fertilizes ovules of other plants. When individuals cross with *K* mates (*K* is large) and each plant produces *g* pollen grains and *m* ovules, the fitness of the resident allele is

$$w_t = w_{tt} = m[2s(1 - \delta) + p(1 - s) + Kp(1 - s) \cdot g/Kg]$$
$$= m[2s(1 - \delta) + 2p(1 - s)]$$

To find a parental ESS for the *t* locus, we examine the fate of a mutant dominant allele *T* that causes a decrease in the selfing frequency to $s - \Delta$. The fitness of the mutant allele,

$$w_T = w_{Tt} = m[2(s - \Delta)(1 - \delta) + p(1 - s + \Delta) + p(1 - s)]$$

Then,

$$w_T - w_t = m\Delta[p - 2(1 - \delta)]$$

The rare mutant can invade the population when $w_T - w_t > 0$, or

$$p > 2(1 - \delta) \tag{5}$$

In this simple case, the extent of inbreeding depression, the frequency of pollinator-mediated outcrossing, and the cost of meiosis jointly determine whether complete outcrossing or selfing is advantageous. When outcrossing is assured, $p = 1$ and from Eq. (5),

$$\delta > \tfrac{1}{2} \qquad (6)$$

The cost of meiosis is outweighed and self-incompatibility is selected when inbreeding depression exceeds one-half. As T spreads, the difference between the fitness of T (in TT plants) and t (in tt plants) remains as above, and the heterozygote contributes equal numbers of both alleles to its progeny. Hence the conditions for the continued spread and fixation of T remain as in Eq. (5). In this simple situation, the parental ESS switches from complete crossing to complete selfing, depending on pollinating conditions and the degree of inbreeding depression.

The collective fitness rule gives the same ESS more simply. According to the rule, the allele T is favored if the sum of its effects, weighted by their frequencies, is positive. A decrease in selfing $(-\Delta)$ produces a gain in outcrossed ovules and a loss to the parent of both pollen and ovule contributions to previously selfed progeny. Here

$$\sum e_i g_i = \frac{1}{2} m[p\Delta - \Delta(1 - \delta) - \Delta(1 - \delta)]$$

Thus, $\sum e_i g_i > 0$ if $p > 2(1 - \delta)$, as in Eq. (5).

As in the case of parental investment, the optimal strategy of a gene in the parental style can be explained equally well in terms of kin selection. Calculations of inclusive and collective fitness are equivalent because all individuals affected by the action are offspring of the actor, the parent $(r_i = g_i)$.

Now consider conditions under which self-incompatibility is selected when the strength of the reaction is altered by genes acting in the *haploid pollen*. Comparing an allele that causes a frequency of selfing s with an allele that reduces selfing to $s - \Delta$, the collective fitness rule gives conditions for selection in terms of the sum of effects on outcrossed and selfed ovules and selfing pollen. The condition for a stronger pollen reaction is

$$\sum e_i g_i = m\left[\frac{1}{2} p\Delta - \frac{1}{2}\Delta(1 - \delta) - \Delta(1 - \delta)\right] > 0$$

or

$$p > 3(1 - \delta) \qquad (7)$$

When $p = 1$,

$$\delta > 2/3. \qquad (8)$$

Conditions for the selection of self-incompatibility are more stringent for pollen modifiers than for modifiers expressed in the style. A comparison of the derivations of Eqs. (5) and (7) shows that this is because a pollen gene is present alone in the haploid tissue and passes to all descendants of the pollen. Hence it bears the *full* cost of the lost pollen contribution from ovules which are not self-fertilized. In contrast, a stylar allele is passed to only half the parent's progeny. Consequently, it *shares* the opportunity cost with the other allele in the diploid genotype.

Hamilton's rule gives the same conditions for a pollen modifier as the genetic and collective fitness models. When $\Sigma \ e_i r_i > 0$, $m[\frac{1}{2}p\Delta - \frac{1}{2}\Delta(1 - \delta) - \Delta(1 - \delta)] > 0$, or $p > 3(1 - \delta)$, as in Eq. (7). Hamilton's rule is valid here because the ratios of the numbers of gene copies experiencing the benefit from increased outcrossing and the losses from nonselfed ovules and pollen ($\frac{1}{2}:\frac{1}{2}:1$ in the collective fitness model) are identical to the relative relatednesses of the three components to the pollen actor ($\frac{1}{2}:\frac{1}{2}:1$). Thus, genetic relatedness accurately weights the three effects so they can be summed, as in the derivation of Eq. (7). Although it is correct, the explanation provided by the calculation of the inclusive fitness of a pollen modifier is less straightforward than that provided by a consideration of the lineal contributions involved in collective fitness.

In the light of these calculations, we may ask which, if any, of the five offered explanations of self-incompatibility adequately describe the phenomenon. The two parental selection hypothesis and pollen altruism can describe part of the situation, but none of these hypotheses incorporates the actions of both stylar and pollen genes. Hence each of these perspectives (including a version of parental selection in Lloyd, 1989) is insufficient for a full explanation of the behavior of pollen *and* styles. There is some conflict between the pollen and stylar genes [when $3(1 - \delta) > p > 2(1 - \delta)$]. This is simultaneously a conflict between sexes (male versus female, as suggested by Willson, 1983), generations (parent versus offspring, as suggested by Beach and Kress, 1980), and life history stages (diploid sporophyte versus haploid gametophyte). The conflict arises from the meiotic segregation of genes in the haploid pollen but not in the diploid style, and might best be described as being between postmeiotic and premeiotic stages of life history. The source of the conflict is the same as that which causes the conflict over the amount of parental investment considered above.

The conflict occurs over only a narrow range of inbreeding depression ($1/2 < \delta < 2/3$ when $p = 1$). This zone occupies only one-sixth of the total range; above and below this zone the pollen and stylar genes act in agreement to select either for or against self-incompatibility. Self-incompatibility requires the cooperative action of both stylar and pollen genes, and like parental investment it is better viewed as a joint action of the two participating entities. The cooperation is forthcoming only when the pollen genes, which require more stringent conditions, are selected to express their self-incompatibility reaction. A hypothesis of cooperation between haploid and diploid genes for their mutual benefit provides a better explanation of the facts of self-incompatibility than any of the five previously suggested hypotheses.

Other Disabled Pollen, Seeds, Flowers, and Fruits

We are concerned here with individuals (pollen or seeds) or structures (flowers or fruits) that are reproductively disabled but in which parental expenditure is retained or even increased in some respects over that in their fertile equivalents. The better documented examples that have come to my notice are described below; no doubt many others exist.

Sterile pollen is more common than sterile seeds and fruits. In at least 15 distantly related angiosperm taxa, there are species in which the flowers have two kinds of anthers, described as fertilizing and feeding anthers. Such taxa include the Melastomataceae (Müller, 1881; Renner, 1989), Commelinaceae (Lee, 1961; Mattson, 1976; Simpson et al., 1986), Fabaceae: Caesalpinioideae (Dulberger, 1981; Gottsberger and Silbauer-Gottsberger, 1988), Lecythidaceae (Mori and Prance, 1987, 1990), *Verbascum*, *Lagerstroemia*, and others (reviewed by Vogel, 1978 and Buchmann, 1983). The feeding anthers are often more conspicuous and supposedly function principally or entirely as attractants and rewards, whereas the fertilizing anthers furnish most or all of the pollen that fertilizes ovules.

The differentiation of pollen functions is unambiguous and total in those species in which the pollen grains of the feeding anthers are inviable ("sterile"), e.g., *Tripogandra grandiflora* (Lee, 1961). In other species, the pollen of feeding anthers is viable and capable of fertilizing ovules (e.g., all species of heteranthic Melastomataceae; Renner, 1989). Nevertheless, fertile pollen from feeding anthers may succeed less frequently in fertilization than that of the fertilizing anthers. In *Solanum rostratum*, a fluorescent dye dusted on the fertilizing anthers was redeposited on a significantly higher proportion of stigmas of the flowers subsequently visited than when the dye was dusted on the feeding anthers (Bowers, 1975). A similar differential success is likely in species of Melastomataceae and Fabaceae in which the pollen of fertilizing anthers is presented in a position within a flower that is similar to that of stigmas and is deposited on a part of the bee's body that is not readily groomed. The more accessible and less congruent position of the feeding anthers makes transfer of pollen to a bee's pollen basket more likely and transfer to a stigma less probable (Vogel, 1978; Gottsberger and Silbauer-Gottsberger, 1988; Renner, 1989).

The important distinction between the pollen of feeding and fertilizing anthers is not in their constitutional viability or fertility. What matters is the relative reproductive success of the two types. Pollen from a feeding anther is reproductively disabled just as effectively when it is fertile but is not transferred to a stigma as when it is sterile or inviable. There may well be a continuum in the degree of functional differentiation of "feeding" and "fertilizing" pollen from slight differences in their chances of fertilization to total differentiation with zero fitness of the pollen of the feeding anthers.

A number of plants have seeds or fruits that are sterile but continue to be invested in because they perform ancillary functions of pollination, dispersal, or protection. In monoecious figs, the female flowers in a synconium occur in two modal types (Verkerke, 1989). Both kinds of flowers receive pollen and have fertilized ovules that are capable of producing seeds, but the short-styled flowers are usually parasitized by larvae of the pollinators, agaonid wasps. Female wasps are unable to oviposit in the long-styled flowers, which produce most or all of the seeds. When they mature, the adult wasps transfer pollen from the syconium in which they hatch to other syconia, thus facilitating the pollen fitness of the plant in which they were raised. The seeds of the short-styled flowers are sacrificed conditionally (when parasitized) to the benefit of the pollen in the same synconium.

In all three genera of the gnetophytes (*Gnetum, Welwitschia,* and *Ephedra*), there are animal-pollinated species in which both male and female plants bear ovules that produce pollination drops that provide nectar rewards for pollinators. The ovules of females are fertile but those of males are sterile. The system has apparently evolved from cosexual ancestors that bore fertile pollen and seeds (reviewed by Lloyd and Wells, submitted).

In several *Paeonia* species (Mildbraed, 1954; van der Pijl, 1969; P. N. Johnson, personal communication) and probably a number of other angiosperm species (Kerner, 1898, but not in the English edition), there are conspicuous fleshy but inviable "display seeds" that contrast strongly with other viable seeds in the same fruit. The distribution of the two types of seeds suggests that the display seeds may be unfertilized ovules (Lloyd and Johnston, personal observation). In the 40 species of the araliad genus *Osmoxylon*, one of the three rays of every compound umbel bears sterile flowers that become seedless "pseudofruits" that are larger and more conspicuous than the fertile fruits (Philipson, 1979). The *Osmoxylon* fruits and peony seeds presumably function as attractants and/or rewards for dispersers, but no observations of dispersal have been made.

Umbels of parsnip, *Pastinaca sativa*, often contain parthenocarpic fruits that are similar to normal fruits but lack an embryo and endosperm (Zangerl et al., 1991). When larvae of the parsnip webworm, the principal herbivore of wild parsnip, were offered a choice between normal and parthenocarpic fruits, they exhibited a strong preference for parthenocarpic fruits but grew more slowly on them than on normal fruits. Zangerl et al. (1991) postulated that the parthenocarpic fruits act as decoys that divert herbivores away from fruits containing embryos. There seems to be no pattern in the distribution of parthenocarpic fruits, and they too may arise from unpollinated flowers (A. R. Zangerl, personal communication).

In a number of taxa there are well-developed neuter flowers that increase the ability of an inflorescence to attract pollinators or provide odor or nectar. There are neuter flowers with exaggerated corollas on the margins of compact inflorescences of some *Hydrangea* and *Viburnum* species (Darwin, 1877; Weberling, 1989). In some Compositae, the ray florets are similarly sterile (Dittrich, 1977; Norlindh, 1977; Stuessy, 1977). The centers of the umbels of carrots and some other umbellifers are sometimes provided with a dark spot of one or more sterile flowers or flower rudiments (Darwin, 1877). Eisikowitch (1980) showed experimentally that the dark spot mimics resting insects and increases the rate at which flies visit an inflorescence. Perhaps the most complex expression of disabled flowers occurs in the species of *Parkia* in which there are two kinds of sterile flowers that provide nectar and odor, respectively (Classen-Bockhoff, 1990; Grunmeier, 1990). Other examples of sterile flowers that function as nectaries occur in *Sesamum* and *Harpagophytum* (Goebel, 1905; Weberling, 1989) and a number of legume genera (van der Pijl, 1955). The inflorescences of *Trifolium subterraneum* bore into the soil to bury the seeds, and the upper flowers are transformed into organs that anchor the inflorescence (Weberling, 1989). In the inflorescences of species of *Arum* and some related genera, the bristles that prevent the premature exit of pollinators from the floral chamber are modified neuter male flowers (Proctor and Yeo, 1973; Dahlgren et al., 1985). The female flowers of *Dieffenbachia longispatha* have staminodes that feed beetles (Young, 1986); this may be a case of a sterilized organ being "resurrected" for a new duty.

All types of reproductively disabled eggs, sperm, seeds, pollen, flowers, and fruits mentioned above are exactly parallel to the much-discussed sterile castes of eusocial animals in being dispossessed of some or all of their reproductive functions. They therefore raise problems for the theory of natural selection that are just as acute as that for sterile castes. How does natural selection produce individuals or ogans that are active and metabolically expensive but have little or no fitness? Despite the severity of the problem, the selection of these reproductively disabled individuals and organs other than sterile castes has received almost no attention. The question that has occupied a generation of zoologists investigating the selection of sterile castes—namely, which individuals benefit from a disability and control it—has not been addressed at all for the disabled plant individuals and structures.

The variously disabled animals and plants lose reproductive fitness while that of other family members is enhanced. The situation is the reverse of the selfish offspring begging for more parental investment that were considered previously, because the disabled entities experience a reproductive cost while others obtain a benefit. There is an obvious possibility of kin selection. But as with parental investment, we must first ask whether the disabling strategy is a parental strategy, an offspring strategy, or a compromise. To be more precise, we should ask whether the strategies are caused by genes expressed in parents or their offspring or by a combination of both.

With the much-debated exception of sterile castes, in all other cases of reproductive disabilities it appears that the developmental decision to produce fertile or disabled individuals or structures is made entirely by the parents, often before the gametes or offspring even appear. The nature of feeding versus fertilizing anthers is determined by the morphological positions of the stamens in the flowers. Similarly, the sterility of conspicuous marginal florets, the dark centers of certain umbels, the sterile flowers of *Parkia* and other nectariferous flowers, the bristles of *Arum* inflorescences, and the pseudofruits of *Osmoxylon* are all associated with their positions in inflorescences. The differentiation of female fig flowers is determined by the length of the style, a purely maternal tissue. The development of ovules and pollination drops in male gnetophytes is controlled by the male plants, whereas the reward offered by the staminodes of *Dieffenbachia* is present only in female flowers. The developmental causes of the display seeds in *Paeonia* and the parthenocarpic fruits of parsnip are less certain but they also seem to be based on parental decisions triggered by pollination failures. In the animal examples, helper sperms and eggs are produced in specialized lobes of the testes and ovaries, respectively (Trivers, 1985), again under control of the parent.

Where the differentiation of reproductively able and disabled individuals and structures is controlled by parental tissues, the distinction must be explained as a differentiation strategy of the parents, benefiting them and genes expressed in them. The disabled entities (other than sterile castes) are only passive vehicles of parental strategies. The selection of the disabled plants, like that of the seed envelopes considered previously, is a manifestation of parental manipulation, the mechanism that Alexander (1974) postulated to explain how parent–offspring conflict might be resolved in the parent's favor. This explanation for disabled offspring is therefore identical to that for accessory organs such as nectaries and petals that aid the parent in producing progeny. Indeed, in some cases, such as sterile anthers with rudimentary pollen or the display seeds of peonies without embryos, the distinction between inviable offspring and sterile parental tissues becomes difficult to make.

Whenever disabled individuals are selected as parental strategies, kin selection can be invoked, but, as with the parental investment strategies described above, kin selection is equivalent to a hypothesis based on the lineal fitness of parents. The collective fitness rule can predict the quantitative conditions for parental selection of disabled progeny or structures. Suppose a fraction q of the pollen or seeds have their fitness reduced by a proportion c, while the remainder have their competitive ability boosted a proportion b by the actions of the disabled individuals. Parents gain from this if the total benefits exceed the total costs. Because the costs and benefits are both experienced by the parents, the collective fitness rule reduces to $e_1 > 0$ (see the Appendix). Sterilizing some pollen or seeds has a net advantage when $b(1 - q) > cq$, or

$$b \left(\frac{1 - q}{q} \right) > c \qquad (9)$$

The fraction of disabled individuals is limited because the benefit required to compensaste for the disabled individuals rises as the able individuals become fewer (the term in parentheses).

Unisexuality, a Total Unilateral Disability

The great majority of animal species are dioecious, and hence the loss of male or female fitness by an individual is the loss of its whole fitness. But most plants are cosexual, and individuals can be partially or totally disabled in one sexual function only. There are numerous cases of female sterility (andromonoecy) and fewer cases of male sterility (gynomonoecy) of some flowers of each individual in a population. In many clades, some individuals have entirely lost male function and are functional females. This is often followed by the partial or total loss of female function by the males of the dimorphic populations. The unisexual disabilities that are most parallel to sterile castes and neuter flowers (in that considerable expenditure continues to be invested in disabled structures) are the cases of female mimicry. In these dioecious species, pollen is the only reward for pollinators and females produce large quantities of sterile pollen (Charlesworth, 1984; Willson and Ågren, 1989). In cryptically dioecious species, both sexes produce fully developed hermaphroditic flowers; the selective basis for this is unknown (Kevan et al., 1990).

It has been taken for granted that the unilateral disabilities of functionally male or female plants are parental strategies, but the question should be asked whether any of them are strategies caused by genes expressed in the offspring. Two factors suggest that these disabilities, the source of the spectacular array of sex conditions in plants, are everywhere the result of parental strategies. First, the nonfunctioning of some potential gametes or progeny is always more easily selected by parents than by offspring. As in the case of self-incompatibility considered above, a gene in a parent experiences the benefits and costs of disabled offspring equally, whereas a gene in disabled offspring bears all the costs but shares the benefits with the other allele from the parent. Second, the evidence clearly indicates parental manipulation wherever flower or fruit sterility depends on position, as in the female flowers of gynomonoecious Compositae, or when sterility is decided before the progeny or even the gametes are formed, as in the male flowers of andromonoecious Umbelliferae. Similarly, the male sterility of females of dimorphic species is always achieved by the total sterility of individuals, or at least of whole flowers of anthers, as would be expected of parent-expressed genes. There are no known cases of natural semisterility caused by a haploid pollen-sterilizing gene segregating within an anther in a parent. Nor are they likely to be recorded, as females have a strong outbreeding advantage and are likely to invade a cosexual population only if male sterility is complete and self-fertilization is totally precluded (Lloyd, 1975).

Conclusion

Finding the appropriate subject of selection, the entity whose fitness is maximized, is not always a straightforward matter. The interests of parents and their offspring may differ, as may those of males and females or haploid and diploid individuals. Whenever a mechanism is influenced by the expression of genes in different generations, sexes, or life history stages, the effects of all genes must be taken into consideration. When two or more participants can act as subjects of selection, as happens frequently in reproduction, three outcomes are possible.

First, the participants may have identical selection targets. Selection may, however, still be controlled by one participant, which dictates events by virtue of its position. This is the case for the parental tissues that contribute to seed dispersal and dormancy mechanisms. The mechanisms are constructed and paid for by the parents, but they benefit the parents and offspring at the same time.

Second, the subjects are in conflict part of the time, and one participant controls the outcome by virtue of its greater power or better ability to retaliate. This may happen in the case of parental investment. The conflict is real, but it is restricted to a limited range of parameters and its importance in the total system should not be exaggerated.

Third, and probably most importantly, the different participants experience a mixture of selection conflicts and agreements and both contribute to a combined strategy. Mutually interdependent participants that have *divergent and complementary* roles in an interaction are cooperating partners more than competing adversaries. Neither parental investment, nor seed dispersal and seedling establishment, nor self-incompatibility could be accomplished by actions of one partner alone. In the case of parental investment during seed development, the parent and the seed (or fruit) have access to different, equally valuable information. Each seed signals its stage of development (and thus the resources needed to reach maturity), but only the parent can assess its resources. The begging-response system is not just a sequence of actions by two selfish individuals. It involves a vital exchange of information that benefits both partners. In the postrelease life of seeds, parents largely control events up to germination, but it is the genuine offspring, the embryo, that must take over and establish the next generation. In self-incompatibility, mutual cooperation of the style and pollen components is fundamental to the operation of the system, which depends entirely on the matching of pollen and style specificities. Even when conflicts between the interacting individuals are present in these systems, they play a relatively minor role in the whole mechanism. Mutual cooperation, not conflict, is the dominant theme of all the parent–offspring interactions considered here. Most features of the biology of flowers and seeds conform to parental strategies; conflicts with offspring interests are a subsidiary element and are likely to cause only minor divergences from parental strategies.

The other major aspect of analyses of the selection of reproductive characters that has been considered here concerns the mode of assigning fitness to individuals. Theorists face a choice between calculating the fitness of individuals or genes conventionally through lineal descent to direct descendants or as inclusive fitness, which incorporates the weighted effects on various relatives from the reference point of an actor. As we have seen above, the choice is not an idle one. Calculations based on inclusive fitness effects (e.g., Hamilton's rule) give an accurate description of events only when the relative weightings of the fitness effects on the various participants are specified exactly by the relatednesses of the actor to each affected individual, including itself. This is not so in the case of parental investment in equally competing offspring, where the genotypes of recipients are not determined independently of those of the actors. Hence the recipients do not have a random sample of the common ancestor's genes, an assumption of relatedness calculations based on identity by descent.

We can compare alternative formulations using collective or inclusive fitness to calculate the selection of parental investment and disabled individuals. In both types of family events, there is a difference in the fitness interests of parents and their offspring under some conditions. Offspring are more selfish than their parents over parental investment, and they require more stringent conditions for a disability to be selected. Kin selection explains the conflict by the fact that parents are equally related to all offspring, whereas offspring are less related to their sibs than to themselves. The collective fitness arguments put forward here explain the disparity by considering the interests of genes controlling the behavior of parents and their progeny. Parents are more even-handed over their investment and more inclined to have some offspring disabled because a controlling gene in a parental body is equally likely to be transmitted to any offspring. Hence a parental gene experiences the benefits and costs of a social action equally. In contrast, an offspring-expressed gene is present in all offspring expressing the action but in only half its sibs. Hence it experiences all of the actor effects (benefits in parental care, costs in disabled individuals) but shares the recipient effects (costs or benefits, respectively) with other individuals in the family that carry the competing allele.

The two types of explanation are equivalent in those situations where kin selection models using identity by descent (IBD) measures of relatedness are valid. But even when collective and inclusive fitness calculations give the same answer and are equally valid, one or the other method may provide a simpler formulation of selection and hence be preferred. One reason that kin selection is invoked so frequently is that it often gives simpler insights into social actions than standard calculations of lineal fitness do. This stems in part from the simplification than inclusive fitness rules consider only the differences between contrasted phenotypes and ignore the elements they have in common. Another advantage of kin selection models using IBD measures of relatedness is that the models can be expressed at the level of genes or individuals. This is so because IBD relatedness assumes that genes are transmitted randomly from ancestors to descendants. Hence relatedness can be interpreted as the probability of genes being simultaneously inherited from a common ancestor or as the proportion of genes that actors and recipients share.

In considering social acts among free-living animal relatives, the assumption that genes are transmitted randomly is not likely to be violated. Calculating IBD relatedness coefficients is then much easier than calculating the probabilities of all individuals inheriting a gene from their common ancestor, especially when that ancestor is remote. Maynard Smith (1981) has given a spectacular example of the elegance of kin selection formulations in these circumstances. On the other hand, for the immediate family matters considered above, it was easier to compare the effects of parent- and offspring-expressed genes for parental investment and self-incompatibility in arguments based on collective fitness than on those based on inclusive fitness. Collective fitness considers events the way they happen, by the descent of genes from parents to their offspring, rather than by tracing back and forth between generations as in IBD formulations of inclusive fitness.

Genetic models of lineal fitness, including applications of the collective fitness rule, refer explicitly to the alleles controlling strategies. As a consequence, they are restricted to the specified mode of inheritance. This is compensated for by their greater generality in terms of the kinds of events that can be modelled. Only genetic models can describe social actions in which the actors or recipients do not carry a random set of alleles of the action locus. In these situations, IBD formulations of inclusive fitness are not valid; but the collective fitness rule still works because it incorporates the lineal transmission of genes from a common ancestor.

There is no single universal answer as to whether calculations of collective or inclusive fitness give more useful descriptions of the selection of social acts in plant reproduction. Each phenomenon must be carefully examined. It is clear, however, that kin selection formulations are not always accurate and that they have sometimes been invoked inappropriately in the past. We cannot simply assume that kin selection is the preferred mode for describing the action of natural selection whenever we are dealing with social acts among relatives. In the future, kin selection formulations should be employed more advisedly, only when they give an accurate description of events.

Appendix: The Collective Fitness Rule

The fitness of an allele is defined as the average number of descendants of each copy of the allele that is present in the population exactly one generation later. It follows from the definition that a fitter allele (one with more descendant copies per original copy) will increase in frequency from one generation to the next at the expense of a less fit allele.

Suppose that at a locus there is a null "nonaction" allele and another "action" allele that when expressed in a social act has effects, e_i, on the fitnesses of the ith individual in the population ($i = 1, 2, \ldots n$). The effects are positive or negative and may be experienced by the "actor" expressing the action allele and/or by other individuals, "recipients," who need not be relatives. The affected individuals may differ in their reproductive value (Fisher, 1930), v_i, the relative contributions (unaffected by the social actions) that individuals of different types make to the future gene pool of the population. If g_i copies of the allele are affected by a social act in the ith individual, the net effect of one act on the number of descendants of the action allele in the next generation is

$$\sum_{i=1}^{n} e_i g_i v_i$$

The action allele increases when its fitness is greater than that of the nonaction allele, that is, when

$$\sum_i e_i g_i v_i > 0 \tag{A1}$$

This is the collective fitness rule, which sums up the weighted effects on the fitness of the various carriers of an allele responsible for a social act. In the first instance, the rule describes the effects of social acts on *genes*.

Frequently, the reproductive values of the affected individuals are constant or are of little interest and can be assumed to be constant. Setting v_i = constant gives

$$\sum_i e_i g_i > 0$$

The collective fitness rule provides a simple method for integrating the divergent effects of a social act on the fitnesses of the various interactants. The method, which has not been previously exploited, relies on calculating the relative numbers of copies of an action allele that are present in the various interactants. These numbers are found by describing the transmission of genes from a common ancestor to the individuals affected by a social act. Calculations of collective fitness provide an intuitively obvious alternative to those of inclusive fitness, which obtain the relative numbers of action alleles in interactants by using the actor, rather than the common ancestor, as the initial reference point. The differences between the two approaches and their relative merits are explained for particular examples in the main text.

The rule can be applied to comparisons of phenotypes in models of the selection of nonsocial behaviors. Phenotypic models of selection are valid only when contrasting phenotypes are controlled by genes with constant effects (Lloyd, 1977, 1992). It is therefore possible to examine the selection of phenotypes by enumerating the fitness effects of a controlling allele via the collective fitness rule. When the costs and benefits of a behavior fall on a single type of individual, $i = 1$ and Eq. (A2) reduces to

$$e_1 > 0 \tag{A3}$$

Summing the costs and benefits of particular behaviors can provide the simplest possible type of phenotypic selection model, as in the case of parent-manipulated disabled pollen and seeds discussed in the main text.

Literature Cited

Alexander, R.D. 1974. The evolution of social behaviour. Annu. Rev. Ecol. Syst. 5:325–383.

Beach, J.H., and W.J. Kress. 1980. Sporophyte versus gametophyte: A note on the origin of self-incompatibility in flowering plants. Syst. Bot. 5:1–5.

Bowers, K.A. 1975. The pollination ecology of *Solanum rostratum* (Solanaceae). Am. J. Bot. 62:633–638.

Buchmann, S.L. 1983. Buzz pollination in angiosperms. *In* C.E. Jones and R.J. Little, eds., Handbook of Experimental Pollination Biology. Scientific and Academic Editions, New York, pp. 73–113.

Charlesworth, D. 1984. Androdioecy and the evolution of dioecy. Biol. J. Linn. Soc. 23:333–348.

Charnov, E.L. 1979. Simultaneous hermaphroditism and sexual selection. Proc. Natl. Acad. Sci. USA 76:2480–2484.

Classen-Bockhoff, R. 1990. Pattern analysis in pseudanthia. Plant Syst. Evol. 171:57–88.

Crozier, R. 1982. On insects and insects: Twists and turns in our understanding of the evolution of eusociality. *In* M.D. Breed, C.D. Michener, and H.E. Evans ed., The Biology of Social Insects. Westview Press, Boulder, CO.

Dahlgren, R.M.T., H.T. Clifford, and P.F. Yeo. 1985. The Families of the Monocotyledons. Springer-Verlag, Berlin.

Darwin, C. 1859. On the Origin of Species. Murray, London.

Darwin, C. 1876. The Effects of Cross and Self Fertilisation in the Vegetable Kingdom. Murray, London.

Darwin, C. 1877. The Different Forms of Flowers on Plants of the Same Species. Murray, London.

Dittrich, M. 1977. Cynareae-systematic review. *In* V.H. Heywood, J.B. Harborne, and B.L. Turner, eds., The Biology and Chemistry of the Compositae, Vol. II. Academic Press, London, pp. 999–1015.

Dulberger, R. 1981. The floral biology of *Cassia didymobotrya* and *C. auriculata* (Caesalpiniaceae). Am. J. Bot. 68:1350–1360.

Eisikowitch, D. 1980. The role of dark flowers in the pollination of certain Umbelliferae. J. Nat. Hist. 14:737–742.

Evans, H.E. 1977. Extrinsic and intrinsic factors in the evolution of insect sociality. BioScience 27:613–617.

Fisher, R.A. 1941. Average excess and average effect of a gene substitution. Ann. Eugen. 11:53–63.

Goebel, K. von. 1905. The Organography of Plants. Clarendon Press, Oxford.

Goodwin, P.B. 1978. Phytohormones and fruit set. *In* D.S. Letham, P.B. Goodwin, and T.J.V. Higgins, eds., Phytohormones and Related Compounds: A Comprehensive Treatise, Vol. 2. Elsevier, Amsterdam, pp. 175–204.

Gottsberger, G., and I. Silbauer-Gottsberger. 1988. Evolution of flower structures and pollination in Neotropical Cassiinae (Caesalpiniaceae) species. Phyton **28**:293–320.

Grunmeier, R. 1990. Pollination by bats and non-flying mammals of the African tree, *Parkia bicolor* (Mimosaceae). Mem. N.Y. Bot. Gard. **55**:83–104.

Guilford, T. 1988. The evolution of conspicuous coloration. Am. Nat. 131 Suppl.:S7–21.

Haig, D. 1990. New perspectives on the Angiosperm female gametophyte. Bot. Rev. **56**:236–274.

Haig, D. and M. Westoby. 1987. Inclusive fitness, seed resources and maternal care. *In* J. Lovett Doust and L. Lovett Doust, eds., Plant Reproductive Ecology. Oxford University Press, New York, pp. 60–79.

Hamilton, W.D. 1963. The evolution of altruistic behaviour. Am. Nat. **97**:354–356.

Hamilton, W.D. 1964. The genetical evolution of social behavior, I, II. J. Theor. Biol. **7**:1–51.

Hamilton, W.D. 1987. Discriminating nepotism: Expectable, common, overlooked. *In* D.J.C. Fletcher and C.D. Michener, eds., Kin Recognition in Animals. John Wiley, New York, pp. 417–437.

Harper, J.L., P.H. Lovell, and K.G. Moore. 1970. The shapes and sizes of seeds. Annu. Rev. Ecol. Syst. **1**:327–356.

Kerner, A. 1898. Pflanzenleben, Vol. 2, Ed. 2. Bibliographisches Institut, Leipzig.

Kevan, P.G., D. Eisikowitch, J.D. Ambrose, and J.R. Kemp. 1990. Cryptic dioecy and insect pollination in *Rosa setigera* Michx. (Rosaceae), a rare plant of Carolinian Canada. Biol. J. Linn. Soc. 40:229–243.

Kimura, M. 1959. Conflict between self-fertilisation and outbreeding in plants. Annu. Report. Natl. Inst. Genet. Jpn. **9**:87–88.

Knight, T.A. 1799. An account of some experiments on the fecundation of vegetables. Phil. Trans. R. Soc. London 1799. Part II:195–204.

Law, R., and C. Cannings. 1984. Genetic analysis of conflicts arising during development of seeds in the Angiospermophytes. Proc. R. Soc. London Ser. B **221**:53–70.

Lee, R.E. 1961. Pollen dimorphism in *Tripogandra grandiflora*. Baileya **9**:53–56.

Lee, T.D. 1988. Patterns of fruit and seed reproduction. *In* J. Lovett Doust and L. Lovett Doust, eds., Plant Reproductive Ecology. Oxford University Press, New York, pp. 179–202.

Lively, C.M., and D.G. Lloyd. 1990. The cost of biparental sex under individual selection. Am. Nat. **135**:489–500.

Lloyd, D.G. 1975. The maintenance of gynodiecy and androdioecy in angiosperms. Genetica **45**:325–339.

Lloyd, D.G. 1977. Genetic and phenotypic models of natural selection. J. Theor. Biol. **69**:543–560.

Lloyd, D.G. 1979. Parental strategies of angiosperms. New Zealand J. Bot. **17**:595–606.

Lloyd, D.G. 1987. Selection of offspring size at independence and other size-versus-number strategies. Am. Nat. **129**:800–817.

Lloyd, D.G. 1989. The reproductive ecology of plants and eusocial animals. *In* P.J. Grubb and J.B. Whittaker, eds., Towards a More Exact Ecology. Blackwell, Oxford, pp. 185–208.

Lloyd, D.G. 1993. The Theory of Adaptive Strategies. Oxford University Press, Oxford. In press.

Lloyd, D.G., and M.S. Wells. 1992. Reproductive biology of a primitive angiosperm, *Pseudowintera colorata* (Winteraceae) and the evolution of pollination systems in the Anthophyta. Plant Syst. Evol. In press.

Mattson, O. 1976. The development of dimorphic pollen in *Tripogandra* (Commelinaceae). *In* K. Ferguson and J. Muller, eds., The Evolution of the Exine. Academic Press, New York, pp. 163–183.

Maynard Smith, J. 1981. The evolution of social behavior—a classification of models. *In* King's College Sociobiology Group, ed., Current Problems in Sociobiology. Cambridge University Press, Cambridge, pp. 29–44.

Maynard Smith, J. 1989. Evolutionary Genetics. Oxford University Press, Oxford.

Mildbraed, J. 1954. Die Schausamen von *Paeonia corallina* Retz. Ber. Deutsch. Bot. Gesellschaft **67**:73–74.

Müller H. 1881. Two kinds of stamens with different functions in the same flower. Nature (London) **24**:307–308.

Müller, H. 1883. The Fertilisation of Flowers. Macmillan, London.

Mori, S.A., and G. Prance. 1987. A guide to collecting Lecythidaceae. Ann. Missouri Bot. Gard. **74**:321–330.

Mori, S.A., and G. Prance. 1990. Taxonomy, ecology, and economic botany of the Brazil nut (*Bertholletia excelsa* Humb. & Bonpl.: Lecythidaceae). Adv. Econ. Bot. **8**:130–150.

Norlindh, T. 1977. Arctoteae—systematic review. *In* V.H. Heywood, J.B. Harborne, and B.L. Turner, eds., The Biology and Chemistry of the Compositae, Vol. II. Academic Press, London, pp. 943–959.

Nitsch, J.P. 1952. Plant hormones and the development of fruits. Quart. Rev. Biol. **27**:33–57.

Nitsch, J.P. 1971. Perennation through seeds and other structures. Fruit development. *In* F.C. Steward, ed., Plant Physiology: A Treatise. Academic Press, New York, pp. 413–501.

Parker, G.A., and M.R. Macnair. 1978. Models of parent-offspring conflict. I. Monogamy. Anim. Behav. **26**:97–110.

Philipson, W.R. 1979. Araliaceae—I. Flora Malesiana, Ser. I, Vol. 9, Kluwer, Dordrecht. pp. 1–105.

Proctor, M., and P. Yeo. 1973. The Pollination of Flowers. Collins, London.

Queller, D. 1983. Kin selection and conflict in seed maturation. J. Theor. Biol. 100:153–172.

Queller, D. 1984. Models of kin selection on seed provisioning. Heredity 53:151–165.

Queller, D. 1989. Inclusive fitness in a nutshell. Oxford Surv. Evol. Biol. 6:73–109.

Renner, S. 1989. A survey of reproductive biology in neotropical Melastomataceae and Memecylaceae. Ann. Missouri. Bot. Gard. 76:496–518.

Simpson, B.B., J.L. Neff, and G. Dieringer. 1986. Reproductive biology of *Tinantia anomala*. Bull. Torrey Bot. Club 113:149–158.

Smith, C.C., and S.D. Fretwell. 1974. The optimal balance between size and number of offspring. Am. Nat. 108:499–506.

Stebbins, G.L. 1971. Adaptive radiation of reproductive characteristics in angiosperms. II. Seeds and seedlings. Annu. Rev. Ecol. Syst. 2:237–260.

Stephenson, A.G. 1981. Flower and fruit abortion: Proximate causes and ultimate functions. Annu. Rev. Ecol. Syst. 12:253–279.

Stubblefield, J.W., and E.L. Charnov. 1986. Some conceptual issues in the origin of eusociality. Heredity 57:181–187.

Stuessy, T.F. 1977. Heliantheae—systematic review. *In* V.H. Heywood, J.B. Harborne, and B.L. Turner, eds., The Biology and Chemistry of the Compositae, Vol. II. Academic Press, London, pp. 621–671.

Trivers, R.L. 1972. Parental investment and sexual selection. *In* B. Campbell, ed., Sexual Selection and the Descent of Man 1871–1971. Aldine, Chicago, pp. 136–179.

Trivers, R.L. 1974. Parent-offspring conflict. Am. Zool. 14:249–264.

Trivers, R.L. 1985. Social Evolution. Bemjamin/Cummings, Menlo Park, CA.

Uma Shaanker, R.U., K.N. Ganeshaiah, and K.S. Bawa. 1988. Parent-offspring conflict, sibling rivalry, and brood size patterns in plants. Annu. Rev. Ecol. Syst. 19:177–205.

Uyenoyama, M. 1988. On the evolution of genetic incompatibility systems: Incompatibility as a mechanism for the regulation of outcrossing distance. *In* R.E. Michod and B.R. Levin, eds., The Evolution of Sex. Sinauer, Sunderland, MA, pp. 212–232.

van der Pijl, L. 1955. Some remarks on myrmecophytes. Phytomorphology 5:190–199.

van der Pijl, L. 1969. Principles of Dispersal in Higher Plants. Springer-Verlag, Berlin.

Verkerke, W. 1989. Structure and function of the fig. Experientia 45:612–622.

Vogel, S. 1978. Evolutionary shifts from reward to deception in pollen flowers. *In* A.J. Richards, ed., The Pollination of Flowers by Insects. Academic Press, London, pp. 89–96.

Weberling, F. 1989. Morphology of Flowers and Inflorescences. Cambridge University Press, Cambridge.

West Eberhard, M.J. 1975. The evolution of social behavior by kin selection. Quart. Rev. Biol. 50:1–33.

Westoby, M., and B. Rice. 1982. Evolution of seed plants and inclusive fitness of plant tissues. Evolution 36:713–724.

Willson, M.F. 1983. Plant Reproductive Ecology. John Wiley, New York.

Willson, M.F., and J. Ågren. 1989. Differential floral rewards and pollination by deceit in unisexual flowers. Oikos 55:23–29.

Wilson, E.O. 1971. The Insect Societies. Belknap Press, Cambridge, MA.

Young, H.J. 1986. Beetle pollination of *Dieffenbachia longispatha* (Araceae). Am. J. Bot. 73:931–944.

Zangerl, A.R., M.R. Berenbaum, and J.K. Nitao. 1991. Parthenocarpic fruit in wild parsnip: Decoy defence against a specialist herbivore. Evol. Ecol. 5:136–145.

8

Ecological Models of Plant Mating Systems and the Evolutionary Stability of Mixed Mating Systems

Kent E. Holsinger
University of Connecticut

Introductory biology texts present an oversimplified and distorted view of the ecology and evolution of plant reproduction. Typically, they contain a stylized diagram of a hermaphroditic flower and a discussion implying that flowering plants reproduce only when cross-pollinated. Of course, the real world is far more complicated and interesting than the one found in introductory biology texts. Allard (1975) guessed that nearly a third of flowering plant species are predominantly self-fertilizing, and Stebbins (1974: p. 51) suggested that the evolutionary pathway from cross-fertilization to self-fertilization "has probably been followed by more different lines of evolution in flowering plants than has any other." Understanding why this pathway has been so frequently followed is an important task for plant evolutionists, especially since the mating system is arguably the most important influence on the genetic structure of populations and species (e.g., Hamrick and Godt, 1989). Yet our understanding of the forces determining the evolution of plant mating systems has, in some ways, advanced only a little since Darwin (1859) first discussed the advantages of outcrossing in the *Origin of Species*.

Darwin (1859: pp. 96–97) argued that "close interbreeding diminishes vigour and fertility" and that as a result "no organic being self-fertilises itself for an eternity of generations." He used these observations to suggest that natural selection for outcrossing explains the evolution of complex floral morphologies in peas, *Lobelia fulgens*, and barberry. He even suggested that monoecy and dioecy in trees evolved as mechanisms to avoid geitonogamous self-fertilization, which he assumed to be prevalent in large trees with hermaphroditic flowers. Of course, he later developed these ideas in much more detail in his book on orchids (Darwin, 1862) and in a book devoted entirely to the effects of cross- and self-fertilization in plants (Darwin, 1876), but the principle remains the same. Inbred progeny are less fit than outbred progeny, causing natural selection to favor those forms that outcross. Thus, Darwin (1859) argued that self-fertilization will evolve only when it is necessary to ensure reproduction.

Although many botanists, including Stebbins (1957), Grant (1958), and Baker (1959), have identified potential advantages to selfing other than ensuring reproduction, they have tended to agree with Darwin (1859) that it demands an adaptive explanation (cf. Holsinger, 1988a). Geneticists, on the other hand, have tended to regard outcrossing as the condition that requires an explanation because, as Fisher (1941) showed, a gene that causes complete selfing will rapidly become fixed in a population unless opposed by some selective force. As different as these traditions seem, they share a common theme: that inbreeding depression is critical in determining when selfing will evolve. To botanists, inbreeding depression is important because it is an obstacle that must be overcome before selfing can evolve. To geneticists, it is important because it appears to be the only force that can oppose the spread of a selfing allele. Consequently, inbreeding depression has dominated theoretical and empirical investigations for many years, but recent theoretical work (Holsinger, 1991a) has suggested that understanding the biology of pollination may be equally critical.

In a pair of provocative papers Lande and Schemske (1985) and Schemske and Lande (1985) suggested that only complete selfing and complete outcrossing can be evolutionarily stable. They pointed out that recessive or partly recessive deleterious alleles are rapidly purged from any population undergoing partial selfing. Thus, one of the primary forces opposing the spread of a selfing variant is lessened as that variant increases in frequency, a result that is consistent with most theoretical models that have been analyzed to date (e.g., Fisher, 1941; Nagylaki, 1976; Maynard Smith, 1978; Lloyd, 1979; Wells, 1979; Charlesworth, 1980; Feldman and Christiansen, 1984; Holsinger et al., 1984). As Waller (1986) argued, however, the empirical evidence suggests that mixed mating systems can be evolutionarily stable. In particular, Aide (1986) pointed out that genetically estimated selfing rates span the range from zero to one, with many animal-pollinated taxa having intermediate rates. Both wind-pollinated and animal-pollinated plants appear to have a bimodal distribution of selfing rates (Barrett and Eckert, 1990), but many animal-pollinated plants have selfing rates between 20 and 80%, whereas intermediate selfing rates are far less common in wind-pollinated taxa. This difference in the frequency of intermediate selfing rates could be a sampling artifact (the wind-pollinated taxa included in the sample are almost entirely conifers and weedy grasses), but it might also be related to different functional relationships between the extent of pollen dispersal and selfing in wind- versus animal-pollinated taxa (Holsinger, 1988a).

Models in which inbreeding depression is the only force opposing spread of a selfing variant may be sufficient to explain the apparent bimodal distribution of selfing rates in wind-pollinated taxa, but they cannot explain the existence of mixed mating systems in animal-pollinated ones. In this chapter I review theoretical results on the role of inbreeding depression in mating system evolution, concluding that we cannot yet make empirically testable predictions about it. I also propose an alternative approach to the study of plant mating systems that focuses on factors affecting the success of outcrossed reproduction.

Inbreeding Depression and the Evolution of Plant Mating Systems

Although the genetic basis of inbreeding depression is uncertain (Charlesworth and Charlesworth, 1987), there is no doubt that it is nearly universal. It is probably the most widely documented empirical phenomenon in plant evolutionary genetics (e.g., Knight, 1799; Darwin, 1876; Neal, 1935; Robinson et al., 1954; Schemske, 1983; Schoen, 1983; Kohn, 1988; Dudash, 1990; Holtsford and Ellstrand, 1990). Whether the plants studied are cultivated or wild, outcrossers or selfers, the progeny of selfed matings are often less vigorous and less fertile than those of outcrossed matings. This is especially true in outcrossers. In corn, for example, Neal (1935) found a 30% decrease in seed yield after a single generation of selfing. Inbreeding depression may be smaller in magnitude and more difficult to detect in predominant selfers (e.g., Barrett and Charlesworth, 1991), but it is often present. In tobacco, for example, Robinson et al. (1954) reported that leaf yield was increased by 2% and plant height by 6% after one generation of outcrossing. It is no wonder that most theoretical and experimental investigations of mating-system evolution in plants have concentrated on the role of fitness differences between selfed and outcrossed progeny (i.e., inbreeding depression).

Direct comparisons of progeny performance from controlled crosses are the most common means of measuring inbreeding depression, but Ritland (1990) has recently developed an alternative approach that can be used without making any crosses. Changes in the inbreeding coefficient, F, at marker loci are combined with an estimate of the selfing rate to infer the average fitness of selfed vs. outcrossed progeny. In general, three measurements of F or two measurements of F and one of the selfing rate, spanning two generations, are needed. If the F in adults is constant, however, measurement of the selfing rate and of F in one generation will suffice. Not only does Ritland's method eliminate the need to make many controlled crosses, it ensures that the fitness consequences of selfing are measured in a natural environment and minimizes any artificial disturbance associated with transplanting selfed and outcrossed seed. Unfortunately, enormous sample sizes (on the order of thousands of individuals) may be necessary to reduce standard errors below about 30% of the expected estimate (Ritland, 1990). Because its statistical power is greatest when several multiallelic marker loci are available in a species with a moderate degree of selfing, this method may not be useful for highly selfing or highly outcrossing species.

Regardless of whether the relative performance of selfed and outcrossed progeny is measured directly or inferred from changes in the inbreeding coefficient, however, the objective remains the same: to determine the average fitness effect of inbreeding in the population. Thus, if w_i is the average fitness of selfed progeny and w_o is the average fitness of outbred progeny, inbreeding depression is defined as $1 - w_i/w_o$. I shall refer to this quantity as the *population inbreeding depression*. One could be forgiven for thinking that population inbreeding depression is the only fitness effect of inbreeding, as it is the only effect that is commonly discussed. But individuals may also differ in the extent to which inbreeding lowers the fitness of their progeny, and these differences may have a profound impact on the course of mating system evolution. If $w_{i,k}$ is the fitness of selfed progeny from the kth individual and $w_{o,k}$ is the fitness of outcrossed progeny from that individual, the *sibship inbreeding depression* for individual k is simply $1 - w_{i,k}/w_{o,k}$.

Most models and most experimental investigations of mating system evolution have assumed that these two quantities are equivalent. In particular, most theoretical studies assume that sibship inbreeding depression is independent of the genotype at loci determining the mating system (e.g., Lloyd, 1979; Charlesworth, 1980; Feldman and Christiansen, 1984; Holsinger et al., 1984; Holsinger, 1986), an assumption we now know to be false (Holsinger, 1988b). To see why it is false, we must consider a more explicit genetic scenario. Imagine a population in which there is genetic variation at two separate sets of loci, one set determining fitness and the other determining selfing rates. Genotypes at these two sets of loci will become associated with one another, even if they are unlinked, for two reasons: (1) the tendency for a heterozygous genotype at one locus to be associated with heterozygous genotypes at other loci (identity disequilibrium: Haldane, 1949; Bennett and Binet, 1956; Weir and Cockerham, 1973) and (2) the tendency for mating system alleles that increase variability in offspring fitness to become associated with high fitness genotypes (Charlesworth and Charlesworth, 1990; Charlesworth et al., 1990; Uyenoyama and Waller, 1991a,b,c).

Identity disequilibrium arises because an individual that is heterozygous at one locus is more likely to have been produced by outcrossing than is a randomly chosen individual. Given that this individual is more likely to have been produced through outcrossing than a randomly chosen individual, it is also more likely to be heterozygous at other loci than a randomly chosen individual. In short, there is an excess of multiple heterozygotes and multiple homozygotes in a partially selfing population (Haldane, 1949; Bennett and Binet, 1956; Weir and Cockerham, 1973). Thus genotypes at mating system and fitness loci will tend to be associated, even if there is no gametic disequilibrium between the loci (Holsinger, 1988b). Nevertheless, the associations that develop between fitness and mating system loci are more complex than that.

The fitness of outcrossed progeny is independent of their genotype at the mating system loci, because a single generation of outcrossing eliminates the identity disequilibrium among outcrossed progeny (Weir and Cockerham, 1973; Charlesworth and Charlesworth, 1990). The fitness of selfed progeny, however, *does* depend on their genotype at the mating system loci. To see why, recall that predominantly outcrossing species usually show more severe effects of inbreeding than those that are predominantly selfing (Jain, 1976; Wright, 1977; Charlesworth and Charlesworth, 1987). Similarly, individuals that are predominantly outcrossing will have little inbreeding in their pedigree and show higher levels of individual inbreeding depression than individuals that are predominantly selfing, even if they occur in the same population. Moreover, alleles promoting outcrossing may become associated with either high viability genotypes or low viability genotypes, depending on how inbreeding depression is genetically determined (Uyenoyama and Waller, 1991a,b,c).

Two basic hypotheses have been proposed to explain inbreeding depression. Both attribute inbreeding depression to the increase in homozygosity associated with inbreeding. The overdominance hypothesis is that homozygosity per se is disfavored because heterozygous genotypes are selectively favored. The partial dominance hypothesis is that homozygosity increases the frequency with which deleterious recessive or partially recessive deleterious alleles are expressed (Charlesworth and Charlesworth, 1987). With heterozygous advantage, the average fitness of outcrossed offspring is greater than that of selfed offspring because the outcrossed offspring are more highly heterozygous. Because outcrossed offspring are produced more frequently by individuals with a genotype that promotes outcrossing than by those with a genotype that promotes selfing, alleles promoting outcrossing become associated with high viability genotypes. With recessive or partially recessive deleterious alleles, however, the genotype that outcrosses will carry many deleterious alleles, whereas the genotype that selfs will carry few, causing alleles that promote outcrossing to become associated with low viability genotypes (cf. Holsinger, 1991b). The effect of these associations may be as great as that of the population inbreeding depression itself.

Simple models usually treat sibship inbreeding depression and population inbreeding depression as equivalent (e.g., Lloyd, 1979; Charlesworth, 1980;

Feldman and Christiansen, 1984; Holsinger et al., 1984; Holsinger, 1986). These models suggest that genotypes promoting outcrossing are unconditionally favored if inbreeding depression is >0.5. Conversely, genotypes promoting selfing are unconditionally favored if inbreeding depression is < 0.5. In models that include genotypic associations between viability and mating system loci, however, this threshold value for population inbreeding depression gives the correct prediction only when the population is primarily outcrossing and the difference in selfing rates among genotypes is small (Holsinger, 1988b; Charlesworth and Charlesworth, 1990; Charlesworth et al., 1990). Simulations of a full genetic model (Holsinger, 1988b) have shown, for example, that when inbreeding depression results from the expression of recessive lethals a genotype that selfs 60% of the time will invade an outcrossing population, even if the population inbreeding depression is 70%.

These results make it clear that differences in *sibship* inbreeding depression are at least as important as the *population* inbreeding depression in determining the fate of alleles at mating system loci. Unfortunately, predicting mating system evolution from differences in sibship inbreeding depression requires more information on the nature of selection at fitness loci than we now have. We must know whether the overdominance hypothesis, the partial dominance hypothesis, or some combination of the two is the most accurate description for the genetic basis of inbreeding depression; we must know the average selection coefficient against deleterious alleles and the average fitness advantage of heterozygotes; and we must know the rate at which mutation introduces new genetic variation in fitness. We could, in principle, restrict our studies to primarily outcrossing populations and assume that only variants with a small effect on the mating system ever arise. The threshold value of 0.5 for population inbreeding depression would then determine whether partially selfing variants can invade, but even this makes sense only if the partial selfers are as successful as outcross-pollen parents as are the outcrossers. If the partial selfers are less successful than the outcrossers, the threshold of 0.5 is not even the appropriate standard for comparison.

A Mass-Action Approach to Plant Mating Systems

To see why the reproductive success of partial selfers as pollen parents should matter, recall Fisher's (1941) argument for the spread of an allele causing selfing. An outcrosser will have, on average, two gametes represented in the next generation: one as a seed parent and one as a pollen parent. But a selfer will have three gametes represented in the next generation: two in its selfed progeny and one as a pollen parent of outcrossed progeny of other individuals in the population. The 50% advantage of selfing arises from the assumption that selfing has no effect on reproductive success through outcross pollen. To the extent that selfing does reduce that component of reproductive success, the advantage of selfing is also reduced. If it eliminates that component of reproductive success, the advantage of selfing is also eliminated (Nagylaki, 1976; Wells, 1979; Charlesworth, 1980; Feldman and Christiansen, 1984; Holsinger et al., 1984). The fractional reduction in reproductive success through outcross pollen that occurs because of selfing is known as the "discounting rate" (Holsinger et al., 1984).

Why should we think that selfing has any effect on reproductive success through outcross pollen? The obvious answer is that a plant can self only if it does not export all of its pollen. To the extent that it reduces the amount of pollen it exports in order to self, it reduces its reproductive success through outcross pollen. Of course, the fraction of pollen grains produced that are used in selfing may be miniscule. Thomson and Thomson (1989), for example, have estimated that <0.5% of the pollen removed from anthers of *Erythronium grandiflorum* participates in self-pollination. The rest is either wasted or used in cross-pollination. This would appear to suggest that selfing has only a minor effect on reproductive success through outcross pollen in *E. grandiflorum*. To conclude this, however, would be to miss an important point. Whether or not a partially selfing variant would be favored in *E. grandiflorum* depends not on the fraction of pollen it uses in selfing, but on whether the resulting decrease in its rate of pollen export leads to an even greater increase in its selfing rate. If it does, the decreased reproductive success through outcross pollen is more than compensated by the additional reproductive success through selfed progeny, and the selfing variant will increase in frequency. If not, it will be eliminated (Nagylaki, 1976; Charlesworth, 1980; Holsinger et al., 1984).

Recognizing how important the functional relationship between selfing rate and discounting rate is, another important point becomes clear. Geneticists especially, but botanists as well, have tended to think of self-*fertilization* as the thing that evolves, the thing that there is genetic variation for. What there is really genetic variation for, of course, are reproductive syndromes that either promote or discourage self-*pollination*. The rate of self-fertilization is the fraction of ovules produced by an individual that are self-fertilized, whereas the rate of self-pollination is the fraction of self pollen on receptive stigmas. The rate of self-fertilization is obviously related to the rate of self-pollination, but they need not be equal. Most self-incompatible plants experience some degree of self-pollination, as evidenced by intramorph pollen flow in heterostylous plants (e.g., Crosby, 1949; Ornduff, 1976; Piper et al., 1984; Vekemans et al., 1990) and an abundance of ungerminated pollen grains on stigmas of many naturally pollinated self-incompatible plants (Plitmann and Levin, 1990). Even in self-compatible plants, phenomena like cryptic self-incompatibility (Bateman, 1956; Bowman, 1987) and competition between self and outcross pollen to fertilize available ovules (Glover and Barrett, 1986) may precluce a strict equality between the fraction of self pollen grains on the stigma and the fraction of ovules self-fertilized.

More importantly, the fraction of pollen grains on a stigma that are self grains (i.e., the rate of self-pollination) depends not only on the number of self grains deposited there, but on the number of outcross grains received. The number of outcross grains received in turn depends on the number of plants from which an individual can expect to receive pollen, the fraction of their pollen that these individuals export, and the efficiency with which exported pollen is captured. If we assume that the fraction of self-fertilized ovules depends on the relative amount of self and outcross pollen on stigmas, then the selfing rate of an individual can be written as a simple function of the rate of pollen export (Fig. 8.1).

Calculating the amount of self pollen a plant receives is a relatively simple matter. Let $1 - \delta$ be the fraction of the pollen produced by an individual that is exported (i.e., δ is the discounting rate). If each plant in the population produces N pollen grains, then $N\delta$ of them remain on the plant that produced them. Of course, not all of this pollen is effective in self-fertilization. To reflect this fact, let ε be the fraction of self pollen that is effective. Then the total number of effective self pollen grains received by

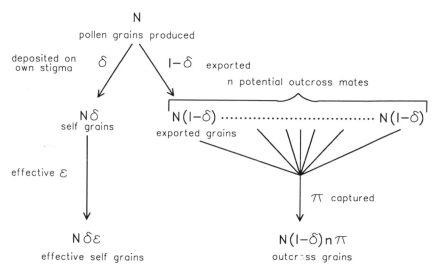

Figure 8.1. Schematic diagram of the processes included in the mass-action model. Refer to text for details and definitions of the model parameters.

any plant is $N\delta\varepsilon$. Calculating the amount of outcross pollen a plant receives is only a little more difficult. Each plant in the population exports $N(1 - \delta)$ pollen grains, but only plants that grow close to one another are likely to exchange pollen. Let n be the average number of potential outcross mates that any plant has. Then the total number of pollen grains that an individual could capture, the outcross pollen pool, is $N(1 - \delta)n$. Of course, no plant can capture all of the outcross pollen available, so let π be the fraction of all available outcross pollen than any plant captures. Then the number of outcross pollen grains received by any plant is $N(1 - \delta)n\pi$. Thus, the selfing rate of an individual is simply

$$\sigma = \frac{\text{number of self}}{\text{number of self} + \text{number of outcross}}$$

$$= \frac{N\delta\varepsilon}{N\delta\varepsilon + N(1 - \delta)n\pi}$$

$$= \frac{\delta}{\delta + (1 - \delta)n(\pi/\varepsilon)}$$

If genotypes differ in their discounting rate, the selfing rate of genotype A_iA_j can be written as

$$\sigma_{ij} = \frac{\delta_{ij}}{\delta_{ij} + (1 - \delta_{11}x_{11} - \delta_{12}x_{12} - \delta_{22}x_{22})n(\pi/\varepsilon)} \tag{1}$$

where δ_{ij} is the discounting rate of genotype A_iA_j, x_{ij} is the frequency of genotype A_iA_j in the population, and the remaining parameters are as above (Holsinger, 1990, 1991a; cf. Gregorius et al., 1987).

Notice that high rates of self-fertilization are compatible with very small amounts of pollen being used for selfing, provided that the success rate of outcross pollen relative to that of self pollen is sufficiently low. Suppose, for example, that 0.5% of the pollen a plant produces is used for selfing. If $n\pi/\varepsilon = 0.001$, then the selfing rate in the population is over 80%. Similarly, low rates of self-fertilization are compatible with very large amounts of pollen being used for selfing, if the success rate of outcross pollen relative to that of self pollen is sufficiently high. For example, if 50% of the pollen a plant produces is used for selfing, the selfing rate might be as low as 10% if $n\pi/\varepsilon = 5$.

Equation (1) embodies in mathematical form an assumption that I have called the "mass-action assumption," because it is analogous to mass-action physical models in which the rate of a process is determined by the frequencies of its parts. It is similar to a formulation used by many other authors for studying the dynamics of self-incompatibility systems (e.g., Finney, 1952; Charlesworth and Charlesworth, 1979; Uyenoyama, 1988). This simple description of plant mating systems leads to two predictions that are easily tested: (1) outcrossing rates will be higher in populations where outcrossers have many potential mates than in those where they have few, and (2) in species with a mating system polymorphism, the selfing rate of any particular genotype will increase as the frequency of the more highly selfing genotypes increases. In short, if the mass-action assumption is correct, selfing rates should be both density and frequency dependent. There is some support for both predictions.

The relationship between population density and selfing rates has not been extensively studied in self-compatible species, but it does appear that selfing rates in low-density populations are higher than in high-density populations. The selfing rate in low-density stands of *Pinus ponderosa*, for example, is about 15%, whereas it is only about 4% in high-density stands (Farris and Mitton, 1984). Similarly, population size and selfing rate are negatively correlated in Jamaican populations of *Eichhornia paniculata* (Barrett and Husband, 1990), and the selfing rate in low-density populations of *Cuphea lutea* is 20–30% lower than in high-density populations (Krueger and Knapp, 1991). The relationship between selfing rates and the frequency of different mating system morphs has also been little studied, but again the data are consistent with the mass-action assumption, suggesting that selfing rates are highest in populations with a high frequency of the more highly selfing morph. For example, the nonradiate morph of *Senecio vulgaris* selfs more frequently than the radiate morph, and an increase in the frequency of the nonradiate morph from 0.297 to 0.614 in an Edinburgh population was accompanied by an increase in the selfing rate of the radiate morph from 0.643 to 0.955 (Marshall and Abbott, 1984). Similarly, white-flowered morphs are associated with increased rates of selfing in *Ipomoea purpurea*: the selfing rate of the white morph increased from 3% in experimental populations composed of equal proportions of white and colored morphs to 16% in populations composed of 75% white morphs (Epperson and Clegg, 1987). Clearly, much more experimental work is required before we can say with confidence whether the mass-action assumption is correct, but the preliminary evidence is encouraging.

This discussion began with the assertion that the functional relationship between the rate of self-fertilization and the discounting rate could have a major impact on the evolution of plant mating systems. If we accept for the moment that the mass-action assumption is an accurate description of the mating process in self-compatible plants, what precisely are the implications? Three important predictions have emerged from the analysis of a model incorporating this assumption (Holsinger, 1991a): (1) outcrossing may be favored even in the absence of inbreeding depression; (2) complete selfing is evolutionarily stable only if reproduction is pollen limited; and (3) mixed mating systems are evolutionarily stable whenever selfing can evolve. Each of these predictions is a direct result of the frequency-dependent ordering of reproductive success associated with a mass-action description of plant mating systems. None of them requires the fitness of selfed progeny to be lower than that of outcrossed progeny.

Consider, for example, invasion of a completely selfing population by a partial outcrosser. Complete selfing can occur only if the population is fixed for a genotype that exports no pollen. Thus, another way to describe a partial outcrosser is to say that it exports some of the pollen that it produces. A genotype that exports some pollen will, when rare, have the same selfing rate as the resident selfer that exports no pollen, because it receives no outcross pollen. Unlike the resident selfer, however, it can also donate gametes to the outcrossed progeny of other individuals through its pollen. Thus, a pollen exporter will actually have a higher mean reproductive success than a complete selfer; hence, complete selfing is never evolutionarily stable. This is similar to the scenario for the evolution of dispersal described by Hamilton and May (1977), in which dispersal evolves to avoid local competition. The argument I have just presented, however, depends on one critical assumption: that the number of seeds produced by the pollen exporter is the same as the number produced by the nonexporter. If reproduction is pollen limited, this assumption is violated and complete selfing may be evolutionarily stable (Holsinger, 1991a).

If complete selfing can occur only when the population is fixed for a genotype that exports no pollen, complete outcrossing can occur only if the population is fixed for a genotype that exports all of its pollen. A partially selfing variant is one that does not export all of its pollen, and it will invade a completely outcrossing population only if it gains more by selfing than it loses by decreasing its contribution to the outcrossed progeny of other individuals. Specifically, it will invade only if the selfing rate it experiences is greater than its discounting rate. When rare, its selfing rate is simply

$$\sigma = \frac{\delta}{\delta + n\pi/\varepsilon}$$

implying that it will invade only when $\delta < 1 - n\pi/\varepsilon$ (Holsinger, 1991a). Note that $n\pi$ measures not only the amount of outcross pollen that a plant receives but also the success rate of exported pollen grains. Specifically, $n\pi$ is the fraction of exported pollen grains that land on the stigmas of another plant. If the success rate of exported pollen exceeds that of non-exported pollen (i.e., $n\pi > \varepsilon$), this inequality can never be satisfied, since $\delta > 0$ for a partially selfing variant. Thus, such a variant will always be eliminated. If, on the other hand, the success rate of exported pollen is less than that of nonexported pollen, a partially selfing variant can invade. In short, selfers may have a lower mean reproductive success than outcrossers whenever the success rate of exported pollen is high, either because the number of potential outcross mates is high or because pollinator service is very reliable. Another way of saying this is that outcrossing is favored when the chances of successful outcrossed reproduction are high.

Given that self-fertilization can evolve when the chances of outcrossed reproduction are low and complete selfing is evolutionarily stable only when reproduction is pollen limited, it should not be surprising that mixed mating systems are evolutionarily stable whenever selfing can evolve (Holsinger, 1991a). This conclusion holds in the absence of a difference in fitness between selfed and outcrossed progeny; including inbreeding depression, at least in the form of population inbreeding depression, appears to have little effect on these results. Preliminary analyses of a model including inbreeding depression suggest that the primary effect of inbreeding depression is to lower the evolutionarily stable selfing rate (Holsinger, 1990). The prediction that mixed mating systems can be evolutionarily stable stands in stark contrast to the results implicit in almost every other model of mating system evolution in plants, in which only complete outcrossing or complete selfing can be evolutionarily stable (e.g., Fisher, 1941; Nagylaki, 1976; Maynard Smith, 1978; Lloyd, 1979; Wells, 1979; Charlesworth, 1980; Feldman and Christiansen, 1984; Holsinger et al., 1984). In fact, Lande and Schemske (1985) and Schemske and Lande (1985) made this implication explicit. According to their view, rare cases of plants that have a mixed mating system merely represent transient stages in the evolution of complete selfing from complete outcrossing.

As Waller (1986) argued, however, the empirical evidence suggests that a plant with a mixed mating system may represent a stable endpoint of the evolutionary process. The best examples are plants like *Impatiens capensis* (Schemske, 1978), *Lamium amplexicaule* (Lord, 1979, 1982), and *Viola canina* (Darwin, 1877) in which an individual produces developmentally distinct chasmogamous (open-pollinated) and cleistogamous (self-pollinated) flowers (Lord, 1984), but there are many others. In *Clarkia tembloriensis*, for example, intermediate selfing has been maintained in one population for at least 20 years (Vasek and Harding, 1976; Holtsford and Ellstrand, 1989), and the morphology associated with this syndrome is found in herbarium specimens collected nearly a century ago. Moreover, other populations of *Clarkia tembloriensis* are almost entirely outcrossing, and the partially selfing ones are confined to the margin of its range, suggesting that the change in mating system may represent an adaptation to reduced pollinator service. Similarly, Rick et al. (1977) reported a regular cline in selfing rates of *Lycopersicon pimpinellifolium*, ranging from

90–100% in southern Peru to 30–40% in northern Peru, and Schoen (1982) found heritable variation for morphological features associated with selfing in *Gilia achilleifolia* and documented a cline in selfing rates from 4% in southern California to 85% in central California.

Although other theoretical models have predicted the existence of stable mixed mating systems, they have done so only for a small range of parameter values in four cases: when self-fertilization reduces the probability of successful reproduction (Lloyd, 1979; Holsinger, unpublished), when both selfing and biparental inbreeding are occurring (Uyenoyama, 1986), when selfed progeny are at a disadvantage in dispersal (Holsinger, 1986), or when inbreeding depression is a result of overdominant selection (Campbell, 1986; Holsinger, 1988b; Uyenoyama and Waller, 1991b). Thus, the mass-action approach is unusual in predicting that mixed mating systems are evolutionarily stable whenever selfing can evolve, but that alone cannot make it a useful approach to the analysis of plant mating systems. It is useful only to the extent that it can help us to understand the evolutionary dynamics of the mating system in real plants.

Applications of the Mass-Action Approach

The mating system polymorphism in *Senecio vulgaris* has been the subject of nearly a decade of study (e.g., Marshall and Abbott, 1980, 1982, 1984; Abbott, 1985, 1986; Ross and Abbott, 1987; Abbott and Irwin, 1988; Abbott et al., 1988), and it provides a useful system for the study of mating system evolution. Many populations are polymorphic for flower type, a trait that is controlled by alternative alleles at a single locus (Trow, 1912), and differences in selfing rate are associated with differences in flower type. Radiate plants (those with flowering heads bearing both ray and disc flowers) often outcross as much as 13–20%, but nonradiate plants (those with heads bearing only disc flowers) rarely outcross more than 1% (Marshall and Abbott, 1982, 1984). The polymorphism appears to have evolved in Britain during the mid-1800s (Stace, 1977) through introgression of the ray-floret allele into *S. vulgaris* from the introduced *S. squalidus* (Ingram et al., 1980; Marshall and Abbott, 1980). Over the past century the radiate variant has spread to many parts of Britain, where it forms large, polymorphic populations. There are, however, no large populations that are monomorphic for the radiate morph, and some populations have remained polymorphic for over 90 years.

Many maximum-likelihood techniques are now available for estimating selfing rates in natural populations (e.g., Clegg, 1980; Ritland, 1983, 1988; Brown, 1989), and it is easy to construct a similar maximum-likelihood model to estimate the parameters of the mass-action model. As with other maximum-likelihood techniques for mating system inference, these estimates are based on the distribution of genotypes in a progeny array. Complete details of the statistical procedures will be published elsewhere (Holsinger and Abbott, 1992), and a brief description will suffice for now.

Let $P_i\{n\}$ be the probability that n individuals can donate pollen to individual i. $P_i\{n\}$ can, in principle, be any discrete probability distribution (e.g., Poisson or negative binomial), but for definiteness, let us assume that it is Poisson. Given the maternal genotype, the conditional probability that an offspring is of a particular genotype is easily calculated from the selfing rate and the allele frequency in pollen (e.g., Clegg, 1980; Ritland, 1983). The selfing rate, σ, is given by Eq. (1) and the frequency of the ray-floret allele in the pollen is

$$p = [(1 - \delta_{11})x_{11} + (1 - \delta_{12})x_{12}/2]/(1 - \delta)$$

$$1 - \delta = 1 - \delta_{11}x_{11} - \delta_{12}x_{12} - \delta_{22}x_{22}$$

Let $Q_i\{n_{11}, n_{12}, n_{22}\}$ be the probability of obtaining the observed sample array from maternal plant i given p, the genotype of i, n, and the population genotype frequencies. Then the probability of obtaining the observed progeny array from plant i is $\Sigma P_i\{n\}Q_i\{n_{11}, n_{12}, n_{22}\}$. The likelihood of the entire sample is the product of these individual likelihoods across all maternal plants. We obtain estimates for π/ε, for n, and for rates of pollen export by maximizing this function with respect to those parameters, and confidence limits are determined either from the curvature of the likelihood surface or from bootstrap estimates (e.g., Syzmura and Barton, 1986; Holsinger, 1987).

Applying this estimation procedure to data from two natural populations of *Senecio vulgaris* reveals that the estimated rates of pollen discounting are stable through the season in Edinburgh and that the nonradiate morph retains far more of its pollen than the radiate morph in both the Edinburgh and Rhosllanerchrugog populations (Table 8.1). (A likelihood-ratio test shows that the difference in discounting rates between the radiate and nonradiate morphs is significant at the 0.001% level in both populations.) In addition, the rate of pollen discounting in radiate plants is close to zero in both the Edinburgh population and the Rhosllanerchrugog population, but the rate of pollen discounting in nonradiate plants is greater in the Edinburgh population than in the Rhosllanerchrugog population. The discounting rates in each of the samples examined is >80%, indicating that more than 80% of the pollen produced by nonradiate plants never leaves the plant on which it was produced.

A likelihood-ratio test for goodness-of-fit between the data and expectations derived from the model estimates (Mood et al., 1974) shows no

Table 8.1. Mass-Action Parameter Estimates for *Senecio vulgaris* Assuming Genotype-Independent Pollen Receipt[a]

Parameter estimates	Edinburgh			Rhosllanerchrugog July 1979
	April 1980	May 1980	August 1980	
$\delta_{radiate}$	0.000577	0.000000135	0.0148	0.00661
$\delta_{nonradiate}$	0.921	0.937	0.959	0.804
n	0.219	1.02	1.25	3.26
π/ε	0.00380	0.00738	0.0156	0.0560
Prediction	Fixation of nonradiate	Fixation of nonradiate	Fixation of nonradiate	Fixation of nonradiate

[a]$\delta_{radiate}$ is the discounting rate of the radiate morph; $\delta_{nonradiate}$ is the discounting rate of the nonradiate morph; n is the average number of potential outcross mates for any given plant in the population; and π/ε is effectiveness of outcross pollen capture relative to the success rate of self pollen. The line labeled prediction refers to the evolutionary consequences expected given the parameter estimates in the column above. These expectations were determined by iterating the model in Holsinger (1991a) with the parameter values given and determining the stable population configuration.

statistically significant departure from model expectations in samples from Edinburgh. A statistically significant departure from model expectations was found in the July 1979 sample from Rhosllanerchrugog. The poor fit in this population may be a result of spatial variation in genotype frequencies, because the maximum-likelihood procedure assumes that genotype frequencies are equal in all parts of the sample population, but Abbott (personal communication) noted that morphs in this population were spatially aggregated. A more serious difficulty for the mass-action approach is that the parameter estimates presented in Table 8.1 fail to explain the evolution and maintenance of the mating system polymorphism. Specifically, the ray-floret allele is eliminated for each set of parameters, regardless of its initial frequency.

How can it be that the mass-action model provides a reasonably accurate description of the mating system in *Senecio vulgaris*, yet fails to predict the evolutionary dynamics of the mating system polymorphism? There are at least three possibilities: (1) the amount of pollen a plant receives depends on its genotype because of differences in the frequency of pollinator visits; (2) the rates of pollen export are frequency dependent because pollinator preferences are frequency dependent; and (3) the survivorship of radiate plants is different from that of nonradiate plants either because of pleiotropic effects of the ray-floret allele or because of the effect of alleles at associated loci. All these possibilities are currently under investigation, but only the first can be assessed with the data currently available.

The basic mass-action model can be extended to genotype-dependent pollen receipt by allowing π, the fraction of the outcross-pollen pool captured by the stigmas of an individual plant, to depend on genotype. Thus,

$$\sigma_{ij} = \frac{\delta_{ij}}{\delta_{ij} + (1 - \delta_{11}x_{11} - \delta_{12}x_{12} - \delta_{22}x_{22})n(\pi_{ij}/\varepsilon)} \tag{2}$$

where π_{ij} is the fraction of the pollen exported by a plant's n neighbors that an individual of genotype A_iA_j captures and the remaining parameters are as in Eq. (1). The parameters of this model can be estimated using the same maximum-likelihood procedure described earlier, substituting Eq. (2) for Eq. (1) in the calculation of selfing rates. Not surprisingly, the model with genotype-dependent pollen receipt fits the data better than the model lacking it, although the improved fit is not statistically significant in any of the populations by a likelihood-ratio criterion (Table 8.2). But even this improved model fails to predict the maintenance of the mating system polymorphism. Thus, it appears likely that either frequency dependence in the mating system parameters or differential fitnesses associated with morph type are required to explain the evolution and maintenance of the polymorphism (cf. Ross and Abbott, 1987). In fact, the discounting rate of the nonradiate morph appears to be positively frequency dependent, and this may be sufficient to allow the polymorphism to become established (Holsinger and Abbott, 1992).

Ecology and Plant Mating Systems

The analysis of inbreeding depression has dominated experimental and theoretical analyses of plant mating systems for nearly 200 years, and rightly

Table 8.2. Mass-Action Parameter Estimates for *Senecio vulgaris* Assuming Genotype-Dependent Pollen Receipt[a]

Parameter estimates	Edinburgh			Rhosllanerchrugog July 1979
	April 1980	May 1980	August 1980	
δ_{radiae}	0.126	0.000473	0.000973	0.000188
$\delta_{nonradiate}$	0.934	0.943	0.959	0.807
n	0.220	1.47	1.25	3.79
π_{radiae}/ε	1.40	0.440	0.00105	7.82
$\pi_{nonradiate}/\varepsilon$	0.00442	0.00510	0.0151	0.0426
Prediction	Fixation of nonradiate	Fixation of nonradiate	Fixation of nonradiate	Fixation of nonradiate

[a]See footnote a to Table 8.1 for symbols and explanation.

so. Darwin, Grant, Stebbins, Baker, and other botanists saw it primarily as a barrier that must be overcome if selfing was ever to evolve. They thought that selfing could evolve only when it was adaptively advantageous (e.g., through ensuring reproduction or through producing locally adapted progeny that breed true for a well-adapted genotype). Population geneticists, on the other hand, saw it as the only force sufficient to oppose the reproductive advantage associated with selfing. In its absence, nothing could explain the prevalence of outcrossing. Although the role they attributed to it was very different in each case, both botanists and geneticists regarded inbreeding depression as a critical factor determining when selfing could evolve (cf. Holsinger, 1988a).

The mass-action approach described here is a hybrid of these traditions. It borrows the insight from population genetics that selfing may evolve simply because it is associated with a reproductive advantage, but it acknowledges the intuition from botany that ecological circumstances also play an important role in determining when selfing evolves. Botanists and geneticists alike have recognized for many years that selfing sometimes evolves simply because there is no other way for the plant to reproduce. It is, after all, better to have reproduced by selfing than never to have reproduced at all. But ecology can play a role much different from that. Any ecological variable that affects the amount of outcross pollen a plant receives (e.g., population density, rate of pollinator visitation, or average rate of pollen export) also affects whether selfers have a reproductive advantage at all.

Population geneticists have attributed a reproductive advantage to selfers because they assumed that an increase in the selfing rate can be accomplished without affecting a plant's ability to contribute to the pool of successful outcrossed male gametes. A plant's selfing rate depends, however, not only on how much pollen it deposits on its own stigmas, but on how much pollen it receives from its neighbors. In dense populations or those in which pollinator service is reliable, the density of outcross pollen on stigmas may be so great that to increase the selfing rate by 10% would require that over 20% of the pollen a plant produces be used for selfing. The decreased reproductive success through outcrossed male gametes may not be compensated by additional reproductive success through selfed male gametes, and selfing will be selected against. Notice that without a reproductive advantage for selfing, outcrossing may be favored even in the absence of inbreeding depression.

What role is there for inbreeding depression in understanding mating system evolution in this scenario? The mass-action models described here allow us to determine when selfers have a reproductive advantage and when they do not. If selfed progeny are less fit than outcrossed progeny, as is almost universally the case, knowing that selfers have a reproductive advantage tells us only part of the story. The much talked about threshold of $s = 0.5$ (e.g., Holsinger, 1988b; Charlesworth and Charlesworth, 1990; Charlesworth et al., 1990; Uyenoyama and Waller, 1991a,b,c), for example, is no longer the appropriate standard for determining when selfing will evolve in a mass-action model (Holsinger, 1990; unpublished). Nevertheless, the relative fitness of selfed and outcrossed progeny will be an important factor determining whether selfing evolves when the probability of successful outcrossed reproduction is low. Twenty years of research on the magnitude, causes, and consequences of inbreeding depression has added substantially to our understanding of plant mating systems and their evolution. The time has come to spend an equal effort on understanding the ecological processes that determine when outcrossed reproduction is likely to be successful.

Acknowledgments

Greg Anderson, Spencer Barrett, Rob Colwell, Michelle Dudash, Steve Pacala, Carl Schlichting, and Robert Wyatt provided valuable comments on an earlier version of this paper. This research was supported, in part, by grants from the University of Connecticut Research Foundation and the National Science Foundation (BSR-9107330).

Literature Cited

Abbott, R.J. 1985. Maintenance of a polymorphism for outcrossing frequency in a predominantly selfing plant. *In* J. Haeck and J. Woldendorp, eds., Structure and Partitioning of Plant Populations. II. Phenotypic and Genotypic Variation in Plant Populations. North-Holland Publ. Co., Amsterdam, pp. 277–286.

Abbott, R.J. 1986. Life-history variation associated with the polymorphism for capitulum type and outcrossing rate in *Senecio vulgaris* L. Heredity **56**:381–391.

Abbott, R.J., and J.A. Irwin. 1988. Pollinator movements and the polymorphism for outcrossing at the ray floret locus in groundsel, *Senecio vulgaris* L. Heredity **60**:295–298.

Abbott, R.J., J.C. Horrill, and G.D.G. Noble. 1988. Germination behavior of the radiate and non-radiate morphs of groundsel, *Senecio vulgaris* L. Heredity **60**:15–20.

Aide, T.M. 1986. The influence of wind and animal pollination on variation in outcrossing rates. Evolution **40**:434–435.

Allard, R.W. 1975. The mating system and microevolution. Genetics **79**:115–126.

Baker, H.G. 1959. Reproductive methods as factors in speciation in flowering plants. Cold Spring Harbor Symp. Quant. Biol. **24**:177–191.

Barrett, S.C.H., and D. Charlesworth. 1991. Effects of a change in the level of inbreeding on the genetic load. Nature (London) **352**:522–524.

Barrett, S.C.H., and C.G. Eckert. 1990. Variation and evolution of mating systems in seed plants. *In* S. Kawano, ed., Biological Approaches and Evolutionary Trends in Plants. Academic Press, London, pp. 230–254.

Barrett, S.C.H., and B.C. Husband. 1990. Variation in outcrossing rates in *Eichhornia paniculata*: The role of demographic and reproductive factors. Plant Species Biol. **5**:41–55.

Bateman, A.J. 1956. Cryptic self-incompatibility in *Cheiranthus cheiri* L. Heredity **10**:257–261.

Bennett, J.H., and F.E. Binet. 1956. Association between Mendelian factors with mixed random mating and selfing. Heredity **10**:51–56.

Bowman, R.N. 1987. Cryptic self-incompatibility and the breeding system of *Clarkia unguiculata* (Onagraceae). Am. J. Bot. **74**:471–476.

Brown, A.H.D. 1989. Genetic characterization of plant mating systems. *In* A.H.D. Brown, M.T. Clegg, A.L. Kahler, and B.S. Weir, eds., Plant Population Genetics, Breeding, and Genetic Resources. Sinauer, Sunderland, MA, pp. 145–162.

Campbell, R.B. 1986. The interdependence of mating structure and inbreeding depression. Theor. Pop. Biol. **30**:232–244.

Charlesworth, B. 1980. The cost of sex in relation to mating system. J. Theor. Biol. **84**:655–671.

Charlesworth, D., and B. Charlesworth. 1979. The evolution and breakdown of S-allele systems. Heredity **43**:41–55.

Charlesworth, D., and B. Charlesworth. 1987. Inbreeding depression and its evolutionary consequences. Annu. Rev. Ecol. Syst. **18**:273–268.

Charlesworth, D., and B. Charlesworth. 1990. Inbreeding depression with heterozygote advantage and its effect on selection for modifiers changing the outcrossing rate. Evolution **44**:870–888.

Charlesworth, D., M.T. Morgan, and B. Charlesworth. 1990. Inbreeding depression, genetic load, and the evolution of outcrossing rates in a multilocus system with no linkage. Evolution **44**:1469–1489.

Clegg, M.T. 1980. Measuring plant mating systems. BioScience **30**:814–818.

Crosby, J.L. 1949. Selection of an unfavourable gene-complex. Evolution **3**:212–230.

Darwin, C.D. 1859. On the Origin of Species by Means of Natural Selection, or Preservation of Favoured Races in the Struggle for Life. John Murray, London.

Darwin, C.D. 1862. On the Various Contrivances by which British and Foreign Orchids Are Fertilised by Insects, and on the Good Effects of Intercrossing. John Murray, London.

Darwin, D.D. 1876. The Effects of Cross- and Self-Fertilisation in the Vegetable Kingdom. John Murray, London.

Darwin, C.D. 1877. The Different Forms of Flowers on Plants of the Same Species. John Murray, London.

Dudash, M.R. 1990. Relative fitness of selfed and outcrossed progeny in a self-compatible, protandrous species, *Sabatia angularis* L. (Gentianaceae): A comparison in three environments. Evolution **44**:1129–1139.

Epperson, B.K., and M.T. Clegg. 1987. Frequency-dependent variation for outcrossing rate among flower-color morphs of *Ipomoea purpurea*. Evolution **41**:1302–1311.

Farris, M.A., and J.W. Mitton. 1984. Population density, outcrossing rate, and heterozygous superiority in Ponderosa pine. Evolution **38**:1151–1154.

Feldman, M.W., and F.B. Christiansen. 1984. Population genetic theory of the cost of inbreeding. Am. Nat. **123**:642–653.

Finney, D.J. 1952. The equilibrium of a self-incompatible polymorphic species. Genetics **26**:36–64.

Fisher, R.A. 1941. Average excess and average effect of a gene substitution. Ann. Eugen. **11**:53–63.

Glover, D.E., and S.C.H. Barrett. 1986. Variation in the mating system of *Eichhornia paniculata* (Spreng.) Solms (Pontederiaceae). Evolution **40**:1122–1131.

Grant, V. 1958. The regulation of recombination in plants. Cold Spring Harbor Symp. Quant. Biol. **23**:337–363.

Gregorius, H.-R., M. Ziehe, and M.D. Ross. 1987. Selection caused by self-fertilization. I. Four measures of self-fertilization and their effects on fitness. Theor. Pop. Biol. **31**:91–115.

Haldane, J.B.S. 1949. The association of characters as a result of inbreeding and linkage. Ann. Eugen. **15**:15–23.

Hamilton, W.D., and R.M. May. 1977. Dispersal in stable habitats. Nature (London) **269**:578–581.

Hamrick, J.L., and M.J. Godt. 1989. Allozyme diversity in plant species. *In* A.H.D. Brown, M.T. Clegg, A.L. Kahler, and B.S. Weir, eds., Plant Population Genetics, Breeding, and Genetic Resources. Sinauer, Sunderland, MA, pp. 43–64.

Holsinger, K.E. 1986. Dispersal and plant mating systems: The evolution of self-fertilization in subdivided populations. Evolution **40**:405–413.

Holsinger, K.E. 1987. Gametophytic self-fertilization in homosporous plants: Development, evaluation, and application of a statistical method for evaluating its importance. Am. J. Bot. **74**:1173–1183.

Holsinger, K.E. 1988a. The evolution of self-fertilization in plants: Lessons from population genetics. Oecol. Plant. **9**:95–102.

Holsinger, K.E. 1988b. Inbreeding depression doesn't matter: The genetic basis of mating system evolution. Evolution **42**:1235–1244.

Holsinger, K.E. 1990. The population genetics of mating system evolution in homosporous plants. Am. Fern J. **80**:148–155.

Holsinger, K.E. 1991a. Mass-action models of plant mating systems: The evolutionary stability of mixed mating systems. Am. Nat. (in press).

Holsinger, K.E. 1991b. Inbreeding depression and the evolution of plant mating systems. Trends Ecol. Evol. **6**:307–308.

Holsinger, K.E., and R.J. Abbott. 1992. In preparation.

Holsinger, K.E., M.W. Feldman, and F.B. Christiansen. 1984. The evolution of self-fertilization in plants: A population genetic model. Am. Nat. **124**:446–453.

Holtsford, T.P., and N.C. Ellstrand. 1989. Variation in outcrossing rate and population genetic structure of *Clarkia tembloriensis* (Onagraceae). Theor. Appl. Genet. **78**:480–488.

Holtsford, T.P., and N.C. Ellstrand. 1990. Inbreeding effects in *Clarkia tembloriensis* (Onagraceae) populations with different natural outcrossing rates. Evolution **44**:2031–2046.

Ingram, R., J. Weir, and R.J. Abbott. 1980. New evidence concerning the origin of inland radiate groundsel, *Senecio vulgaris* L. var. *hibernicus* Stone. New Phytol. **84**:543–546.

Jain, S.K. 1976. The evolution of inbreeding in plants. Annu. Rev. Ecol. Syst. **7**:469–495.

Knight, T. 1799. Experiments on the fecundation of vegetables. Phil. Trans. R. Soc. London **89**:195–204.

Kohn, J.R. 1988. Why be female? Nature (London) **355**:431–433.

Krueger, S.K., and S.J. Knapp. 1991. Mating systems of *Cuphea laminuligera* and *Cuphea lutea*. Theor. Appl. Genet. **82**:221–226.

Lande, R., and D.W. Schemske. 1985. The evolution of self-fertilization and inbreeding depression. Evolution **39**:24–40.

Lloyd, D.G. 1979. Some reproductive factors affecting the selection of self-fertilization in plants. Am. Nat. **113**:67–79.

Lord, E.M. 1979. The development of cleistogamous and chasmogamous flowers in *Lamium amplexicaule* L. (Labiatae): An example of heteroblastic inflorescence development. Bot. Gaz. **140**:29–50.

Lord, E.M. 1981. Cleistogamy: A tool for the study of floral morphogenesis, function, and evolution. Bot. Rev. **47**:421–449.

Lord, E.M. 1982. Floral morphogenesis in *Lamium amplexicaule* L. (Labiatae) with a model for the evolution of the cleistogamous flower. Bot. Gaz. **143**:63–72.

Lord, E.M. 1984. Cleistogamy: A comparative study of intraspecific floral variation. *In* R.A. White and W.C. Dickison, eds., Contemporary Problems in Plant Anatomy. Academic Press, New York, pp. 451–494.

Marshall, D.F., and R.J. Abbott. 1980. On the frequency of introgression of the radiate (T_r) allele from *Senecio squalidus* L. into *Senecio vulgaris* L. Heredity **48**:227–235.

Marshall, D.F., and R.J. Abbott. 1982. Polymorphism for outcrossing frequency at the ray floret locus in *Senecio vulgaris* L. I. Evidence. Heredity **48**:227–235.

Marshall, D.F., and R.J. Abbott. 1984. Polymorphism for outcrossing frequency at the ray floret locus in *Senecio vulgaris* L. II. Confirmation. Heredity **52**:331–336.

Maynard Smith, J. 1978. The Evolution of Sex. Cambridge University Press, Cambridge.

Mood, A.M., F.A. Graybill, and D.C. Boes. 1974. Introduction to the Theory of Statistics, 3rd ed. McGraw-Hill, New York.

Nagylaki, T. 1976. A model for the evolution of self-fertilization and vegetative reproduction. J. Theor. Biol. **58**:55–58.

Neal, N.P. 1935. The decrease in yielding capacity in advanced generations of hybrid corn. J. Am. Soc. Agron. **27**:666–670.

Ornduff, R. 1976. The reproductive system of *Amsinckia grandiflora*. Syst. Bot. **1**:57–66.

Piper, J.G., B. Charlesworth, and D. Charlesworth. 1984. A high rate of self-fertilization and increased seed fertility of homostyle primroses. Nature (London) **310**:50–51.

Plitmann, U., and D.A. Levin. 1990. Breeding systems in the Polemoniaceae. Plant Syst. Evol. **170**:205–214.

Rick, C.M., J.F. Fobes, and M. Hollé. 1977. Genetic variation in *Lycopersicon pimpinellifolium*: Evidence for evolutionary change in mating systems. Plant Syst. Evol. **127**:139–170.

Ritland, K.R. 1983. Estimation of plant mating systems. *In* S.D. Tanksley and T.J. Orton, eds., Isozymes in Plant Genetics and Plant Breeding, Part A. Elsevier, Amsterdam, pp. 289–302.

Ritland, K.R. 1988. Joint maximum-likelihood estimation of genetic and mating system structure using open-pollinated progenies. Biometrics **42**:23–43.

Ritland, K.R. 1990. Inferences about inbreeding depression based on changes of the inbreeding coefficient. Evolution **44**:1230–1241.

Robinson, H.F., T.J. Mann, and R.E. Comstock. 1954. An analysis of quantitative variability in *Nicotiana tabacum*. Heredity **8**:365–376.

Ross, M.D., and R.J. Abbott. 1987. Fitness, sexual asymmetry, functional sex and selfing in *Senecio vulgaris* L. Evol. Trends Plants **1**:21–28.

Schemske, D.W. 1978. Evolution of reproductive characters in *Impatiens* (Balsaminaceae): The significance of cleistogamy and chasmogamy. Ecology **59**:596–613.

Schemske, D.W. 1983. Breeding system and habitat effects on fitness components in three neotropical *Costus* (Zingiberaceae). Evolution **37**:523–539.

Schemske, D.W., and R. Lande. 1985. The evolution of self-fertilization and inbreeding depression in plants. II. Empirical observations. Evolution **39**:41–52.

Schoen, D.J. 1982. The breeding system of *Gilia achilleifolia*: Variation in floral characteristics and outcrossing rate. Evolution **36**:352–360.

Schoen, D.J. 1983. Relative fitness of selfed and outcrossed progeny in *Gilia achilleifolia*. Evolution **37**:292–301.

Stace, C.A. 1977. The origin of radiate *Senecio vulgaris* L. Heredity **39**:383–388.

Stebbins, G.L. 1957. Self-fertilization and population variability in the higher plants. Am. Nat. **91**:337–354.

Stebbins, G.L. 1974. Flowering Plants: Evolution Above the Species Level. Harvard University Press, Cambridge, MA.

Syzmura, J.M., and N.H. Barton. 1986. Genetic analysis of a hybrid zone between the fire-bellied toads, *Bombina bombina* and *B. variegata*, near Cracow in southern Poland. Evolution **40**:1141–1159.

Thomson, J.D., and B.A. Thomson. 1989. Dispersal of *Erythronium grandiflorum* pollen by bumblebees: Implications for gene flow and reproductive success. Evolution **43**:657–661.

Trow, A.H. 1912. On the inheritance of certain characters in the common groundsel, *Senecio vulgaris* L., and its segregates. J. Genet. **2**:239–276.

Uyenoyama, M.K. 1986. Inbreeding and the cost of meiosis: The evolution of selfing in populations practicing biparental inbreeding. Evolution **40**:388–404.

Uyenoyama, M.K. 1988. On the evolution of genetic incompatibility systems. II. Initial increase of strong gametophytic self-incompatibility under partial selfing and half-sib mating. Am. Nat. **131**:700–722.

Uyenoyama, M.K., and D.M. Waller. 1991a. Coevolution of self-fertilization and inbreeding depression. I. Genetic modification in response to mutation-selection balance at one and two loci. Theor. Pop. Biol. (in press).

Uyenoyama, M.K., and D.M. Waller. 1991b. Coevolution of self-fertilization and inbreeding depression. II. Symmetric overdominance in viability. Theor. Pop. Biol. (in press).

Uyenoyama, M.K., and D.M. Waller. 1991c. Coevolution of self-fertilization and inbreeding depression. III. Homozygous lethal mutations at multiple loci. Theor. Pop. Biol. (in press).

Vasek, F.C., and J. Harding. 1976. Outcrossing in natural populations. V. Analysis of outcrossing, inbreeding, and selection in *Clarkia exilis* and *Clarkia tembloriensis*. Evolution **30**:403–411.

Vekemans, X., C. Lefebvre, L. Belalia, and P. Meerts. 1990. The evolution and breakdown of the heteromorphic incompatibility system of *Armeria maritima* revisited. Evol. Trends Plants **4**:15–23.

Waller, D.M. 1986. Is there disruptive selection for self-fertilization? Am. Nat. **128**:421–426.

Weir, B.S., and C.C. Cockerham. 1973. Mixed self- and random-mating at two loci. Genet. Res. **21**:247–262.

Wells, H. 1979. Self-fertilization: Advantageous or disadvantageous? Evolution **33**:252–255.

Wright, S. 1977. Evolution and the Genetics of Populations. Vol. 3. Experimental Results and Evolutionary Deductions. University of Chicago Press, Chicago, IL.

9

Experimental Studies of Mating-System Evolution: The Marriage of Marker Genes and Floral Biology

Spencer C.H. Barrett, Joshua R. Kohn, and Mitchell B. Cruzan

University of Toronto

Introduction

Analysis of the causes and consequences of mating-system evolution is a major focus of research in plant population biology since shifts in mating pattern have important consequences for population genetic structure (Brown, 1979; Hamrick and Godt, 1990) and patterns of evolutionary diversification (Stebbins, 1974; Jain, 1976; Charlesworth, 1992). Empirical studies of the genetic consequences of contrasting mating systems have provided clear evidence of their role as primary determinants of the amounts and organization of genetic diversity within and among populations (e.g., Layton and Ganders, 1984; Holtsford and Ellstrand, 1989; Costich and Meagher, 1992). The evolutionary fate of genes that alter the mating system can therefore have large consequences for the future evolution of the species. Predicting the course of mating-system evolution is a complex problem, however, because of the numerous genetic, environmental, demographic, and historical factors that can influence mating patterns (Schemske and Lande, 1985; Brown et al., 1990; Barrett and Eckert, 1990).

During the past 15 years, a rich theoretical literature on the evolution of plant mating systems has developed. A variety of genetic and phenotypic models incorporating variation in mating, fertility, allocation patterns, differences in fitness between selfed and outcrossed offspring, and contrasting demographic, life history, and environmental variables have been used to explore the conditions that might favor the evolution of different levels of self- and cross-fertilization (e.g., Lloyd, 1979a; Wells, 1979; Ross and Gregorius, 1983; Holsinger et al., 1984; Lande and Schemske, 1985; Campbell, 1986; Uyenoyama, 1986; Charlesworth and Charlesworth, 1990; Iwasa, 1990; Holsinger, 1991; Yahara, 1992). The formulation of theoretical models of mating-system evolution has benefited from a recognition that floral traits influencing mating patterns may be subject to sexual selection and the optimal allocation of resources to female and male function (Charnov, 1982; Willson and Burley, 1983; Queller, 1983; Bell, 1985; Lyons et al., 1989). In addition, models have been developed to account for the evolution of particular reproductive systems such as heterostyly (Charlesworth and Charlesworth, 1979a; Muenchow, 1982; Lloyd and Webb, 1992a,b), self-incompatibility systems (Charlesworth and Charlesworth, 1979b; Uyenoyama, 1988), gynodioecy and dioecy (Lloyd, 1975; Charlesworth and Charlesworth, 1978; Givnish, 1982; Ross, 1982; Muenchow and Grebus, 1989), cleistogamy (Schoen and Lloyd, 1984), and more generally the evolution of combined versus separate sexes (Charlesworth and Charlesworth, 1981; Charnov, 1982; Lloyd, 1982). This work has provided the necessary conceptual framework for microevolutionary studies of mating systems based on models of individual selection. The approach contrasts with earlier largely intuitive group selection arguments involving "immediate fitness" and "long-term flexibility" that were proposed to account for the evolution of mating systems (Darlington, 1939; Stebbins, 1958; Grant, 1958).

Although these conceptual developments represent considerable progress for the study of mating-system evolution, it is clear that theoretical models have advanced more rapidly than the collection of relevant empirical data. For many of the important parameters in models of mating-system evolution (e.g., inbreeding depression, pollen discounting, biparental inbreeding, allocation patterns), few data are available from natural populations. As a result, their relative importance as factors influencing mating-system evolution are often difficult to evaluate.

Many generalizations concerning the evolutionary forces influencing mating-system evolution have been inferred from comparative studies of closely related taxa (Stebbins, 1957; Raven, 1979; Wyatt, 1988). In addition, a variety of ecological and geographical correlates have been identified as important for understanding particular mating-system shifts such as the evolution of self-fertilization (Baker, 1959; Jain, 1976; Lloyd, 1980; Barrett, 1989) and dioecy (Bawa, 1980; Thomson and Brunet, 1990; Weller and Sakai, 1990; Barrett, 1992). Paradoxically, similar conditions, such as bottlenecks, low density, and uncertain or altered pollinator regimes, have often been invoked to account for the evolution of selfing as well as mechanisms that promote outcrossing (e.g., dioecy, gynodioecy, and heterostyly). Because of the complex interplay between ecological, genetic, and historical factors, a strictly correlative approach is unlikely to enable isolation of the specific selective forces governing mating-system change. Moreover, for particular mating systems (e.g., predominant self-fertilization) the evolutionary processes responsible for the establishment of genes modifying mating patterns may be difficult to infer from present-day processes, because the genetic and ecological conditions that favored such changes are transient (Uyenoyama and Antonovics, 1987). To prevent the gulf between theoretical and empirical work on mating-system evolution from widening further, more experimental studies are required, particularly on species with broad variation in mating patterns.

A prerequisite for the development of biologically realistic models of mating-system evolution is accurate information on the mating process, the ecological and demographic context in which it occurs, and the fitness consequences of various mating strategies. During the past decade, considerable progress has been made in the use of isozyme markers for measuring a variety of mating-system parameters (Clegg, 1980; Ritland, 1983; Brown, 1990). Quantitative estimates of outcrossing rates using single loci and the mixed-mating model have given way to new estimation procedures based on multilocus approaches including the effective selfing model, the analysis of correlated matings, estimates of male fertility, and the paternity of outcrossed progeny. Electrophoretic techniques provide the reproductive biologist with a powerful tool for obtaining quantitative information on mating patterns in plant populations. The marriage of marker gene studies with detailed ecological observations on floral biology is required to determine how floral morphology and physiology influence the frequency of various classes of mating events. For example, information on ecological influences causing different modes of self- and cross-pollination in natural populations is needed to evaluate particular models of the evolution of self-fertilization (Lloyd, 1979a; Schoen and Brown, 1992). Future marker gene studies at increasingly more refined scales (e.g., at the population, individual, and flower levels) are likely to provide the background information for evolutionary investigations on the selective forces maintaining mating-system variation.

Experimental studies using genetic markers can be particularly revealing in plant species that are polymorphic for floral traits. Fitness comparisons of genotypes with similar genetic backgrounds, but that differ in traits influencing reproductive success, provide opportunities for investigating the potential selective forces operating on mating systems. For example, work on flower color polymorphisms in experimental populations of *Ipomoea purpurea* (L.) Roth. (Clegg and Epperson, 1988) and *Raphanus raphanistrum* L. (Stanton et al., 1989) have provided important information on the spread of mating-system modifier genes and on variation in male reproductive success, respectively. Unfortunately, relatively few species are known with discrete genetic polymorphisms for mating behavior within natural populations (although see Abbott, 1985). The best known involve the outcrossing and selfing morphs in mixed populations of heterostylous and homostylous plants (Crosby, 1949; Bodmer, 1960; Ganders, 1975; Barrett, 1979; Piper et al., 1984) and the coexistence of female (unisexual) and hermaphroditic (bisexual) morphs in gynodioecious species (Lloyd, 1976; Gouyon and Couvet, 1987; Kohn, 1989). These polymorphic sexual systems provide a rare opportunity to investigate the relative fitness of different morphs under various ecological conditions. In addition, manipulative studies of the floral biology of polymorphic populations provide data that can be used in a predictive manner to understand the evolution and maintenance of these complex sexual systems.

In this chapter, we review experimental approaches used to study the genetic and environmental factors influencing mating-system variation in tristylous *Eichhornia paniculata* (Spreng.) Solms-Laubach (Pontederiaceae). We begin by outlining the features of this species that have made it a useful experimental system for analyzing mating-system problems and then describe what is known about variation in mating patterns of natural populations. The results from a series of experimental studies are presented next to illustrate the complex forces acting on the mating system and how the interaction of genetic and environmental factors can modify mating patterns, even in a species with major-gene control of floral traits governing its breeding system.

The Experimental System

Eichhornia paniculata is an emergent aquatic of seasonal pools, marshes, ditches, and low-lying pastures in N.E. Brazil and the Caribbean islands of Jamaica and Cuba. Single isolated localities are also reported from W. Brazil, Ecuador, and Nicaragua (Barrett, 1988). When moisture is available for extended periods, plants of *E. paniculata* behave as perennials; however, most populations are annual owing to uncertain moisture levels and desiccation of habitats. Populations range in size from isolated individuals to monospecific stands containing several thousand plants. Most commonly, however, they are small (<100 individuals) and short-lived due to drought and habitat destruction (Barrett et al., 1989). Genetic estimates of effective population size (N_e) in 10 populations from N.E. Brazil ranged from 3.4 to 70.6 with a mean of 15.8, a value approximately 10% of the census number (Husband and Barrett, 1992a). Individuals display considerable phenotypic plasticity depending on site conditions with plants varying in height from 10 to 150 cm. *Eichhornia paniculata* does not regenerate by clonal propagation under field conditions, but plants can be cloned under glasshouse culture by severing young axillary shoots from the main shoot.

Flowers of *E. paniculata* are showy, mauve-blue, zygomorphic, and borne on inflorescences that usually contain 5–100 flowers that last for up to 12 days (Richards and Barrett, 1984). The anthesis period of individual flowers is 6–8 hr, with the major pollinators in N.E. Brazil being long-tongued solitary bees (*Florilegus festivus* Smith and *Ancyloscelis* spp.) and butter-flies (Husband and Barrett, 1992b). Other floral visitors include pollen-collecting Meliponid bees and, in Jamaica, *Apis mellifera* L. and occasional syrphids. Seeds of *E. paniculata* are small (0.15 mg), produced in large numbers per capsule (50–120), and shed 10–14 days after pollination. They are dispersed locally by water, over longer distances apparently by water-fowl, and germinate synchronously when rains follow the marked dry sea-son that is typical of the parts of Brazil and Jamaica in which the species occurs.

Eichhornia paniculata exhibits a wide range of mating patterns that are associated with the evolutionary breakdown of tristyly to semihomostyly (Barrett, 1985a,b). It was the discovery of this variation, and its associated diversity in floral morphology, that motivated our interest in using the species as an experimental system for mating-system analysis. Clearly, a major prerequisite for studies of this type is evidence that the mating system varies within and between populations and that this variation is associated with heritable differences in floral traits. Populations of *E. paniculata* often contain a mixture of outcrossing and selfing variants. Differences between the floral forms have a genetic basis with variation largely under major-gene control (S.C.H. Barrett, unpublished data). The outcrossing forms represent the conventional long-, mid-, and short-styled morphs of a tris-tylous system (hereafter L, M, and S morphs). The selfing variants are modified plants of the M morph (hereafter selfing variants) with between 1 and 3 elongated short-level stamens (Seburn et al., 1990; Barrett and Harder, 1991). Anthers of elongated stamens lie close to mid-level stigmas, making autonomous self-pollination of flowers possible. The different floral phenotypes in *E. paniculata* are easily recognizable under field conditions, thereby enabling surveys of their relative frequency (Barrett et al., 1989).

Eichhornia paniculata has a number of other features that facilitate both observational and experimental studies. Populations are conspicuous, eas-ily studied under field conditions, and abundant, particularly in N.E. Bra-zil. The plant is diploid, highly self-compatible, has low levels of seed abortion (Morgan and Barrett, 1989), is easily crossed and cultured, flowers in 2–3 months from seed, blooms year-round under glasshouse conditions, and is polymorphic at 10–15 isozyme loci (Glover and Barrett, 1987), several of which can be assayed at the seed stage.

One of our major objectives in studying *E. paniculata* has been to obtain a comprehensive picture of the variation in mating systems of populations from different parts of the geographical range (Glover and Barrett, 1986; Barrett and Husband, 1990; Husband and Barrett, 1991b). This has been achieved by sampling open-pollinated families from a large number of populations from throughout the range of the species in N.E. Brazil and on the island of Jamaica. Estimates of population size, plant density, and frequencies of floral morphs have also been made in each population. Multilocus outcrossing rates (*t*) were estimated for each population using isozyme markers and the estimation procedure of Ritland (1990). Figure 9.1 illustrates the range of outcrossing rates among populations of varying size and morph structure. Estimates of *t* range from 0.002 to 1.0, the widest range of values reported for any plant species (reviewed by Barrett and Eckert, 1990). Populations from Jamaica are predominantly selfing (mean *t* = 0.21, range 0.002–0.68, *N* = 11), whereas those from N.E. Brazil are more highly outcrossing, although several populations exhibit selfing rates comparable to those measured in Jamaica (mean *t* = 0.74, range 0.19–1.00, *N* = 43). Multiple regression and partial correlation analyses revealed that 60% of variation in *t* can be explained by style morph diversity and the frequency of selfing variants within populations (Barrett and Husband,

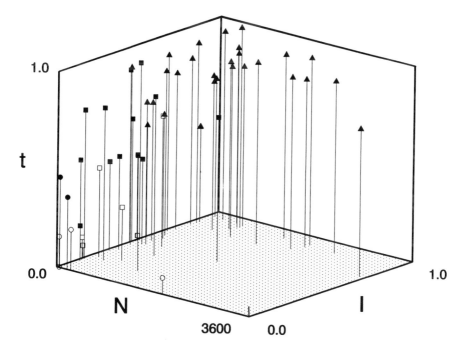

Figure 9.1. Variation in multilocus outcrossing rate (*t*) among 54 populations of *Eichhornia paniculata* in N.E. Brazil (closed symbols) and in Jamaica (open symbols) in relation to their size (*N*) and style morph diversity (*I*). Trimorphic, dimorphic, and monomorphic populations are represented by triangles, squares, and circles, respectively. Values of *I* are based on a normalized index of diversity and range from 1.0, equal frequencies of the floral morphs, to 0, floral monomorphism (after Barrett and Husband, 1990, and unpublished data).

1990). Population size and plant density also account for a significant portion of variation in t, particularly in Jamaica, where style morph diversity is low owing to a predominance of selfing variants and absence of the S morph from the island.

The wide range of population-level outcrossing rates found in *E. paniculata* raises a number of questions concerning the details of the mating process within populations and the ecological and evolutionary processes producing mating-system variation. What factors account for high outcrossing rates in some populations and the occurrence of predominant self-fertilization in others? Which aspects of the floral biology of populations have an important impact on mating patterns? How does population morph structure influence mating patterns and how important is the ecological and demographic context in which mating occurs for mating-system evolution? To address these issues, we have undertaken a series of investigations involving field, garden, and glasshouse experiments. The underlying approach in these studies has been largely reductionistic, involving an attempt to decompose the mating process into its elementary causative agents by experimental means. Although this mechanistic approach suffers from the valid criticism that ecological realism may be sacrificed, we believe that progress in reproductive ecology requires a broader understanding of the specific details of floral biology and its influence on mating patterns. This is more likely to be achieved by adopting an experimental approach in which particular stages of the mating process are analyzed to determine the possible mechanisms governing the dynamics of mating-system change.

Mating in Trimorphic Populations

Trimorphic populations of *Eichhornia paniculata* exhibit moderate to high levels of outcrossing (Fig. 9.1: mean $t = 0.83$, range 0.5–0.96, $N = 27$). Nevertheless, some self-fertilization is evident in most populations, particularly those of small size. This is consistent with the view that demographic factors are likely to influence mating patterns in self-compatible plants (Ganders, 1975; Lloyd, 1980; Schemske and Lande, 1985; Barrett and Eckert, 1990; Holsinger, 1991). Small populations may receive less reliable pollinator service resulting in increased levels of self-pollination. Attempts to correlate pollinator levels and mating patterns have proven difficult (Lloyd, 1965; Rick et al., 1977; Wyatt, 1986) and an investigation of pollinator visitation to 16 populations of *E. paniculata* in N.E. Brazil failed to detect a positive relationship between pollinator densities and population size (Husband and Barrett, 1992b). Visitation levels were, however, more variable among small populations, a pattern consistent with the hypothesis that small populations are more likely to experience uncertain pollinator service. The difficulties in establishing relationships between ecological factors and mating patterns by correlative approaches illustrate the need for experimental studies in this area.

Floral Trimorphism and the Maintenance of Outcrossing

The high outcrossing rates measured in most trimorphic populations of *E. paniculata* are unusual for a predominantly annual, self-compatible species and are more characteristic of long-lived species (Barrett and Eckert, 1990). Controlled pollinations of *E. paniculata* have demonstrated that the species is highly self- and intramorph compatible (Barrett, 1985a; Barrett et al., 1989; Kohn and Barrett, 1992a). Moreover, individuals usually produce many flowers that open synchronously, permitting both intrafloral and geitonogamous pollen transfer. What factors counteract these effects to maintain high outcrossing rates? Several reproductive mechanisms could potentially be involved: (1) the reciprocal arrangement of stamens and styles (reciprocal herkogamy) in the floral morphs of this heterostylous species, (2) differential pollen germination and pollen tube growth of self versus outcross pollen, and (3) differential survivorship of selfed versus outcrossed embryos or plants. Studies of embryo abortion in *E. paniculata* (Morgan and Barrett, 1989; Toppings, 1989) found low levels of ovule abortion (<10%), and a glasshouse study found no differences in germination, seedling growth, and survival of selfed and outcrossed progeny (Toppings, 1989). Hence, none of these factors seems likely to have inflated our measures of outcrossing rate to any great extent. To investigate the relative importance of the remaining two factors, several experimental studies were undertaken.

Reciprocal Herkogamy

We used experimental garden populations of different floral morph structure to investigate the role of reciprocal herkogamy in promoting outcrossing in *E. paniculata*. Our goal was to determine whether mating patterns and measures of fertility differed between trimorphic populations and those containing a single morph. Controlled crosses were used to produce plants that were homozygous for alternative alleles at the triallelic *Aat-3* locus and polymorphic at two additional electrophoretic marker loci (*Pgi-2* and *Acp-1*). Trimorphic populations composed of 12 plants of each morph were assembled in which style morphs were homozygous for different alleles at the *Aat-3* locus. This allowed joint estimation of the levels of intermorph and intramorph mating and outcrossing rate (Kohn and Barrett, 1992a). For logistical reasons, these manipulations could not be undertaken under field conditions in N.E. Brazil. Instead the experiments were conducted in Ontario where populations were serviced by novel pollinators

[*Bombus fervidus* (Fabricius), *B. impatiens* Cresson, and *B. vagans* F. Smith]. Thus, an important issue in interpreting the results of these studies was the extent to which the use of alien pollinators may have modified mating patterns from those occurring in the native habitat.

Both outcrossing rates and seed-set were higher for all three morphs in trimorphic than in monomorphic populations (Fig. 9.2), but the magnitude of the differences varied significantly among the morphs. The higher seed-sets were apparently due to increased pollen deposition on stigmas in trimorphic populations. Outcrossing rates in experimental trimorphic populations did not vary among style morphs or replicates. The mean outcrossing rate (0.81) was similar to that obtained for the population from which experimental plants originated ($t = 0.87$; M.R. Dudash and S.C.H.

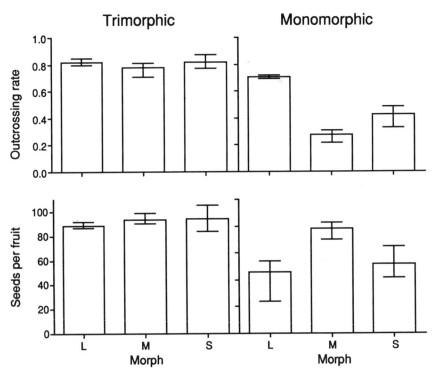

Figure 9.2. Mean morph-specific outcrossing rate (t) and seed-set per fruit in experimental trimorphic and monomorphic garden populations of *Eichhornia paniculata*. Bars represent the range of mean values from three replicates of each treatment. Trimorphic populations were composed of 12 plants per morph and monomorphic populations contained 36 plants of a single morph (after Kohn and Barrett, 1992a).

Barrett, unpublished data). This indicates that, despite the artificial conditions of our experimental garden, the influence of floral trimorphism on outcrossing rates does not appear to have been altered to any great extent. The similar outcrossing rates observed in field and garden populations challenge the widely held assumption that a high degree of pollinator specificity is required for the functioning of heterostylous systems.

The observation that seed-set increased in trimorphic populations may be of general significance for models of the evolution of heterostyly (reviewed by Barrett, 1990). If the ancestral condition in heterostylous groups resembled the L morph, as suggested by Lloyd and Webb (1992a), then the large increase in seed-set compared to the relatively small increase in outcrossing rate (Fig. 9.2), suggests that the selective basis for the evolution of reciprocal herkogamy may have been increased pollen transfer, rather than higher levels of outcrossing. The validity of this interpretation depends, however, on which morph most closely resembles the ancestral condition, as the other two morphs displayed large increases in outcrossing rate when in trimorphic populations.

If outcrossing events in experimental trimorphic populations were random, expected rates of intermorph mating would be 69% because 24 of the 35 non-self pollen donors in each array were of a different morph than the recipient. Of the outcrossed seeds assayed from experimental trimorphic populations, however, 95% resulted from intermorph fertilizations (Fig. 9.3). Rates of intermorph mating, estimated from a different trimorphic population in N.E. Brazil, indicated that a large fraction (92%) of outcrossed progeny also resulted from intermorph mating (Fig. 9.3). The high rates of intermorph mating found in both experimental and natural populations indicate that mechanisms favoring intermorph over intramorph mating occur in *E. paniculata*. Further experiments were conducted to investigate what factors might be involved.

Postpollination Mechanisms

In the garden experiments described above, the high levels of intermorph mating could result from both high levels of intermorph pollen transfer and postpollination discrimination among pollen types. Reciprocal herkogamy alone can increase the frequency of intermorph pollen transfer (Ganders, 1974; Barrett and Glover, 1985; Lloyd and Webb, 1992b); however, in species studied to date, illegitimate pollen grains typically outnumber legitimate grains in most stigmatic pollen loads (Ganders, 1979). A preliminary study (Glover and Barrett, 1986) indicated that, although there may be the potential for discrimination between selfed and outcrossed pollen, there was little evidence of discrimination among outcross pollen types. Such discrimination would be a prerequisite for postpollination pro-

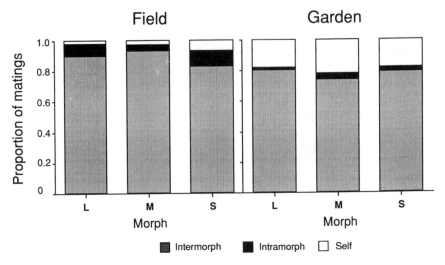

Figure 9.3. Partitioning of mating events in field and garden populations of *Eichhornia paniculata*. For each floral morph (L, M, and S), intermorph, intramorph, and self matings were estimated from open-pollinated progeny arrays using genetic markers. Field population data are from Barrett et al. (1987), and garden population data are from Kohn and Barrett (1992a).

cesses to affect levels of intermorph mating. To assess the potential for postpollination discrimination among pollen types, experiments were conducted using plants derived from the same population in Brazil as the plants used in the experimental garden populations (Cruzan and Barrett, 1992 and unpublished data). The following questions were addressed: (1) Does pollen from different anther levels germinate or grow at different rates when deposited on each of the three style morphs? (2) When a mixture of pollen is deposited on stigmas, does the siring ability of pollen types in the mixture depend on style morph?

If postpollination processes contribute to the observed levels of intermorph mating, we might predict that legitimate pollen (i.e., pollen from anthers equivalent in height to the stigma) should germinate or grow faster than illegitimate pollen (i.e., pollen from the other two anther levels) in all three style morphs. This expectation was not upheld for pollen germination: there was little difference in the germination of pollen from different anther levels on any of the three style morphs. On the other hand, pollen tube growth did depend on style morph. In each morph the number of legitimate pollen tubes reaching the ovary after 3 hr was significantly higher than for both illegitimate types. These patterns could be due to either differential growth rate or failure of pollen tubes (i.e., attrition: see Cruzan, 1989) in the style.

To determine the potential for differences in pollen tube growth rate to affect siring success, mixtures of genetically marked pollen were applied to stigmas of each style morph at different densities. The pollen density classes (high and low) were used to explore the possibility that mating patterns may be influenced by the total amount of pollen deposited on stigmas. Patterns of pollen siring ability reflected the observed differences in pollen tube growth rate; legitimate pollen always obtained more fertilizations than both illegitimate pollen types (Fig. 9.4). The relative competitive ability of pollen in each recipient style morph did not depend on its source (i.e., self, intramorph, or intermorph), but only on the anther level from which it originated. The total amount of pollen applied to stigmas had a large effect on the frequency of fertilizations by legitimate pollen (Fig. 9.4). Higher pollen loads resulted in a significant increase in the frequency of legitimate fertilization in the L and M morphs. In the S morph, however, there was no increase in the frequency of legitimate fertilization for larger stigma pollen loads.

These experiments suggest that postpollination mechanisms involving differential pollen tube growth contribute to the high levels of outcrossing and intermorph mating observed in *E. paniculata*. Further studies are required, however, to determine the relative roles of reciprocal herkogamy and pollen competitive ability in promoting these mating patterns under different demographic and environmental conditions.

The data on pollen competitive ability in *E. paniculata* closely parallel the patterns of pollen tube growth and seed-set observed in related tristylous *Pontederia* species with strong trimorphic incompatibility systems (e.g., Barrett and Anderson, 1985; Scribailo and Barrett, 1991). This suggests that the functional basis of differences in pollen competitive ability in *E. paniculata* results from a weak (or cryptic) trimorphic incompatibility system. The predominantly annual life cycle of *E. paniculata* and ephemeral aquatic habitats it occupies may have favored a more flexible reproductive system than the strong self-incompatibility observed in *Pontederia* species, which tend to occur in more permanent aquatic habitats. The ability to produce large numbers of seeds under a variety of ecological conditions, while also being able to take advantage of high levels of pollinator activity by producing mostly outcrossed progeny, was suggested as the primary factor maintaining cryptic incompatibility in *Clarkia unguiculata* (Bowman, 1987) and has been called the "best-of-both-worlds" strategy by Becerra and Lloyd (1992). Whereas cryptic incompatibility in *E. paniculata* may allow for the maintenance of populations by assuring high seed output, the lack of strong incompatibility barriers combined with uncertain pollinator service likely contributes toward the diversity of morph frequencies and selfing rates found in populations throughout N.E. Brazil (Fig. 9.1).

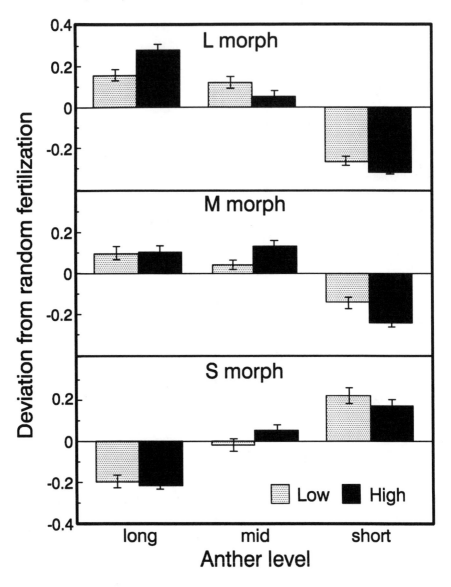

Figure 9.4. Fertilization success of pollen from long-, mid-, and short-level anthers when applied in equal mixtures to stigmas of the L, M, and S morphs of *Eichhornia paniculata*. Pollen mixtures were applied at two densities (high and low), and genetic markers were used to determine the fertilization success of the different pollen types. A value of zero indicates random fertilization. Error bars indicate two standard errors of the mean (M.B. Cruzan and S.C.H. Barrett, unpublished data).

Pollen Precedence and Correlated Mating

Field observations of pollinator visitation indicate that flowers of *E. paniculata* commonly receive multiple visits from insect pollinators. Because flowers are short-lived and pollen types differ in competitive ability, it was of interest to determine whether the probability of fertilization depends on the sequential order of pollination (Epperson and Clegg, 1987a) and whether pollen precedence differed among the morphs. To investigate these possibilities, experiments involving the application of genetically marked pollen mixtures and electrophoretic screening of seed progeny were used (Graham and Barrett, 1990 and unpublished data).

Early-arriving pollen experienced a fertilization advantage, but the magnitude of this advantage varied with floral morph and whether pollen was self or outcross (Fig. 9.5). Time delays in the order of pollination >2–3 hr virtually guaranteed that all fertilizations resulted from early-arriving pollen, irrespective of pollen type. For shorter time intervals, however, outcross pollen experienced a greater fertilization advantage than self pollen when each pollen type was applied first. This effect was most pro-

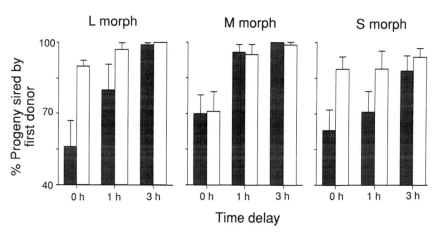

Figure 9.5. Pollen precedence in the L, M, and S morphs of *Eichhornia paniculata*. Self or outcross pollen was applied in equal amounts either 0, 1, or 3 hr after the first pollination. Hatched bars indicate treatments where self pollen was applied first and outcross pollen second; open bars represent treatments where outcross pollen was applied first and self pollen second. Self pollen types used from each morph were L morph, mid-level anthers; M morph, long-level anthers; and S morph, mid-level anthers. Outcross pollen was always from the same anther level as the stigma (legitimate pollen). The mean percentage of progeny sired by the first donor was obtained using genetic markers. Error bars indicate two standard errors of the mean (after Graham and Barrett, 1990 and unpublished data).

nounced in the L and S morphs. In the M morph, there was no significant difference between self and outcross pollen in the degree of precedence shown by early-arriving pollen. This suggests that the M morph is more likely to experience self-fertilization if outcross pollen delivery is delayed owing to reduced levels of pollinator service.

The results of these experiments suggest that individual maternal plants are unlikely to experience random outcrossing under field conditions. Depending on the schedule of pollen receipt and the composition of the pollen load, seeds are likely to be sired by a limited number of male parents. Analysis of mating patterns in trimorphic populations provides evidence in support of this suggestion (Barrett et al., 1987; Morgan and Barrett, 1990). Data on the segregation of both style morphs and allozymes in open-pollinated families indicated heterogeneous patterns, a result consistent with nonrandom outcrossing. Furthermore, estimates of the correlation of outcrossed paternal parentage using Ritland's (1989) correlated mating model, indicated that 32% of progeny sampled from within fruits of *E. paniculata* are full sibs. Studies of several other animal-pollinated species have also reported significant levels of correlated mating (Schoen and Clegg, 1984; Schoen, 1985, 1988; Brown et al., 1986; Dudash and Ritland, 1991), suggesting that it may be a general feature of such plants. Presumably, stigmatic pollen loads of animal-pollinated plants rarely contain gametes sampled from the entire population of potential male parents but instead are composed of pollen from a restricted subset of male donors owing to fertility variation, phenological differences, pollen carryover, and population substructure (Marshall, 1990). Depending on the schedule of pollen arrival, postpollination phenomena such as pollen precedence may serve to restrict further the subset of males that is successful in siring seeds.

Functional Gender

In recent years, several workers have proposed that heterostyly can be viewed from the perspective of sexual selection and sex allocation theory (Wilson, 1979; Beach and Bawa, 1980; Casper and Charnov, 1982; Casper, 1992). Following this view the floral morphs may exhibit gender specialization and gain differential reproductive success through male and female function (Lloyd, 1979b; Hicks et al., 1985; Nicholls, 1987). Unfortunately, because of the difficulties in measuring male reproductive success in hermaphroditic plant populations, there have been few reliable estimates of functional gender based on the use of genetic markers (although see Ennos and Dodson, 1987; Broyles and Wyatt, 1990).

The experimental design used in the garden studies described above has enabled us to measure the functional gender of floral morphs in *E. paniculata*. In trimorphic populations, plants of each morph were homozygous for one of three alleles at the *Aat-3* locus. Thus, a seed heterozygous at this locus indicated an intermorph mating and the identity of the paternal morph could be determined unambiguously. By combining data on male reproductive success with measures of seed fertility, it was possible to estimate the functional gender of each floral morph using the formula:

$$G_i = \frac{f_i}{(2 - d_i)f_i + f_j p_{ji} + f_k p_{ki}}$$

where f_i is the seed production of morph i, d_i is the proportion of seeds of morph i sired by intermorph pollen, and p_{ji} is the proportion of seeds of morph j sired by morph i. Data on functional gender revealed a significant difference between the L and S morphs in each replicate population. The L morph was more female and the S morph more male (mean G_i ± 2 standard errors: L = 0.56 ± 0.04, M = 0.51 ± 0.03, S = 0.45 ± 0.03). Although these differences may not appear large, a deviation of ±0.05 from the value of 0.5 (the value for a hermaphrodite that gains equal reproductive success through male and female function) implies that individuals of that morph gain approximately 20% more fitness through one sex function than the other. Thus, small differences in gender may have significant consequences for optimal allocation to male versus female structures. Interestingly, the difference in gender between the L and S morphs in *E. paniculata* is of the type described by models of the evolution of dioecy from heterostyly (Beach and Bawa, 1980; but see Muenchow and Grebus, 1989). That is, in dioecious taxa derived from distylous ones, it is thought that the L morph became the female and the S morph became the male.

The difference in gender between the L and S morphs was primarily the result of their strikingly different abilities to sire seeds of the M morph. In 1989, when these experiments were first conducted, the S morph was three times more likely than the L morph to sire seeds produced by the M morph. This mating asymmetry, while not as large, was also evident when experiments were repeated in 1990 (Fig. 9.6). What factors could account for the differences in male reproductive success of the L and S morphs when acting as paternal parents to the M morph? Two hypotheses could account for these differences. The first stems from Webb and Lloyd's (1986) suggestion that the female organs of approach herkogamous plants, as in the L morph, may interfere with pollen removal from flowers by pollinators. This "pollen–stigma interference" hypothesis can be tested by comparing the male reproductive success of L plants with styles intact versus removed. The second hypothesis proposes that the difference in siring success of the L and S morphs results from differences in the frequency of illegitimate mating. Pollen from long-level anthers of the S morph

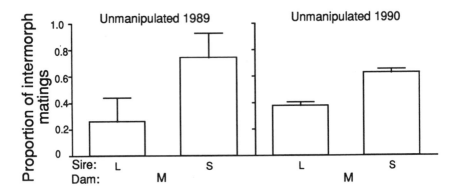

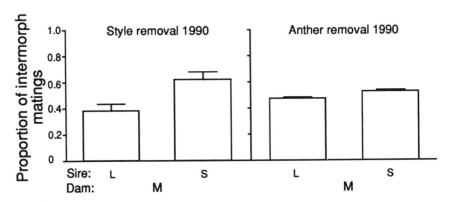

Figure 9.6. Proportion of seeds from the M morph sired through intermorph mating by the L and S morphs of *Eichhornia paniculata* in experimental garden populations in 1989 and 1990. In 1990 two experiments were conducted with trimorphic populations (see text for details); styles were intact versus removed on the L morph, and illegitimate anthers of the L and S morphs were intact or removed (after Kohn and Barrett, 1992b).

may fertilize more ovules of the M morph than pollen from short-level anthers of the L morph due to differences in transfer efficiency and/or siring ability of the two pollen types. This hypothesis can be tested by the removal of short-level anthers from the L morph and long-level anthers from the S morph, leaving only mid-level anthers on the L and S morphs. If the mating asymmetry is largely a result of illegitimate mating, then it should disappear following the anther removal treatment.

Experimental manipulations of the floral biology of trimorphic arrays were used to test the two hypotheses in 1990 (Kohn and Barrett, 1992b). The experiments gave unequivocal results (Fig. 9.6). In the first experiment there were no significant differences in the siring ability of L plants with styles intact or removed, and the mating asymmetry between L and S plants persisted. In the second experiment there were no significant differences in the siring abilities of the L and S morphs, indicating that the asymmetries observed in previous experiments resulted from differential contribution of illegitimate pollen types to male reproductive success. The experiments on pollen competitive ability described above indicated large differences in the siring ability of pollen from long-level versus short-level anthers when present in mixtures on stigmas of the M morph (Fig. 9.4). Pollen from long-level anthers achieved more fertilizations than pollen from short-level anthers, a pattern consistent with the data on gender asymmetry. Several lines of evidence, however, also suggest that pollen transfer from long-level anthers exceeds that from short-level anthers (see Kohn and Barrett, 1992b). Further studies are therefore required to determine the relative contributions of pollen transfer and/or siring ability of illegitimate pollen to the higher male reproductive success of the L morph in comparison with the S morph.

Evolution of Self-Fertilization

Many populations of *E. paniculata* experience moderate to high levels of self-fertilization (Fig. 9.1). The primary cause of increased rates of selfing is the spread of mating-system modifier genes among plants of the M morph in dimorphic and monomorphic populations. Selfing M variants predominate in Jamaica and occur in varying frequencies in nontrimorphic populations from different parts of the geographical range in N.E. Brazil. A survey of the frequency of selfing variants in populations of different morph structure in N.E. Brazil clearly indicates their nonrandom distribution (Table 9.1). The variants occur only rarely in trimorphic populations, whereas they are abundant in many dimorphic and monomorphic populations. Morph-

Table 9.1. The Frequency of Selfing Variants of the M Morph in *Eichhornia paniculata* Populations of Contrasting Morph Structure in N.E. Brazil[a]

Measure	Trimorphic	Dimorphic	Monomorphic
N (populations)	118	42	7
Percentage with selfing variants	8.5	51.2	85.7
Mean frequency of variants in pupulations occupied	0.05	0.37	1.00
Mean population size	146.3	38.2	10.2

[a]The average population sizes are based on the geometric means of population samples (after Barrett et al., 1989 and Husband and Barrett, 1992b).

specific estimates of outcrossing rate in dimorphic populations of *E. paniculata* have demonstrated dramatic differences in levels of selfing between the L and variant M morphs. For example, in three Jamaican populations the mean value of *t* for the L and variant M morphs was 0.82 and 0.12, respectively (Barrett et al., 1989). This suggests that selfing variants are likely to experience a significant transmission advantage in comparison with the L morph.

The potential transmission advantage of selfing variants of the M morph may be further augmented in some ecological circumstances by maternal differences in seed fertility. Where pollinator service is unreliable, the facility for autonomous self-pollination in variant M plants can potentially provide them with reproductive assurance. The geographically marginal distribution of selfing variants, and their occurrence in populations that tend to be smaller and less dense than trimorphic populations, suggests that founder events and periods of uncertain pollinator service may have been of primary importance for the evolution of self-fertilization in *E. paniculata* (Barrett et al., 1989). The association between selfing and low-density, marginal conditions is a recurrent theme in the mating-system literature (Baker, 1955; Jain, 1976; Lloyd, 1980), but few empirical studies have provided clear evidence bearing on this relationship. This is because selfing and outcrossing plants are usually geographically or ecologically segregated. Consequently, their performances in particular environments are difficult to compare. Because heterostylous and homostylous plants can coexist within populations, however, comparisons of seed fertility can provide direct evidence in support of the reproductive assurance hypothesis (Barrett, 1979; Piper et al., 1986). Comparisons of this type in dimorphic populations of *E. paniculata* have demonstrated large differences in the maternal fertility of selfing and outcrossing morphs, particularly in Jamaican populations (Barrett et al., 1989). Hence, depending on the ecological and demographic conditions in which the variant M morph occurs (see below), it can benefit from increased genetic transmission through both high selfing rates and elevated seed fertility. These advantages may help explain the predominance of selfing variants in Jamaica and their occurrence in many nontrimorphic populations in N.E. Brazil.

Floral Instability and Modes of Self-Fertilization

Patterns of floral variation in selfing variants of the M morph of *E. paniculata* are complex. The variation results from both genetic and non-genetic causes and is manifested at a number of levels, including between populations, between genotypes within populations, and between flowers of individual plants (Richards and Barrett, 1992). The most conspicuous variation involves the elongation of filament length in short-level stamens.

An unusual feature of this modification is the discontinuous nature of the elongation patterns. The most common variant, particularly in N.E. Brazil, has a single stamen in the mid-level position, with the remaining two stamens largely unmodified. Another variant occurring in Jamaica and only rarely in N.E. Brazil has all three "short-level stamens" adjacent to mid-level stigmas. This type of phenotype is referred to as a "semihomostyle" in the literature on heterostyly (Ornduff, 1972). Inheritance studies of the modified M variants indicate recessive gene control of filament elongation. Crosses between variants from different parts of the geographical range in N.E. Brazil give rise to progenies composed of M plants lacking stamen modifications. This indicates complementary gene action resulting from the occurrence of different recessive modifiers in the populations (C.B. Fenster and S.C.H. Barrett, unpublished data).

A curious feature of the genetic modifications to stamen position in the M morph is that in some plants not all flowers within inflorescences exhibit elongated filaments (Barrett, 1985a; Seburn et al., 1990). As a result, inflorescences are frequently composed of both unmodified and modified flowers. Because the former are incapable of autonomous self-pollination, genotypes displaying floral instability are likely to produce both selfed and outcrossed progeny, if plants are visited by pollinators. This type of variation within an inflorescence is particularly evident in dimorphic Brazilian populations that contain both unmodified and variant M plants and presumably represents an early stage in the evolution of self-fertilization. Unmodified flowers occur very rarely in variant M plants from Jamaica, where selfing rates are considerably higher and populations have probably experienced a much longer history of inbreeding (Barrett, 1985b; Husband and Barrett, 1991).

To investigate the influence of genetic, developmental and environmental factors on stamen modification in *E. paniculata*, experimental studies were undertaken of cloned genotypes of the M morph grown under various environmental conditions (Barrett and Harder, 1991). Significant position effects were detected among genotypes with variant flowers. Flowers with modified short-level stamens were most frequently produced on later inflorescence branches in the flowering sequence and at proximal flower positions within an inflorescence branch. These patterns, however, were complex and varied among populations, genotypes, and experimental treatments. Stamen modification increased in clones grown under water stress or at high temperature, demonstrating a significant environmental component to floral instability.

The occurrence of floral instability in *E. paniculata* may have ecological and evolutionary significance. Genotypes producing modified and unmodified flowers are likely to display mixed mating systems. In this respect they resemble taxa that produce cleistogamous and chasmogamous flowers, but the position effects are far more subtle and the contrast between the flower types in terms of selfing rate may not be as extreme (Lord, 1981; Ellstrand et al., 1984). Another parallel with cleistogamous taxa concerns the strong environmental component to the formation of selfing flowers (Waller, 1980). Stressful growing conditions increase the frequency of selfing flowers. Because the habitats in which the species occurs in N.E. Brazil are subject to frequent droughts and desiccation, the ability of selfing variants of the M morph to adjust the frequency of flowers capable of autonomous self-pollination may have contributed to the spread of this morph in populations, particularly those in marginal habitats.

The mode of self-pollination (sensu Lloyd, 1979a) in variant M plants displaying floral instability is of relevance to models concerned with the evolution of predominant self-fertilization. Models that focus primarily on the joint evolution of selfing and inbreeding depression generally predict that mixed mating systems should be evolutionarily unstable, with selection favoring either complete outcrossing or complete selfing (Lande and Schemske, 1985; Charlesworth and Charlesworth, 1990). On the other hand, models that incorporate details of floral biology suggest that intermediate levels of self-fertilization can be evolutionarily stable, particularly if pollination conditions limit seed-set and progeny from selfing are viable (Lloyd, 1979a; Schoen and Lloyd, 1984; Schoen and Brown, 1992). Under conditions of uncertain pollinator service, the mode of self-fertilization in unstable variant M plants is likely to vary from flower-to-flower on a plant. Unmodified flowers may primarily experience competing selfing, in which a mixture of selfed and outcrossed offspring is produced. In contrast, in flowers with stamen modification all ovules within a flower may be self-fertilized. This is particularly likely where pollinators are infrequent and time delays in the delivery of outcross pollen give a competitive advantage to self pollen already deposited on stigmas through prior contact with anthers (see Fig. 9.5). A mixture of whole-flower and part-flower selfing will result in intermediate levels of self-fertilization (see Schoen and Brown, 1991).

Whether mixed mating remains evolutionarily stable or not will depend on levels of pollinator service. Schoen and Brown (1992) have shown that whenever flowering seasons contain a mixture of temporal phases, both favorable and unfavorable for pollinator activity, and where whole-flower and part-flower selfing of the type found in E. paniculata occurs, stable mixed mating systems should evolve. The rarity of predominantly selfing populations in N.E. Brazil and their abundance in Jamaica may reflect contrasting levels of pollinator activity and differences in the modes of self-fertilization in the two regions (Barrett and Husband, 1990).

Constraints on the Spread of Selfing Variants

Whereas selfing variants of *E. paniculata* occur commonly in dimorphic and monomorphic populations, they occur infrequently in trimorphic populations (Table 9.1). This nonrandom distribution with respect to morph structure raises the question: What evolutionary forces constrain the spread of selfing variants of the M morph in trimorphic populations? Three hypotheses could potentially explain the overall rarity of selfing variants in trimorphic populations: (1) the recessive genes governing mating-system modification are unlikely to spread in large outcrossing populations since they remain unexpressed in the heterozygous condition (Haldane's Sieve Hypothesis); (2) inbreeding depression in the progeny of selfing variants is large enough to prevent the "automatic selection" of selfing genes (Inbreeding Depression Hypothesis); and (3) the selfing variants have no transmission advantage in trimorphic populations due to the influence of morph structure on mating-system parameters (Context-Dependent Transmission Hypothesis). To evaluate the relative importance of these three hypotheses, several lines of inquiry were pursued. Below, we summarize observations and experiments that we have conducted to address why selfing variants rarely occur in tristylous populations.

Haldane's Sieve

In large outcrossing populations the chance of a new favorable recessive gene (as opposed to one with dominant expression) spreading is very small. Such mutations are more likely to be lost through drift than to reach high enough frequencies to occur in the homozygous state and be exposed to selection (Haldane, 1924, 1927). This principle, known as "Haldane's sieve" (Turner, 1981; Charlesworth, 1992), results in a strong bias against the evolution of newly evolved traits governed by recessive genes, despite the fact that most mutations that arise in the laboratory with large phenotypic effects are recessive (Fisher, 1931). Because mating-system modification in *E. paniculata* is largely governed by recessive genes, the overall rarity of selfing variants in trimorphic populations may be partly explained by the operation of Haldane's sieve.

Two related factors, however, can reduce the bias against the selection of recessive genes. In small populations genetic drift may occasionally lead to the fixation of recessive genes, even those with detrimental effects (Pollack, 1987). More importantly, for populations experiencing inbreeding, through partial self-fertilization or biparental inbreeding, recessive genes are more likely to occur in the homozygous condition. If the genes increase fitness, then they are likely to increase in frequency and become fixed. Recent models by Charlesworth (1992) on the rate of fixation of favorable and deleterious mutations in partially self-fertilizing populations predict that moderate rates of selfing will greatly increase the probability of fixation of favorable recessive mutations, even in large populations. Favorable recessive genes are much less likely to become fixed than dominant genes in outcrossing populations, but there is little difference between the two in highly selfing populations.

Episodes of small population size and inbreeding may account for the irregular occurrence of selfing variants in trimorphic populations of *E. paniculata*. Whereas population sizes tend to be larger in trimorphic than nontrimorphic populations (Table 9.1), annual censuses indicate dramatic fluctuations in numbers, irrespective of morph structure. Crashes of population size in trimorphic populations may provide occasional opportunities for the exposure and spread of recessive mating-system modifier genes as a result of inbreeding. Moderate levels of self-fertilization occur in many small trimorphic populations (Fig. 9.1) and estimates of biparental inbreeding [$(t_{ML} - t_{SL})/t_{ML}$; Waller and Knight, 1989] averaged 0.30, suggesting that matings between related individuals are common (Barrett and Husband, 1990 and unpubl. data). Haldane's sieve may explain the rarity of selfing variants in certain parts of the geographical range of *E. paniculata* in N.E. Brazil (e.g., Ceará and Bahia States), where historically large, highly outcrossed populations may have been maintained because of more extensive habitats suitable for the species. In other areas of N.E. Brazil (e.g., Pernambuco and Alagoas States), however, habitat fragmentation and small population sizes may have provided greater opportunities for the exposure and spread of mating-system modifier genes.

Inbreeding Depression

Most models of mating-system evolution in plants incorporate inbreeding depression, the reduced fitness of selfed offspring in comparison with outcrossed offspring, as the major force opposing the spread and fixation of selfing variants (e.g., Lande and Schemske, 1985; Charlesworth and Charlesworth, 1987 and references therein). The maintenance of high genetic loads in historically large, outcrossing populations serves to restrict the spread of selfing variants because deleterious recessive genes, normally sheltered in the heterozygous condition, reduce the fitness of selfed progeny. Where genetic loads are reduced by bottlenecks, or pollinator failure, however, selfing variants can spread because fitness differences between selfed and outcrossed offspring are not of sufficient magnitude to prevent the automatic selection of selfing genes. Models of the joint evolution of mating systems and inbreeding depression generally predict that levels of inbreeding depression should decline with increasing levels of self-fertilization. However, few empirical studies have investigated the relationships between mating patterns and inbreeding depression (see Holtsford and Ellstrand, 1990) and little is known about the genetic basis of inbreeding depression (reviewed by Charlesworth and Charlesworth, 1987; Barrett and Kohn, 1991).

What are the relationships between mating patterns and inbreeding depression in *E. paniculata* and could fitness differences between selfed and outcrossed offspring account for the failure of selfing variants to spread in trimorphic populations? To investigate the relationships between mating patterns and inbreeding depression, fitness comparisons were undertaken of selfed and outcrossed progeny grown under a variety of experimental conditions. Eleven populations of *E. paniculata*, with outcrossing rates spanning the full range encountered in the species, were used in these studies (Toppings, 1989). Fitness differences between selfed and outcrossed offspring were largest in trimorphic populations with high outcrossing rates, intermediate in nontrimorphic populations with mixed mating systems, and absent altogether in Jamaican populations with high levels of self-fertilization. This overall pattern was confirmed in a second experiment involving a comparison of the effects of continued inbreeding on two populations representing the extremes of mating-system variation in the species (Barrett and Charlesworth, 1991). Five generations of selfing followed by random outcrossing had no effects on fitness components in a monomorphic selfing population from Jamaica. In contrast, in an outcrossing trimorphic population from N.E. Brazil, fitness components declined on selfing but showed a dramatic increase when plants in the S_5 generation were randomly intercrossed (Fig. 9.7). The large difference in flower production between the S_5 and C_5 generations in comparison to the difference between the S_1 and C_1 generations (Fig. 9.7) was most likely the result of purging of recessive or partially recessive deleterious alleles within the inbred lines. The experimental results were in close agreement with theoretical expectations of a model of inbreeding depression based on mutation-selection balance at many unlinked loci with a high mutation rate to moderately deleterious, partially recessive alleles (Barrett and Charlesworth, 1991). The results support the partial dominance (mutational) hypothesis for genetic load (see Charlesworth and Charlesworth, 1987) and suggest that purging of partially recessive mutations accounts for a significant component of heterosis.

Although these experimental results provide useful comparative data on the relative patterns of inbreeding depression in populations with contrasting mating systems, they cannot determine whether inbreeding depression limits the spread of selfing variants in trimorphic populations. First, because the studies were conducted under glasshouse rather than field conditions, they may provide little information about the magnitude of inbreeding depression under field conditions (reviewed by Charlesworth and Charles-

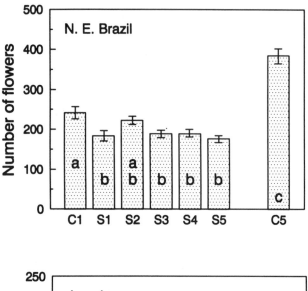

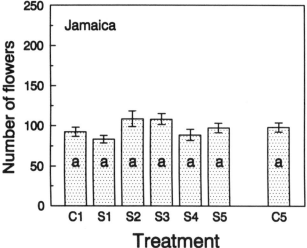

Figure 9.7. The effects of five generations of selfing on flower production in populations of *Eichhornia paniculata* with contrasting mating systems. The population from N.E. Brazil was trimorphic and highly outcrossing, whereas the population from Jamaica was monomorphic and predominantly selfing. Selfed ($S_1–S_5$) and outcrossed generations (C_1 and C_5) were compared in a single multigeneration experiment under uniform greenhouse conditions. The Brazilian population was represented by 30 lines and the Jamaican population by 10 lines. Error bars represent two standard errors of the mean, and letters indicate statistically significant ($p < 0.05$) differences (after Barrett and Charlesworth, 1991).

worth, 1987; and see Schoen, 1983; Schemske, 1983; Kohn, 1988; Dudash, 1990). Such information is required to evaluate whether selfing variants would be likely to experience a transmission advantage in trimorphic populations. Second, even though inbreeding depression was found to be lower in more inbred populations (Toppings, 1989), it cannot be determined whether the loss of load (e.g., due to small population size) preceded the spread of selfing variants or whether load was purged as a consequence of the spread of selfing variants.

Transmission Characteristics of Selfing Variants

The effects of population morph structure on gamete transmission by variant M plants was investigated by J.R. Kohn and S.C.H. Barrett (unpublished data) using genetic markers and garden populations that were trimorphic, dimorphic (S morph absent), or monomorphic (L and S morphs absent). In each replicated population male and female reproductive success of unmodified and variant M plants were measured. Male reproductive success was partitioned into self and outcross pollen components. It was hypothesized that, in trimorphic populations, the elongated short-level stamen of variant M plants might result in a reduced ability to fertilize plants of the S morph. This effect could counter the increased transmission through selfing caused by the stamen modification. In dimorphic and monomorphic populations, the absence of the S morph would eliminate this cost.

Fruit- and seed-set by unmodified and variant M plants were not significantly different among any of the morph structure treatments. Variant M plants had higher mean selfing rates than unmodified M plants in each treatment, and mean selfing rates of unmodified and variant M plants increased with the progressive loss of morphs (Fig. 9.8). In trimorphic populations, variant M plants suffered reduced transmission through outcross pollen donation relative to unmodified M plants. This effect was largely due to an approximate 50% reduction in the ability of variant M plants to sire seeds of the S morph relative to unmodified M plants. Thus, in experimental trimorphic populations, the cost of increased selfing was reduced reproductive success through outcross pollen donation. As a result, total transmission of gametes by unmodified and variant M plants were similar in trimorphic populations.

In dimorphic and monomorphic populations, the modification increased both self and outcross pollen donation. Apparently, the occurrence of a single anther at the mid-level caused variant M plants to sire a greater proportion of seeds of unmodified M plants than vice versa. Thus, in dimorphic and monomorphic populations, variant M plants were favored over unmodified M plants through a large increase in gamete transmission

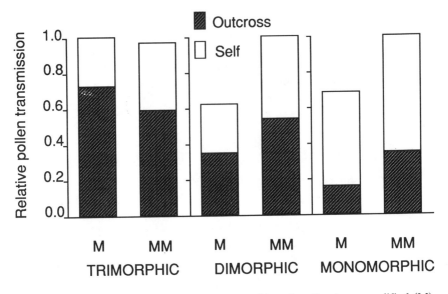

Figure 9.8. Relative transmission of gametes through pollen by unmodified (M) and variant (MM) plants of the mid-styled morph in experimental trimorphic, dimorphic (S morph absent), and monomorphic (L and S morphs absent) garden populations of *Eichhornia paniculata*. All populations contained a total of 36 plants. Estimates of male fertility through self and cross pollen donation are based on electrophretic assays of five seeds from each of two fruits per plant. In trimorphic populations there was no difference in transmission by unmodified (M) and variant (MM) plants of the mid-styled morph. In populations lacking the S morph, transmission by variant M plants was significantly greater ($F_{1,8} = 22.2$, $p < 0.01$) than transmission by unmodified plants (J.R. Kohn and S.C.H. Barrett, unpublished data).

by pollen (Fig. 9.8). These results demonstrate the large role that the morph structure of populations can play in altering the effects that a given floral morphology has on the mating system.

In trimorphic populations in N.E. Brazil, variant M plants may not benefit from the transmission advantage normally associated with selfing. In addition to possible reduced transmission through cross pollen donation, the altered morphology of M variants may not increase their selfing rate relative to unmodified M plants as much in trimorphic as in nontrimorphic populations. Since trimorphic populations tend to be larger and contain plants at higher density, M variants may experience less increase in selfing due to greater availability of outcross pollen. A study of outcrossing rates in two trimorphic populations containing variant and unmodified M plants provided data that are consistent with this view (M.R. Dudash and S.C.H.

Barrett, unpublished data). Outcrossing rates of the two types of M plants were high and not significantly different within each population (B46: M morph, $t = 0.79$; variant M, $t = 0.72$; B58: M morph, $t = 1.0$; variant M, $t = 0.95$). The high outcrossing rates may result from delivery of large amounts of legitimate pollen early in the anthesis period of flowers, allowing little opportunity for precedence of self pollen (Fig. 9.5). Such an effect, combined with effective postpollination discrimination among pollen types, would serve to maintain high levels of outcrossing even in a selfing variant capable of autonomous self-pollination. Thus, the ecological context may alter the transmission of genes for floral modification by affecting relative female function (seed-set) and relative male function through both self and outcross pollen donation.

The available data from experiments with *E. paniculata* do not enable us to reject any of the three hypotheses proposed to explain the overall rarity of selfing variants in trimorphic populations. The hypotheses are not mutually exclusive, and it seems likely that some elements of each hypothesis (i.e., genetic basis of selfing genes, inbreeding depression, transmission characteristics of selfing variants) will determine whether selfing variants spread in populations. Perhaps the only way to identify unequivocally the selective forces influencing the frequencies of selfing variants in trimorphic populations is to introduce variants into populations occurring under different ecological circumstances and then follow their fate using demographic genetic approaches. Field experiments have been used to investigate both local adaptation in plant populations (Antonovics, 1976; Schemske, 1984; Waser and Price, 1985) and the fitness consequences of different mating patterns (Clay and Antonovics, 1985; Waser and Price, 1989; Schmitt and Gamble, 1990), but no field studies to date have followed the fate of mating-system variants introduced into natural populations. Such experiments, while logistically demanding, would provide powerful insights into the mechanisms governing mating-system change. The major drawback of using this approach in *E. paniculata* is the short-lived nature of many populations, which makes meaningful chronological data on changes in floral morph frequency difficult to obtain. Whereas colonizing species with short life cycles can provide excellent experimental material for answering certain types of questions, the ephermeral nature of many of the habitats they occupy can limit opportunities for long-term studies of the ecological basis of evolutionary change.

Conclusions

Populations of *Eichhornia paniculata* exhibit a wide range of outcrossing rates associated with the dissolution of the tristylous genetic polymorphism and the evolution of self-fertilization. By coupling marker gene analysis of mating-system parameters with information on the demography and floral biology of populations, we have made some progress in unraveling the complex forces influencing mating-system evolution in this species. We have been able to identify the stages of the mating process on which selection is most likely to operate and have determined how pre- and post-pollination mechanisms influence mating behavior. Outcrossing rates in *E. paniculata* are strongly influenced by various aspects of the pollination biology of populations (e.g., levels of pollinator service, timing of self and outcross pollen arrival, and size and composition of pollen loads). Moreover, mating patterns in *E. paniculata* are strongly correlated with the morph structure of populations and, in particular, the frequency of selfing variants.

An important finding from our experimental manipulations of morph structure in garden populations is that the mating patterns of floral morphs are strongly influenced by the morph composition of the array. The transmission characteristics of selfing variants varied with population structure, and outcrossing rates and seed-set in unmodified plants (L, M, S morphs) differed between trimorphic and monomorphic populations. This context-dependent mating behavior has also been demonstrated in experimental populations of *Ipomea purpurea* polymorphic for flower color (Epperson and Clegg, 1987b and unpublished data). Collectively the two studies provide empirical support for mating-system models that assume frequency-dependent influences on selfing rate (Charlesworth and Charlesworth, 1981; Gregorius et al., 1987; Holsinger, 1991). Not only were selfing rates influenced by morph structure, but male reproductive success also varied, depending on the frequency and occurrence of other floral phenotypes within populations. Empirical evidence for frequency-dependent male reproductive success has not been previously demonstrated in plant populations, although it has been inferred from studies of temporal variation in phenotypic gender of diclinous taxa (Thomson and Barrett, 1981; Ross, 1990).

Although our experimental studies of *E. paniculata* have provided useful insights into the potential selective mechanisms governing mating-system evolution, they cannot by themselves provide definitive answers about the causes of mating-system change. This can be achieved only by long-term field studies of the ecological genetics of natural populations (Ford, 1964; Endler, 1986). Field studies can be particularly revealing if they involve experimental manipulations of both the population and the environment in which it occurs. Using such an approach, it is possible to identify not only how selection occurs, but also why it operates in the way it does (Wade and Kalisz, 1990). There have been few attempts to measure selection on floral traits influencing mating patterns in plant populations (but see Campbell, 1989; Schemske and Horvitz, 1989) and only limited work involving the experimental manipulation of floral traits and their reproductive consequences (Ganders, 1974; Queller, 1983; Barrett and Glover, 1985; Bell, 1985; Cruzan et al., 1988). Future work concerned with the selective basis of mating-system evolution could profitably focus on experimental field studies of taxa with wide intraspecific variation in outcrossing rates.

The relative roles of floral morphology, pollen–pistil interactions, and ecological factors in determining mating patterns in populations need to be determined for individual species. If the mating system is primarily controlled by traits under genetic control, then genetic models of mating-system evolution can guide our experimental efforts. On the other hand, data from field and experimental populations reported here indicate that ecological variation can have large effects on mating patterns. To the extent that this is true, more attention will need to be paid to the development and testing of models that incorporate both ecological and genetic factors.

Acknowledgments

We thank C.G. Eckert and K.E. Holsinger for comments on the manuscript, W. Cole for technical assistance, B.C. Husband, S.W. Graham, M.R. Dudash, and C.B. Fenster for permission to cite unpublished data, and A.H.D. Brown, B. Charlesworth, K.E. Holsinger, and D.J. Schoen for providing unpublished manuscripts. Research funds and postdoctoral support for JRK and MBC were provided by operating grants to SCHB from the Natural Sciences and Engineering Research Council of Canada.

Literature Cited

Abbott, R.J. 1985. Maintenance of a polymorphism for outcrossing frequency in a predominantly selfing plant. *In* J. Haeck and J. Woldendorp, eds., Structure and Functioning of Plant Populations, II. Phenotypic and Genotypic Variation in Plant Populations. North Holland, Amsterdam, pp. 277–286.

Antonovics, J. 1976. The nature of the limits to natural selection. Ann. Missouri Bot. Gard. **63**:224–247.

Baker, H.G. 1955. Self-compatibility and establishment after "long-distance" dispersal. Evolution **9**:347–348.

Baker, H.G. 1959. The contribution of autecological and genecological studies to our knowledge of the past migration of plants. Am. Nat. **13**:255–272.

Barrett, S.C.H. 1979. The evolutionary breakdown of tristyly in *Eichhornia crassipes* (Mart.) Solms (water hyacinth). Evolution **33**:499–510.

Barrett, S.C.H. 1985a. Ecological genetics of breakdown in tristyly. *In* J. Haeck and J.W. Woldendorp, eds., Structure and Functioning of Plant Populations. II:

Phenotypic and Genotypic Variation in Plant Populations. North-Holland, Amsterdam, pp. 267–275.

Barrett, S.C.H. 1985b. Floral trimorphism and monomorphism in continental and island populations of Eichhornia paniculata (Spreng.) Solms (Pontederiaceae). Biol. J. Linn. Soc. 25:41–60.

Barrett, S.C.H. 1988. Evolution of breeding systems in Eichhornia (Pontederiaceae): A review. Ann. Missouri Bot. Gard. 75:741–760.

Barrett, S.C.H. 1989. The evolutionary breakdown of heterostyly. In J.H. Bock and Y.B. Linhart, eds., The Evolutionary Ecology of Plants. Westview Press, Boulder, CO, pp. 151–169.

Barrett, S.C.H. 1990. The evolution and adaptive significance of heterostyly. Trends in Ecol. Evol. 5:144–148.

Barrett, S.C.H. 1992. Gender variation in Wurmbea dioica (Liliaceae) and the evolution of dioecy. J. Evol. Biol., in press.

Barrett, S.C.H., and J.M. Anderson. 1985. Variation in expression of trimorphic incompatibility in Pontederia cordata L. (Pontederiaceae). Theor. Appl. Genet. 70:355–362.

Barrett, S.C.H., and D. Charlesworth. 1991. Effects of a change in the level of inbreeding on the genetic load. Nature (London) 352:522–524.

Barrett, S.C.H., and C.G. Eckert. 1990. Variation and evolution of mating systems in seed plants. In S. Kawano (ed.), Biological Approaches and Evolutionary Trends in Plants. Academic Press, London, pp. 229–254.

Barrett, S.C.H., and D.E. Glover. 1985. On the Darwinian hypothesis of the adaptive significance of tristyly. Evolution 39:766–774.

Barrett, S.C.H., and L.D. Harder. 1991. Floral variation in Eichhornia paniculata (Spreng.) Solms (Pontederiaceae) II. Effects of development and environment on the formation of selfing flowers. J. Evol. Biol., in press.

Barrett, S.C.H., and B.C. Husband. 1990. Variation in outcrossing rates in Eichhornia paniculata: The role of demographic and reproductive factors. Plant Species Biol. 5:41–55.

Barrett, S.C.H., and J.R. Kohn. 1991. Genetic and evolutionary consequences of small population size in plants: Implications for conservation. In D.A. Falk and K.E. Holsinger, eds., Genetics and Conservation of Rare Plants, Oxford University Press, New York, pp. 3–30.

Barrett, S.C.H., A.H.D. Brown, and J.S. Shore. 1987. Disassortative mating in tristylous Eichhornia paniculata (Pontederiaceae). Heredity 58:49–55.

Barrett, S.C.H., M.T. Morgan, and B.C. Husband. 1989. The dissolution of a complex genetic polymorphism: The evolution of self-fertilization in tristylous Eichhornia paniculata (Pontederiaceae). Evolution 43:1398–1416.

Bawa, K.S. 1980. Evolution of dioecy in flowering plants. Annu. Rev. Ecol. Syst. 11:15–39.

Beach, J.H., and K.S. Bawa. 1980. Role of pollinators in the evolution of dioecy from distyly. Evolution 34:1138–1142.

Becerra, J.X., and D.G. Lloyd. 1992. Competition-dependent abortion of self-pollinated flowers of *Phormium tenax* (Agavaceae)—a second action of self-incompatibility at the whole flower level? Evolution, in press.

Bell, G. 1985. On the function of flowers. Proc. R. Soc. London Ser. B **224**:223–265.

Bodmer, W.F. 1960. Genetics of homostyly in populations of *Primula vulgaris*. Phil. Trans. R. Soc. London Ser. B **242**:517–549.

Bowman, R.N. 1987. Cryptic self-incompatibility and the breeding system of *Clarkia unguiculata* (Onagraceae). Am. J. Bot. **74**:471–476.

Brown, A.H.D. 1979. Enzyme polymorphism in plant populations. Theor. Pop. Biol. **15**:1–42.

Brown, A.H.D. 1990. Genetic characterization of plant mating systems. *In* A.H.D. Brown, M.T. Clegg, A.L. Kahler, and B.S. Wier, eds., Plant Population Genetics, Breeding, and Genetic Resources. Sinauer, Sunderland, MA, pp. 145–162.

Brown, A.H.D., J.E. Grant, and R. Pullen. 1986. Outcrossing and paternity in *Glycine argyrea* by paired fruit analysis. Biol. J. Linn. Soc. **29**:283–294.

Brown, A.H.D., J.J. Burdon, and A.M. Jarosz. 1990. Isozyme analysis of plant mating systems. *In* D.E. Soltis and P.S. Soltis, eds., Isozymes in Plant Biology. Dioscorides Press, Portland, OR, pp. 73–86.

Broyles, S.B., and R. Wyatt. 1990. Paternity analysis in a natural population of *Asclepias exaltata*: Multiple paternity, functional gender, and the "pollen-donation hypothesis." Evolution **44**:1454–1468.

Campbell, D.R. 1989. Measurements of selection in a hermaphroditic plant: Variation in male and female pollination success. Evolution **43**:318–334.

Campbell, R.B. 1986. The interdependence of mating structure and inbreeding depression. J. Theor. Biol. **30**:232–244.

Casper, B.B. 1992. The application of sex-allocation theory to heterostylous plants. *In* S.C.H. Barrett, ed., Evolution and Function of Heterostyly. Springer-Verlag, Berlin, pp. 209–223.

Casper, B.B., and E.L. Charnov. 1982. Sex allocation in heterostylous plants. J. Theor. Biol. **96**:143–149.

Charlesworth, B. 1992. Evolutionary rates in partially self-fertilizing species. Am. Nat., in press.

Charlesworth, B., and D. Charlesworth. 1978. A model for the evolution of dioecy and gynodioecy. Am. Nat. **112**:975–997.

Charlesworth, D., and B. Charlesworth. 1979a. A model for the evolution of distyly. Am. Nat. **114**:467–498.

Charlesworth, D., and B. Charlesworth. 1979b. The evolution and breakdown of *S*-allele systems. Heredity **43**:41–55.

Charlesworth, D., and B. Charlesworth. 1981. Allocation of resources to male and female functions in hermaphrodites. Biol. J. Linn. Soc. **15**:57–74.

Charlesworth, D., and B. Charlesworth. 1987. Inbreeding depression and its evolutionary consequences. Annu. Rev. Ecol. Syst. **18**:237–268.

Charlesworth, D., and B. Charlesworth. 1990. Inbreeding depression with heterozygote advantage and its effect on selection for modifiers changing the outcrossing rate. Evolution **44**:870–888.

Charnov, E.L. 1982. The Theory of Sex Allocation. Princeton University Press, Princeton, NJ.

Clay, K., and J. Antonovics. 1985. Demographic genetics of the grass *Danthonia spicata*: Success of progeny from chasmogamous and cleistogamous flowers. Evolution **39**:205–210.

Clegg, M.T. 1980. Measuring plant mating systems. BioScience **30**:814–818.

Clegg, M.T., and B.K. Epperson. 1988. Natural selection of flower color polymorphisms in morning glory populations. *In* L.D. Gottlieb and S.K. Jain, eds., Plant Evolutionary Biology. Chapman & Hall, London, pp. 255–273.

Costich, D.E., and T.R. Meagher. 1992. Genetic variation in *Ecballium elaterium*: Breeding system and geographic distribution. J. Evol. Biol., in press.

Crosby, J.L. 1949. Selection of an unfavourable gene-complex. Evolution **3**:212–230.

Cruzan, M.B. 1989. Pollen tube attrition in *Erythronium grandiflorum*. Am. J. Bot. **76**:562–570.

Cruzan, M.B., and S.C.H. Barrett. 1992. Contribution of cryptic incompatibility to the mating system of *Eichhornia paniculata* (Pontederiaceae). Evolution, in press.

Cruzan, M.B., P.R. Neal, and M.F. Willson. 1988. Floral display in *Phyla incisa*: Consequences for male and female reproductive success. Evolution **42**:505–515.

Darlington, C.D. 1939. Evolution of Genetic Systems. Cambridge University Press, Cambridge.

Dudash, M.R. 1990. Relative fitness of selfed and outcrossed progeny in a self-compatible, protandrous species, *Sabatia angularis* L. (Gentianaceae): A comparison of three environments. Evolution **44**:1129–1139.

Dudash, M.R., and K.M. Ritland. 1991. Multiple paternity and self-fertilization in relation to floral age in *Mimulus guttatus* (Scrophulariaceae). Am. J. Bot., **78**:1746–1753.

Ellstrand, N.C., E.M. Lord, and K.J. Eckard. 1984. The inflorescence as a metapopulation of flowers: Position-dependent differences in function and form in the cleistogamous species *Collomia grandiflora* Dougl. ex Lindl. (Polemoniaceae). Bot. Gaz. **145**:329–333.

Endler, J.A. 1986. Natural Selection in the Wild. Princeton University Press, Princeton, NJ.

Ennos, R.A., and R.K. Dodson. 1987. Pollen success, functional gender and disassortative mating in an experimental plant population. Heredity **58**:119–126.

Epperson, B.K., and M.T. Clegg. 1987a. First-pollination primacy and pollen selection in the morning glory, *Ipomoea purpurea*. Heredity **58**:5–14.

Epperson, B.K., and M.T. Clegg. 1987b. Frequency-dependent variation for out-crossing rate among flower color morphs of *Ipomoea purpurea*. Evolution **41**:1302–1311.

Fisher, R.A. 1931. The evolution of dominance. Biol. Rev. **6**:345–368.

Ford, E.B. 1964. Ecological Genetics. Chapman & Hall, London.

Ganders, F.R. 1974. Disassortative pollination in the distylous plant *Jepsonia heterandra*. Can. J. Bot. **52**:2401–2406.

Ganders, F.R. 1975. Mating patterns in self-compatible populations of *Amsinckia* (Boraginaceae). Can. J. Bot. **53**:773–779.

Ganders, F.R. 1979. The biology of heterostyly. N.Z. J. Bot. **17**:607–635.

Givnish, T.J. 1982. Outcrossing versus ecological constraints in the evolution of dioecy. Am. Nat. **119**:849–865.

Glover, D.E., and S.C.H. Barrett. 1986. Variation in the mating system of *Eichhornia paniculata* (Spreng.) Solms (Pontederiaceae). Evolution **40**:1122–1131.

Glover, D.E., and S.C.H. Barrett. 1987. Genetic variation in continental and island populations of *Eichhornia paniculata* (Pontederiaceae). Heredity **59**:7–17.

Gouyon, P.H., and D. Couvet. 1987. A conflict between two sexes, females and hermaphordites. *In* S.C. Stearns, ed., The Evolution of Sex and Its Consequences. Birkhauser Verlag, Basel, pp. 245–261.

Graham, S.W., and S.C.H. Barrett. 1990. Pollen precedence in *Eichhornia paniculata*: A tristylous species. Am. J. Bot. **77**:54–55 (abstract).

Grant, V. 1958. The regulation of recombination in plants. Cold Spring Harbor Symp. Quant. Biol. **23**:337–363.

Gregorius, H.R., M. Ziehe, and M.D. Ross. 1987. Selection caused by self-fertilization. 1. Four measures of self-fertilization and their effects on fitness. Theor. Pop. Biol. **31**:91–115.

Haldane, J.B.S. 1924. A mathematical theory of natural and artificial selection. Part I. Trans. Cambridge Phil. Soc. **23**:19–41.

Haldane, J.B.S. 1927. A mathematical theory of natural and artificial selection. Part V. Selection and mutation. Proc. Cambridge Phil. Soc. **23**:838–844.

Hamrick, J.L., and M.J. Godt. 1990. Allozyme diversity in plant species. *In* A.H.D. Brown, M.T. Clegg, A.L. Kahler, and B.S. Weir, eds., Plant Population Genetics, Breeding, and Genetic Resources. Sinauer, Sunderland, MA, pp. 43–63.

Hicks, D.J., R. Wyatt, and T.R. Meagher. 1985. Reproductive biology of distylous partridgeberry, *Mitchella repens*. Am. J. Bot. **72**:1503–1514.

Holsinger, K.E. 1991. Mass action models of plant mating systems: The evolutionary stability of mixed mating systems. Am. Nat., **138**:606–622.

Holsinger, K.E., M.W. Feldman, and F.B. Christiansen. 1984. The evolution of self-fertilization in plants: A population genetic model. Am. Nat. **124**:446–453.

Holtsford, T.P., and N.C. Ellstrand. 1989. Variation in outcrossing rate and population genetic structure of *Clarkia tembloriensis* (Onagraceae). Theor. Appl. Genet. **78**:480–488.

Holtsford, T.P., and N.C. Ellstrand. 1990. Inbreeding effects in *Clarkia temblo-riensis* (Onagraceae) populations with different natural outcrossing rates. Evolution **44**:2031–2046.

Husband, B.C., and S.C.H. Barrett. 1991. Colonization history and population genetic structure of *Eichhornia paniculata* in Jamaica. Heredity **66**:287–296.

Husband, B.C., and S.C.H. Barrett. 1992a. Effective population size and genetic drift in tristylous *Eichhornia paniculata* (Pontederiaceae). Evolution, in press.

Husband, B.C., and S.C.H. Barrett. 1992b. Pollinator visitation in populations of tristylous *Eichhornia paniculata* in northeastern Brazil. Oecologia, in press.

Husband, B.C., and S.C.H. Barrett. 1992c. Multiple origins of self-fertilization in tristylous *Eichhornia paniculata* (Pontederiaceae): Inferences from style morph and isozyme variation. J. Evol. Biol., submitted.

Iwasa, Y. 1990. Evolution of selfing rate and resource allocation models. Plant Species Biol. **5**:19–30.

Jain, S.K. 1976. The evolution of inbreeding in plants. Annu. Rev. Ecol. Syst. **7**:69–95.

Kohn, J.R. 1988. Why be female? Nature (London) **335**:431–433.

Kohn, J.R. 1989. Sex ratio, seed production, biomass allocation, and the cost of female function in *Cucurbita foetidissima* HBK (Cucurbitaceae). Evolution **43**:1424–1434.

Kohn, J.R., and S.C.H. Barrett. 1992a. Experimental studies on the functional significance of heterostyly. Evolution, in press.

Kohn, J.R., and S.C.H. Barrett. 1992b. Floral manipulations reveal the cause of male fitness variation in experimental populations of *Eichhornia paniculata* (Pontederiaceae). Func. Ecol., in press.

Lande, R., and D.W. Schemske. 1985. The evolution of self-fertilization and inbreeding depression in plants. I. Genetic models. Evolution **39**:24–40.

Layton, C.R., and F.R. Ganders. 1984. The genetic consequences of contrasting breeding systems in *Plectritis* (Valerianaceae). Evolution **38**:1308–1325.

Lloyd, D.G. 1965. Evolution of self-compatibility and racial differentiation in *Leavenworthia* (Cruciferae). Contrib. Gray Herb. **195**:3–133.

Lloyd, D.G. 1975. The maintenance of gynodioecy and androdioecy in angiosperms. Genetica **45**:325–339.

Lloyd, D.G. 1976. The transmission of genes via pollen and ovules in gynodioecious angiosperms. Theor. Pop. Biol. **9**:199–216.

Lloyd, D.G. 1979a. Some reproductive factors affecting the selection of self-fertilization in plants. Am. Nat. **113**:67–79.

Lloyd, D.G. 1979b. Evolution towards dioecy in heterostylous populations. Plant Syst. Evol. **131**:71–80.

Lloyd, D.G. 1980. Demographic factors and mating patterns in angiosperms. *In* O.T. Solbrig, ed., Demography and Evolution in Plant Populations. Blackwell, Oxford, pp. 67–88.

Lloyd, D.G. 1982. Selection of combined versus separate sexes in seed plants. Am. Nat. **120**:571–585.

Lloyd, D.G., and C.J. Webb. 1992a. Evolution of heterostyly. *In* S.C.H. Barrett, ed., Evolution and Function of Heterostyly. Springer-Verlag, Berlin, pp. 151–178.

Lloyd, D.G., and C.J. Webb. 1992b. The selection of heterostyly. *In* S.C.H. Barrett, ed., Evolution and Function of Heterostyly. Springer-Verlag, Berlin, pp. 179–207.

Lord, E.M. 1981. Cleistogamy: A tool for the study of floral morphogenesis, function and evolution. Bot. Rev. **47**:421–449.

Lyons, E.E., N.M. Waser, M.V. Price, J. Antonovics, and A.F. Motten. 1989. Sources of variation in plant reproductive success, and implications for concepts of sexual selection. Am. Nat. **134**:409–433.

Marshall, D.L. 1990. Non-random mating in wild radish, *Raphanus sativus*. Plant Species Biol. **5**:143–156.

Morgan, M.T., and S.C.H. Barrett. 1989. Reproductive correlates of mating system variation in *Eichhornia paniculata* (Spreng.) Solms (Pontederiaceae). J. Evol. Biol. **2**:183–203.

Morgan, M.T., and S.C.H. Barrett. 1990. Outcrossing rates and correlated mating within a population of *Eichhornia paniculata* (Pontederiaceae). Heredity **64**:271–280.

Muenchow, G. 1982. A loss-of-alleles model for the evolution of distyly. Heredity **49**:81–93.

Muenchow, G.E., and M. Grebus. 1989. The evolution of dioecy from distyly: Reevaluation of the hypothesis of the loss of long-tongued pollinators. Am. Nat. **133**:149–156.

Nicholls, M.S. 1987. Pollen flow, self-pollination and gender specialization: Factors affecting seed-set in the tristylous species *Lythrum salicaria* (Lythraceae). Plant Syst. Evol. **156**:151–157.

Ornduff, R. 1972. The breakdown of trimorphic incompatibility in *Oxalis* section Corniculatae. Evolution **26**:52–65.

Piper, J.G., B. Charlesworth, and D. Charlesworth. 1984. A high rate of self-fertilization and increased seed fertility of homostyle primroses. Nature (London) **310**:50–51.

Piper, J.G., B. Charlesworth, and D. Charlesworth. 1986. Breeding system evolution in *Primula vulgaris* and the role of reproductive assurance. Heredity **56**:207–217.

Pollack, E. 1987. On the theory of partially inbreeding finite populations. I. Partial selfing. Genetics **117**:353–360.

Queller, D.C. 1983. Sexual selection in an hermaphroditic plant. Nature (London) **205**:706–707.

Raven, P.H. 1979. A survey of reproductive biology in the Onagraceae. N.Z. J. Bot. **17**:575–594.

Richards, J.H., and S.C.H. Barrett. 1984. The developmental basis of tristyly in *Eichhornia paniculata* (Pontederiaceae). Am. J. Bot. **71**:1347–1363.

Richards, J.H., and S.C.H. Barrett. 1992. Development of heterostyly. *In* S.C.H. Barrett, ed., Evolution and Function of Heterostyly. Springer-Verlag, Berlin, pp. 85–127.

Rick, C.M., J.F. Fobes, and M. Holle. 1977. Genetic variation in *Lycopersicon pimpinellifolium*: Evidence of evolutionary change in mating systems. Plant Syst. Evol. **127**:139–170.

Ritland, K. 1983. Estimation of mating systems. *In* S.D. Tanksley and T.J. Orton, eds., Isozymes in Plant Genetics and Breeding, Part A. Elsevier, Amsterdam, pp. 289–302.

Ritland, K. 1989. Correlated matings in the partial selfer *Mimulus guttatus*. Evolution **43**:848–860.

Ritland, K. 1990. A series of FORTRAN computer programs for estimating plant mating systems. J. Hered. **85**:325–327.

Ross, M.D. 1982. Five evolutionary pathways to subdioecy. Am. Nat. **119**:297–318.

Ross, M.D. 1990. Sexual asymmetry in hermaphroditic plants. Trends Ecol. Evol. **5**:43–47.

Ross, M.D., and H. Gregorius. 1983. Outcrossing and sex function in hermaphrodites: A resource allocation model. Am. Nat. **121**:204–222.

Schemske, D.W. 1983. Breeding system and habitat effects in three neotropical *Costus* (Zingiberaceae). Evolution **37**:523–539.

Schemske, D.W. 1984. Population structure and local selection in *Impatiens pallida* (Balsaminaceae), a selfing annual. Evolution **38**:817–832.

Schemske, D.W., and C.C. Horvitz. 1989. Temporal variation in selection on a floral character. Evolution **43**:461–465.

Schemske, D.W., and R. Lande. 1985. The evolution of self-fertilization and inbreeding depression in plants. II. Empirical observations. Evolution **39**:41–52.

Schmitt, J., and S.E. Gamble. 1990. The effect of distance from the parental site on offspring performance and inbreeding depression in *Impatiens capensis*: A test of the local adaptation hypothesis. Evolution **44**:2022–2030.

Schoen, D.J. 1983. Relative fitness of selfed and outcrossed progeny in *Gilia achilleifolia*. Evolution **37**:291–301.

Schoen, D.J. 1985. Correlation between classes of mating events in two experimental plant populations. Heredity **55**:381–385.

Schoen, D.J. 1988. Mating system estimation via the one pollen parent model with the progeny array as the unit of observation. Heredity **60**:439–444.

Schoen, D.J., and A.H.D. Brown. 1991. Whole- and part-flower self-pollination in *Glycine clandestina* and *G. argyrea* and the evolution of autogamy. Evolution **45**:1651–1664.

Schoen, D.J., and M.T. Clegg. 1984. Estimation of mating system parameters when outcrossing events are correlated. Proc. Nat. Acad. Sci. U.S.A. **81**:5258–5262.

Schoen, D.J., and D.G. Lloyd. 1984. The selection of cleistogamy and heteromorphic diaspores. Biol. J. Linn. Soc. **23**:303–322.

Scribailo, R.W., and S.C.H. Barrett. 1991. Pollen-pistil interactions in tristylous *Pontederia sagittata* (Pontederiaceae). II. Patterns of pollen tube growth. Am. J. Bot. **78**:1662–1682.

Seburn, C.L., T.A. Dickinson, and S.C.H. Barrett. 1990. Floral variation in *Eichhornia paniculata* (Spreng.) Solms (Pontederiaceae). I. Instability of stamen position in genotypes from northeast Brazil. J. Evol. Biol. **3**:103–123.

Stanton, M.L., A.A. Snow, S.N. Handel, and J. Bereczky. 1989. The impact of flower-color polymorphism on mating patterns in experimental populations of wild radish (*Raphanus raphanistrum* L.). Evolution **43**:335–346.

Stebbins, G.L. 1957. Self-fertilization and population variability in the higher plants. Am. Nat. **91**:337–354.

Stebbins, G.L. 1958. Longevity, habitat, and release of genetic variability in the higher plants. Cold Spring Harbor Symp. Quant. Biol. **23**:365–378.

Stebbins, G.L. 1974. Flowering Plants: Evolution above the Species Level. Belknap Press, Cambridge.

Thomson, J.D., and S.C.H. Barrett. 1981. Temporal variation of gender in *Aralia hispida* Vent. (Araliaceae). Evolution **35**:1094–1107.

Thomson, J.D., and J. Brunet. 1990. Hyptheses for the evolution of dioecy in seed plants. Trends Ecol. Evol. **5**:11–16.

Toppings, P. 1989. The significance of inbreeding depression to the evolution of self-fertilization in *Eichhornia paniculata* (Spreng.) Solms (Pontederiaceae). M.Sc. Thesis, University of Toronto.

Turner, J.R.G. 1981. Adaptation and evolution in *Heliconius*: A defense of neoDarwinism. Annu. Rev. Ecol. Syst. **12**:99–121.

Uyenoyama, M.K. 1986. Inbreeding and the cost of meiosis: The evolution of selfing in populations practicing biparental inbreeding. Evolution **40**:388–404.

Uyenoyama, M.K. 1988. On the evolution of incompatibility systems. II. Initial increase of strong gametophytic self-incompatibility under partial selfing and half-sib mating. Am. Nat. **131**:700–722.

Uyenoyama, M.K., and J. Antonovics. 1987. The evolutionary dynamics of mixed mating systems: On the adaptive value of selfing and biparental inbreeding. *In* P.P.G. Bateson and P.H. Klopfer, eds., Perspectives in Ethology, Vol. 7: Alternatives. Plenum Press, New York, pp. 125–152.

Wade, M.J., and S. Kalisz. 1990. The causes of natural selection. Evolution **44**:1947–1955.

Waller, D.M. 1980. Environmental determinants of outcrossing in *Impatiens capensis* (Balsaminaceae). Evolution **34**:747–761.

Waller, D.M., and S.E. Knight. 1989. Genetic consequences of outcrossing in the cleistogamous annual, *Impatiens capensis*. II. Outcrossing rates and genotypic correlations. Evolution **43**:860–869.

Waser, N.M., and M.V. Price. 1985. Reciprocal transplant experiments with *Delphinium nelsonii* (Ranunculaceae): Evidence for local adaptation. Am. J. Bot. **72**:1726–1732.

Waser, N.M., and M.V. Price. 1989. Optimal outcrossing in *Ipomopsis aggregata*: Seed set and offspring fitness. Evolution **43**:1097–1109.

Webb, C.J., and D.G. Lloyd. 1986. The avoidance of interference between the presentation of pollen and stigmas in angiosperms. II. Herkogamy. N.Z. J. Bot. **24**:163–178.

Weller, S.G., and A.K. Sakai. 1990. The evolution of dicliny in *Schiedea* (Caryophyllaceae), an endemic Hawaiian genus. Plant Species Biol. **5**:83–96.

Wells, H. 1979. Self-fertilization: Advantageous or deleterious? Evolution **33**:252–255.

Willson, M.F. 1979. Sexual selection in plants. Am. Nat. **113**:777–790.

Willson, M.F., and N. Burley. 1983. Mate Choice in Plants: Tactics, Mechanisms, and Consequences. Princeton University Press, Princeton, NJ.

Wyatt, R. 1986. Ecology and evolution of self-pollination in *Arenaria uniflora* (Caryophyllaceae). J. Ecol. **74**:403–418.

Wyatt, R. 1988. Phylogenetic aspects of the evolution of self-pollination. *In* L.D. Gottlieb and S.K. Jain, eds., Plant Evolutionary Biology. Chapman & Hall, London, pp. 109–131.

Yahara, T. 1992. Graphical analysis for evolutionary stable mating systems in plants. Evolution, in press.

10

Genetic and Molecular Dissection of Male Reproductive Processes

Bruce Knox, Cenk Suphioglu, Terryn Hough, and Mohan Singh

University of Melbourne

Introduction

Reproduction is essential for living plants to increase their numbers and as such is of primary importance in evolution. Reproduction can now be dissected at the molecular level by isolating genes that regulate the processes involved. The genetic dissection of this system offers new insights into the understanding and manipulation of fertilization and seed-set in flowering plants.

The pollen grain, the male gametophyte of flowering plants, plays a central role in reproduction. The pollen produces and transfers the sperm cells (or their progenitor, the generative cell), into the growing pollen tube to the embryo sac, where fertilization occurs. A feature of such systems is that haploid gene expression occurs during microspore ontogeny, pollen maturation and dehydration, germination, tube growth in the pistil, and sperm–egg interactions. New variants of genes can be expressed and screened during these processes. Haploid:diploid interactions occur at meiosis and during interactions with the tapetum, and offer the potential to explore the extent of diploid-specified gene expression during pollen development.

Relatively few plant genes have been isolated to date. A considerable amount of descriptive molecular biology has been carried out to study tissue-specific gene expression. An important genetic approach is the generation and characterization of developmental mutants, and a number of male-sterile mutants in which anther development is blocked at specific stages of development have begun to be characterized.

Here we focus on the genetic and molecular dissection of processes involved in male reproductive development, especially the isolation of genes that are specifically expressed in pollen. The case of grass pollen allergenic proteins is considered in detail here, as these are examples in which the genes encode products whose biological activity is known. Before reviewing the information on gene expression and isolation, we include an update on the cell biology of the pollen grain in relation to two key features: viability of the pollen grain and organization of the male reproductive cells.

Quantification of Pollen Viability

The functional quality of pollen is very difficult to estimate, either in field populations, in the greenhouse, or during pollen storage. There is no test of what Heslop-Harrison et al. (1984) have termed the essential role of pollen: its fertilizing capacity, which involves the successful transfer of the male reproductive cells to the embryo sac for fertilization. Whereas functional quality can, given time, be estimated by seed-set, there are other complicating factors that regulate seed-set (Heslop-Harrison et al., 1984). Accordingly, other measures of estimating pollen quality have been developed, including a great range of staining tests (Heslop-Harrison et al., 1984; Knox, 1984).

Of these there are three currently employed methods that give consistent, reproducible, and meaningful results (Knox, 1984): (1) the fluorochromatic reaction (FCR) test assesses the integrity of the plasma membranes of the vegetative cell of the pollen grain by fluorescence microscopy and indicates whether the vegetative cell is living or dead; (2) *in vitro* pollen germination tests demonstrate the ability of pollen tubes to grow in an artificial medium following effective hydration; and (3) nuclear magnetic resonance (NMR) estimates the amount of free to bound water content of the pollen grains— a test that depends on the maintenance of the living state.

The FCR test can be difficult to interpret, especially if applied in hypertonic or hypotonic media. Variations in the strength of the fluorescent reactions have been observed in some systems, but, if carefully documented, FCR still provides a rapid test of whether the vegetative cell is living or dead at the time of testing. An excellent example of the application of this method is the demonstration of the deleterious effects of metapleural secretions of the ant *Myrmecia* on pollen quality (Beattie et al., 1985). Metapleural secretion dramatically reduced pollen quality prior to hydration, suggesting that the active component directly interfered with the integrity of the plasma membrane.

In vitro germination tests rely on defining the appropriate medium composition for the pollen type, which can be difficult. The method is largely applicable only to bicellular pollen types, which have less rigid germination requirements. Among tricellular types, special media have been developed for the germination of *Brassica* pollen (Mulcahy and Mulcahy, 1987). The capacity to produce a pollen tube does not depend on newly synthesized mRNA (Mascarenhas, 1975) and thus is to some extent autonomous. This can be seen from mentor pollen experiments with poplar pollen, when the mentor pollen, which had been exposed to lethal doses of gamma radiation, was still able to germinate and pollen tubes grew into the style (Knox et al., 1987, 1988). Both of these measures of viability are destructive, which can be a problem if pollen quantity is limited.

NMR is a nondestructive method and can be carried out using proton ^{1}H NMR to measure water loss from pollen or with phosphorus ^{31}P NMR, which provides information on the amount of organic phosphorus available for metabolism. ^{1}H NMR is not recommended for determining viability, as the pollen is able to rehydrate after desiccation, a phenomenon that is difficult to interpret. ^{31}P NMR is a promising technique for measuring viability, but the pollen sample required is relatively large and the equipment required is expensive or difficult to access.

Ultimately, the objective is to determine the longevity of pollen grains in the field. To date this has been achieved only in the laboratory. Both FCR and NMR tests with *Brassica* pollen indicate that pollen retains its viability for 4 to 5 days (Dumas et al., 1983), whereas the pollen of rye is viable for only a few hours following anthesis (Ladyman and Taylor, 1988). Assessment of pollen viability under field conditions can be performed by FCR, but NMR technology is not suited to these conditions.

Tests of pollen quality can be useful indicators of male fitness in studies of reproductive biology if the limitations of the techniques are taken into account.

The Male Germ Unit

With the development of quantitative techniques for transmission electron microscopy, a new concept concerning the organization of the male reproductive cells emerged—the male germ unit (Dumas et al., 1984). The pair of sperm cells or, in some cases, their progenitor (the generative cell), become linked with the vegetative nucleus, so that a single transmitting unit for fertilization is formed in the pollen tube. Effectively, all of the DNA of male heredity is associated together. Some 30 different genera of angiosperms have been shown to have a male germ unit, either as a stable association during tube growth or an ephemeral association early in tube growth (Knox and Singh, 1987; Knox et al., 1988; Roeckel et al., 1990).

The extent of the sperm cell dimorphism provides some indication of the likely role of the male germ unit. Quantitative differences in volume between the pair of sperm cells have been recorded for *Plumbago* (Russell and Cass, 1983) and *Brassica* (McConchie et al., 1987). In both these tricellular pollen systems, the smaller sperm of the pair (Fig. 10.1a. Sua)

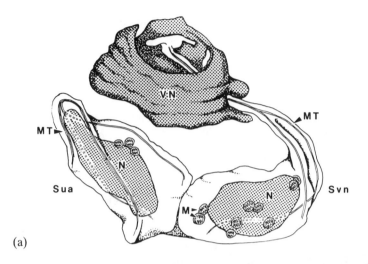

(a)

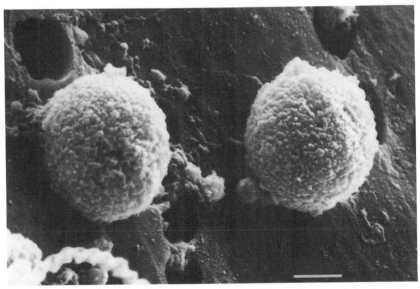

(b)

Figure 10.1. The male germ unit and sperm cells of *Brassica*. (a) A schematic diagram (from *B. campestris*, McConchie et al., 1987) of the paired sperm cells from mature pollen. One sperm cell has a long extension that enters enclaves in the vegetative nucleus (Svn), with longitudinally oriented arrays of microtubules and small spherical mitochondria adjacent to the sperm nuclei. The smaller sperm (Sua) is unattached to the vegetative cell and has few mitochondria. (b) A scanning electron micrograph of isolated sperm cells from pollen tubes of *B. napus* grown *in vitro*, ruptured by osmotic shock and immobilized on a Nuclepore membrane (P.E. Taylor, unpublished). Bar = 1 μm.

has few mitochondria. In *Brassica*, this sperm shows the lowest number of mitochondria (mean of 3.5) recorded in any flowering plant cell (Fig. 10.1a,b). These organelles are much smaller than their vegetative cell counterparts, and because mitochondrial inheritance is maternal, they may already have lost their DNA. In *Plumbago*, the smaller sperm is plastid rich, and Russell (1985) provided some evidence for its role in sperm–egg interactions at double fertilization.

In *Rhododendron* pollen, a bicellular system, the male germ unit commences in the pollen tube when the generative cell wraps long filamentous extensions around the tube nucleus (Taylor et al., 1989). The male germ unit organization in *Rhododendron* tubes has been detected by serial reconstruction of transmission electron micrographs and by image-enhanced fluorescent microscopy. *Rhododendron*, like *Plumbago*, shows biparental inheritance and so possesses sperm cells containing both mitochondria and plastids. Unlike *Plumbago*, *Rhododendron* sperm cells do not show polarization into mitochondria-rich and plastid-rich types. Instead, although sperm cells show dimorphism in size, the larger of the pair has relatively more of both types of organelle (Taylor et al., 1989).

Considerable progress in understanding sperm cell function has come from attempts to isolate sperm-enriched fractions from pollen grains. This can be achieved either by rupturing of the grains by osmotic shock (Russell, 1986; Dupuis et al., 1987) or by physical grinding of pollen (Hough et al., 1986). Sperm cells from bicellular pollen types have been isolated by osmotic shock from pollen tubes (Shivanna et al., 1988). The isolated sperms are natural protoplasts and round off in the isolation medium.

Isolated sperm cells can be used to characterize surface determinants, which are expected to play some role in sperm–egg interactions and *in vitro* fertilization.

In animal sperm, the specificity of cell fusion depends on the major histocompatibility (MHC) factors on the interacting cell surface. Ashida and Scofield (1987) have shown that human sperm can fuse with leukocytes in blood if the leukocytes carry the appropriate MHC factor. Little is known of the surface determinants of plant sperm, and electroporation and fusion with electric currents have been necessary to cause isolated sperm and eggs from plants to fuse (Kranz et al., 1991). The production of monoclonal antibodies to isolated sperm cells provides a versatile tool to study sperm cell determinants and to examine their role in recognition events.

Pollen Gene Expression

Pollen genes are those genes that are newly expressed or whose expression is augmented in pollen grains, and we might expect to find the following types: (1) genes expressed exclusively in pollen; (2) pollen-specific isozymes, isotypes, and variants; and (3) sporophytic genes with overlapping expression in pollen.

The level of complexity of transcripts expressed in pollen has been approached using hybridization kinetic analysis. About 20,000 different mRNA transcripts are expressed in pollen of *Tradescantia* and corn (Willing and Mascarenhas, 1984; Willing et al., 1988). These numbers represent transcripts found in mature pollen, and it is likely that additional genes may be active in earlier stages of pollen development (Mascarenhas, 1989, 1990a,b). Based on colony hybridization with cDNA libraries made to pollen poly(A)$^+$ RNA and hybridized with ^{32}P-labeled cDNAs from pollen and vegetative tissues, about 10% of total sequences expressed in corn pollen were shown to be unique to pollen (Hanson et al., 1989; Stinson et al., 1987). cDNA libraries have been constructed to poly(A)$^+$ RNA from mature pollen of *Brassica campestris* (Theerakulpisut et al., 1989, 1991) and *Lolium perenne* (Knox et al., 1989; Singh et al., 1991). Only 0.5% of sequences represented in mature *Brassica* pollen were unique to pollen. Pollen-specific cDNAs have also been isolated from tomato (McCormick et al., 1987; Twell et al., 1989), and *Oenothera* (Brown and Crouch, 1990).

Two sets of genes exist that differ in regard to their temporal pattern of expression during pollen development (Mascarenhas, 1989, 1990a,b; Wing et al., 1989). The *first or early set* includes housekeeping genes such as actin, which are expressed during microspore development but whose expression diminishes substantially before anthesis. Genes encoding alcohol dehydrogenase (Stinson and Mascarenhas, 1985) and β-galactosidase (Singh et al., 1985) probably belong to this set. Most of the research into pollen genes has involved cDNA libraries constructed from RNA of mature pollen and anther tissues, and as a result most of the genes identified to date fall into the *late set*. These are represented by pollen-specific cDNA clones that are first detected after first microspore mitosis during pollen development in corn and *Tradescantia* (Stinson et al., 1987) and tomato (LAT genes, Ursin et al., 1989). The mRNAs for these genes continue to accumulate during the maturation phases. This accumulation of *late* mRNAs suggests an important role for these genes during the later stages of pollen development and during germination and pollen tube growth.

Isolation of Pollen-Specific Genes by Differential Hybridization

Genes of unidentified function can be isolated by differential hybridization, which involves screening cDNA libraries from pollen against vegetative organs such as the leaf. Clones that are specifically expressed in pollen can be selected for further study. After sequencing, the possible function of these genes can be elucidated by comparison with known sequences in data bases or by use of specific monoclonal antibody probes. The LAT genes of tomato, LAT 56 and LAT 59 cDNAs, possess significant sequence homologies with pectate lyases of the bacterial plant pathogen, *Erwinia* and the eukaryote *Aspergillus* (Wing et al., 1989). This sequence similarity suggests that the LAT genes may encode functional pectate lyase for pectin degradation during pollen germination or growth of the pollen tube down the style.

Another example of sequence homology is the putative amino acid sequence of *Oenothera* pollen cDNA, which has significant homologies with polygalacturonase from tomato fruits (Brown and Crouch, 1990). In a study of pollen-specific genes in *Brassica campestris* (Theerakulpisut et al., 1989, 1991), two cDNA clones showed the same pattern by Northern analysis as late-expressed pollen genes. This is not unexpected, as the clones were isolated from cDNA libraries prepared from mature pollen.

There are practical difficulties in collecting enough material to isolate sufficient poly(A)[+] RNA from early stages of microspore development. Recently, PCR-based methods have been proposed to prepare cDNA libraries from RNA of small numbers of cells (Domec et al., 1990; Murphy et al., 1990). These methods have been used to make cDNA libraries to microspores in early stages of development to isolate early-expressed genes. An early-expressed pollen-specific gene, *Bp4*, from *B. napus* has been isolated by screening the genomic library by hybridization with ^{32}P-labeled cDNAs from pollen and vegetative tissues (Albani et al., 1990). Transcripts did not accumulate in the mature grains.

Tissue specificity of expression in many pollen-specific clones that have been published have been tested by Northern analysis only. This prevents testing cross-hybridization with other anther tissues. For example, the LAT series of cDNA clones of tomato pollen were found by *in situ* hybridization to cross-hybridize with the transcripts in the endothecium and epidermis of the anther wall (Ursin et al., 1989). In the case of *Bcp1*, a cDNA clone isolated from a *B. campestris* pollen cDNA library (Theerakulpisut et al., 1989, 1991), *in situ* hybridization showed that transcripts hybridizing with *Bcp1* cDNA are also present in the tapetum. Thus, this clone shows both haploid and diploid expression in the anther. It is also possible that the genes expressed in the tapetum may be closely related to those isolated from pollen and may be able to cross-hybridize with their pollen counterparts. Alternatively, it is possible that the same gene utilizes a different promoter in pollen and anther tissues (e.g., chalcone synthase: Tunen et al., 1989).

Isolation of Pollen Genes Using Specific Probes

Genes whose products have identified functions that can be expressed in pollen include (1) cell division cycle genes involved in the control of meiosis and mitosis, (2) tapetal genes important in microspore nutrition, (3) genes encoding callose-degrading enzymes needed for microspore release from the tetrad, (4) genes implicated in the maintenance of pollen viability and desiccation protection during dehydration that occurs at maturation, (5) genes encoding pollen enzymes, e.g., β-galactosidase, and (6) genes encoding sperm-specific polypeptides associated with fertilization.

Microspore and pollen development is characterized by a number of cell divisions. Development is initiated by two meiotic divisions, and two mitoses are required to produce the male reproductive cells. Genes involved in controlling these cell division events are obviously important in the development of the male gametophyte (Dickinson, 1987). A novel meiotic protein has recently been isolated that is microsporocyte specific and shares some properties with testis-specific proteins in mammalian systems (Sasaki et al., 1990).

Pollen grains, like seeds, undergo a period of dehydration prior to maturity. Pollen water content declines rapidly, and this change has been correlated with maintenance of viability (Dumas et al., 1983; Ladyman and Taylor, 1988). In the case of seeds, a number of *lea* genes are expressed during the desiccation phase of embryogenesis (Skriver and Mundy, 1990), and these have been isolated and sequenced. The proteins encoded by these genes, dehydrins, have been postulated to be functionally important in desiccation protection. Pollen maturation involves dehydration, thus it will be interesting to look for these genes in a suitable pollen system.

Tapetal cells are in immediate contact with the developing microspores and pollen and are considered to mediate their nutrition, production of exine precursors, and pollencoat materials (Knox, 1984). The initial lesions of male sterility are often first observed in the tapetum (Knox and Heslop-Harrison, 1966; Bino, 1985).

Another important role of the tapetum is secretion of callose-degrading enzyme into the locular fluid at the time of microspore release from the tetrads (Stieglitz, 1977). Developmental expression of the β-glucanase gene appears to be under tight temporal control, as any disturbance in the timing of appearance of this enzyme results in sterile pollen (Frankel et al., 1969). The genes encoding β-glucanase in other tissues, particularly in pathogenesis-related responses, have been isolated and sequenced (Loose et al., 1988; Castresana et al., 1990). No information is yet available, however, on the nature of the tapetal β-glucanase or the gene that encodes this enzyme.

The importance of the tapetum in pollen development has been elegantly demonstrated in joint research by Goldberg and his group at UCLA and Plant Genetic Systems in Belgium (Mariani et al., 1990; Koltunow et al., 1990). The promoter region of a tapetum-specific gene was isolated and spliced to a ribonuclease gene from a fungal source. When *Petunia* and *Brassica* plants were transformed with this lethal construct, the expressed ribonuclease specifically ablated both tapetal and pollen development. These experiments provide the first evidence of the vital role of the tapetum in pollen development. Moreover, these experiments provide a means to create genetically engineered male-sterile lines for hybrid seed production. It remains to be determined, however, what role the polypeptides encoded by these pollen- and tapetum-specific genes play in development.

Pollen Developmental Mutants

The pollen grain is normally haploid, so that any mutations or new variants are expressed in the absence of dominance that is present in the diploid sporophyte. Three biochemical mutants are known in pollen systems (Evans et al., 1990): (1) *waxy*, which controls the nature of starch produced in corn; (2) *Adh*, which controls alcohol dehydrogenase activity in corn; and (3) *gal* which controls β-galactosidase enzymatic activity in pollen of *Brassica campestris*. *Gal* is a naturally occurring mutant that exerts its effect only on pollen grains (Singh and Knox, 1984, 1985a,b). The enzyme-deficient pollen is easily detected by a simple cytochemical method involving an indigogenic substrate. By selfing heterozygous plants, *gal/gal* mutants have been isolated. In these mutants, the pollen is defective for this enzyme, whereas the sporophyte retains normal enzymatic activity. It is unclear whether two different genes are expressed, one in the sporophyte and the other in the male gametophyte. This mutant provides the opportunity to dissect the role of the *gal* enzyme in germinating pollen. *Gal* is the first pollen-wall protein mutant to be identified and should enable structure–function relationships to be determined. The enzyme-deficient pollen is still able to germinate, but it is less effective in male transmission, implying that it plays a role in facilitating pollen tube growth and pollen–stigma interactions.

Pollen-Specific Proteins: The Allergens

As a case history, we will consider the remarkable implications of the molecular cloning of two genes of known function: those encoding major rye-grass pollen proteins that act as allergens in triggering hay fever and asthma.

Pollen of many grass species are highly allergenic. The pollen present in the atmosphere during pollination can provoke an immediate hypersensitive response of allergic asthma in about 20% of the human population (see review by Howlett and Knox, 1984). The symptoms are induced by the production of specific types of human defense molecules of the immune system, especially Immunoglobulin E (IgE). Allergens are antigens that induce formation of specific IgE and are defined as IgE-binding proteins.

The allergens of rye-grass pollen were first isolated using biochemical techniques at Cambridge in 1965 (Johnson and Marsh, 1965). There are at least four groups of allergens, separated according to their relative molecular mass. The principal allergen is a glycoprotein of 27–35 kDa, containing 50% carbohydrate that does not contribute to its allergenicity (Johnson and Marsh, 1966). The allergen, known originally as Group I allergen but now by convention as *Lol p*I, is the immunodominant component (i.e., more than 95% of patients possess specific IgE that binds to this pollen protein). The function of this allergen has only just begun to be understood through molecular analysis.

This has been achieved by isolating and cloning the genes encoding the two major allergens of rye-grass pollen, *Lol p*I and *Lol p*Ib, the latter having been identified recently in our lab.

Molecular Cloning of Rye-Grass Allergens

The deduced amino acid sequence of the full length *Lol p*I clone contains both the N-terminal sequence (Cottam et al., 1986; Perez et al., 1990; Griffith et al., 1991) and internal peptide fragment sequence (Esch and Klapper, 1989) previously reported from protein microsequencing. The entire 97 amino acid sequences of *Lol p*II (Ansari et al., 1989a) and *Lol p*III (Ansari et al., 1989b) have strong homology with the C-terminus of *Lol p*I (Fig. 10.2). Nevertheless, despite strong homology between amino acids 145 to 240 of *Lol p*I and *Lol p*II and III, no serological cross-reactivity has been detected with MAbs raised against *Lol p*I and either *Lol p*II or III (Ansari et al., 1987).

Northern analysis has shown that the *Lol p*I transcripts are only present in pollen of rye-grass. The absence of transcripts in other tissues is consistent with the absence of *Lol p*I antigens, as observed by immunoblotting (Singh et al., 1991). Thus *Lol p*I, like *Lol p*Ib, shows pollen-specific expression. These organ-specific proteins are known allergens, but their natural function in pollen remains to be elucidated.

Targeting of Newly Discovered Allergen to Pollen Amyloplasts

In the process of cloning *Lol p*I, a new major allergenic protein designated *Lol p*Ib has been identified; it is a 31-kDa basic protein. A cDNA

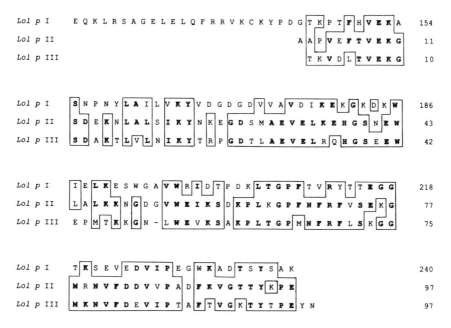

Figure 10.2. Comparison of amino acid sequences of *Lolium perenne* allergen, *Lol p*I, II, and III (Griffith et al., 1991). The standard single letter amino acid code is used; identical residues are shown in bold letters, and together with other similar residues, similarities are indicated by boxes.

clone encoding *Lol p*Ib (12R) has been isolated, sequenced, and characterized (Singh et al., 1992).

Northern analysis of RNA prepared from pollen showed high levels of expression of the 1.2-kb transcripts hybridizing to the cloned allergen gene. Similar hybridizing transcripts were not detectable in RNA from vegetative tissues (Fig. 10.3a). Pollen-specific RNA expression correlated with pollen-specific expression of antigens recognized by monoclonal antibodies (MAbs) 40.1 and 12.3, and IgE antibodies (Fig. 10.3b). Specific antibody binding occurred only when pollen and floral tissues (containing pollen) were used as the protein source.

Using a newly developed anhydrous fixation technique and immunogold probes with specific MAbs (Singh et al., 1992), we have shown *Lol p*I to be located in the cytosol (Fig. 10.4b, small gold particles, 30 nm). In contrast, gold marker particles bind predominantly to the starch granules when *Lol p*Ib-specific MAb is used (Fig. 10.4b, large gold particles, 60 nm). Mature grass pollen (Fig. 10.4a) is filled with starch granules, which are up to 2.5 μm in size and originate in the amyloplasts. Localization of *Lol p*Ib in the starch granules implies that this protein should be transported

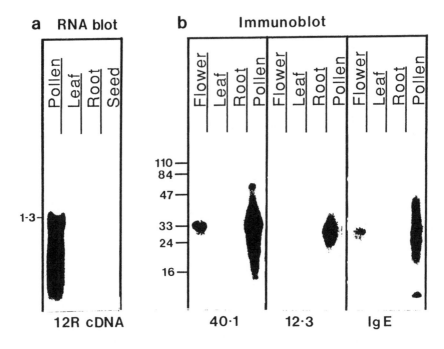

Figure 10.3. Pollen-specificity of the major rye-grass allergen *Lol p*Ib.
(a) Northern blot analysis of total RNA from rye-grass pollen, leaves, roots, and
seeds with 12R cDNA as probe; (b) immunoblot analysis of proteins showing
tissue-specific distribution of *Lol p*I antigens (from Singh et al., 1991).

from the cytosol to the lumen of the amyloplasts during development. For
transport to chloroplasts, the proteins, which are synthesized in the cytosol,
are synthesized as precursors containing a transit peptide sequence that is
cleaved after transport into the organelle. The transit peptides of most
chloroplast-targeted proteins possess a loosely defined cleavage site consensus sequence: (Val/Ile)-X-(Ala/Cys) Ala (Gavel and Von Heijne, 1990).
Arginine is also found in positions -6 to -10. The corresponding motif
in the *Lol p*Ib transit peptide sequence is Ser-Tyr-Ala Ala in this position,
and there are two arginine residues at positions -4 and -9. We conclude
that the *Lol p*Ib molecule is synthesized first as a preallergen in the cytosol
and is transported to the amyloplast for posttranslational modification.

Starch Granules: Atmospheric Carriers of Grass Pollen Allergens

The ecological implications of this finding of a potent allergen in starch
granules are profound. Grass pollen (Fig. 10.5a) is known to rupture by

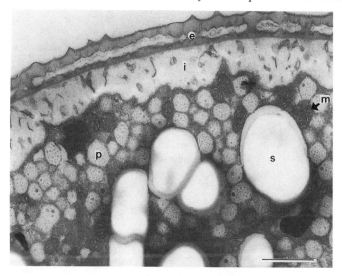

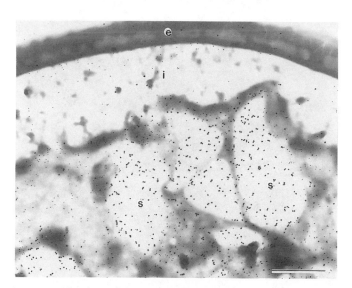

Figure 10.4. Localization of the major allergens of rye-grass pollen, using an anhydrous fixation technique, specific MAbs, and immunogold labeling.
(a) Transmission electron micrograph of thin section of pollen, poststained to show organelles: e, exine; i, intine; s, starch granule; p, polysaccharide particle; m, mitochondrion. (b) Double immunolabeling of unstained thin section. The section was first exposed to MAb 12.3 (specific for *Lol p*Ib), the gold particles silver-enhanced to ~40 nm, and then exposed to MAb 40.1 (specific for *Lol p*I). The gold particles detecting *Lol p*I are present as the small 15 nm markers.
Bar = 1 μm.

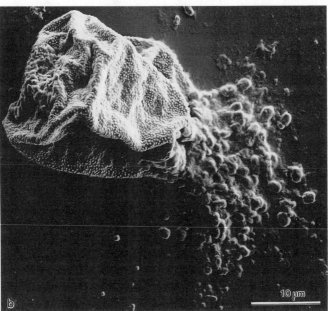

Figure 10.5. Scanning electron micrographs of (a) mature dry rye-grass pollen; (b) similar pollen after rupture by osmotic shock in water (from Suphioglu et al., 1992). Note the release of starch granules from the pore shown in (b).

osmotic shock, and we have shown that this does occur in rye-grass pollen (Fig. 10.5b). Rupture of the pollen ensures the release of 500–1000 allergen-containing starch granules per pollen grain into the atmospheric aerosol. It was predicted that the granules would be released into the atmosphere during rainfall and constitute the source of the previously unexplained micronic particles. Micronic particles are less than 5 μm in diameter, contain grass pollen allergen, and are small enough to enter the airways of susceptible humans to cause asthma. Asthma is known to be a disease of the lungs, and yet grass pollen is too large to penetrate beyond the oral cavity. Could these starch granules become part of the circulating atmosphere and provoke symptoms of asthma? Aerobiological studies in Melbourne last spring confirmed this hypothesis. Volumetric filter samplers detected the presence of starch granules in the atmosphere, with a 50-fold increase occurring on days following significant rainfall (Suphioglu et al., 1991). Starch granules were detected cytochemically by staining filters with IKI. Antibodies specific to the rye-grass allergens were used as probes to detect allergenic particles trapped on the filters (Fig. 10.6). These observations replicate previous observations of micronic particles, which had shown that they were especially frequent on days following rainfall during the grass pollen season (Stewart and Holt, 1985). We conclude that the starch granules are likely to be the unexplained micronic particles.

To confirm the action of starch granules as allergens in triggering asthma, when isolated from whole pollen, we collaborated with Melbourne physicians specializing in allergy and respiratory diseases. Skin prick tests showed that starch granules are as effective as pollen extracts in eliciting an allergic reaction, and lung function tests were positive in every case (Suphioglu et al., 1991). This indicates that the granules are capable of eliciting an IgE-mediated response and are the likely elicitors of the asthmatic reaction. The natural function of the allergen molecule in the starch granules remains unknown.

These kinds of studies illustrate the power and potential of molecular probes in elucidating complex biological interactions.

Grass Pollen Allergens: Immunological Cross-Reactivity and Amino Acid Sequence Homologies

Considerable antigenic and allergenic cross-reactivity has been demonstrated between allergens of different grass species. Immunological relatives, of similar molecular mass, are present in other related and clinically important grass genera (Fig. 10.7). These allergens share antigenic and allergenic determinants with *Lol p*I, as determined by binding studies with

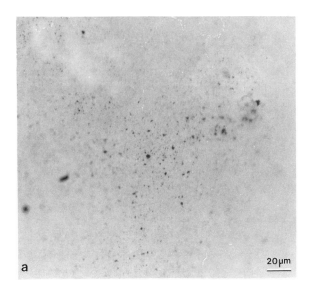

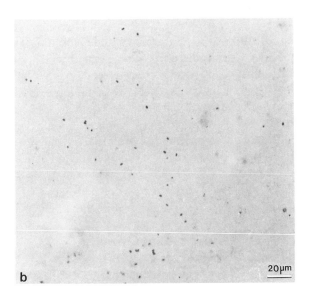

Figure 10.6. Immunological and cytochemical detection of allergens in starch granules from aerobiological samples collected from the atmosphere of Melbourne during the grass pollen season. (a) MAb 40.1 and (b) IKI (iodine/K iodide) staining of starch granules on filters (from Suphioglu et al., 1992).

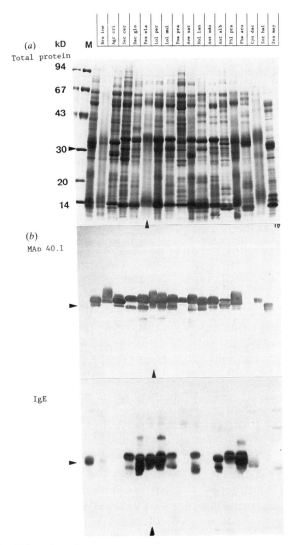

Figure 10.7. Diversity of proteins, antigens, and allergens in rye-grass pollen, compared with a panel of 16 other grasses: *Bromus inermis, Agropyron cristatum, Secale cereale, Dactylis glomerata, Festuca elatior, Lolium perenne, Lolium multiflorum, Poa compressa, Avena sativa, Holcus lanatus, Anthoxanthum odoratum, Agrostis alba, Phleum pratense, Phalaris arundinacea, Cynodon dactylon, Sorghum halepense,* and *Zea mays.* (a) Coomassie blue-stained protein profiles of pollen extracts after SDS-PAGE under reducing conditions; (b) Western blot of similar gel probed with MAb 40.1 specific for *Lol p*I; (c) similar blot probed with specific IgE antibodies from pooled grass pollen-allergic human sera (from Knox and Singh, 1990).

monoclonal antibodies, and specific IgE antibodies (Singh and Knox, 1985a; Knox and Singh, 1990; Singh et al., 1990).

Many other grasses contain allergenic proteins. A number of the allergenic grasses are representatives of the Pooideae, such as *Dactylis glomerata* (Walsh et al., 1990), *Festuca elatior* (Esch and Klapper, 1989), and *Poa pratensis* (Silvanovich et al., 1991). Substantial allergenic cross-reactivity between grass pollens has been demonstrated using an IgE-binding assay, the radioallergosorbent test (RAST: Marsh et al., 1970; Leiferman and Gleich, 1976; Lowenstein, 1978). There is no cross-reactivity, however, with distantly related grasses such as Bermuda grass, *Cynodon dactylon* (Ford and Baldo, 1987). This suggests that patterns of antigenic and allergenic cross-reactivity may follow taxonomic and evolutionary relationships.

*Lol p*Ib has no homology with *Lol p*I, II, or III, but *Lol p*I shows high levels of homology with *Lol p*II and III (Griffith et al., 1991). Another allergen with a high level of homology occurs in Kentucky blue grass, where a basic allergenic protein, M_r 33 k Da, has been identified using MAbs (Silvanovich et al., 1991). The sequence of this allergen, designated *Poa p*IX, has recently been reported, and a short segment of the sequence shows considerable similarities with other allergens. The N-terminal sequence of a newly identified major allergen from timothy grass, *Phl p*V (28), shows 50% similarity (20% identity) with the N-terminal sequence of *Lol p*Ib.

The detection of homologies in amino acid sequence among rye-grass pollen allergens is now possible, because a number of different allergen genes have been isolated and their sequences predicted. These data show that allergens may belong to families of closely related proteins. Whereas their function in pollen remains unknown, the similarity in amino acid sequences suggests that these molecules belong to families of molecules with conserved functions.

The amino acid sequence of a highly conserved antigenic determinant of *Fes e*I, an allergen from *Festuca elatior* corresponding to *Lol p*I, has been obtained from cyanogen bromide-cleaved and trypsin-generated fragments (Esch and Klapper, 1989). This sequence fragment of 28 amino acids is common to the group I allergens of several other grasses, including *Lol p*I, *Poa p*I, *Ant o*I, and *Agr a*I. The possibility exists to use *Lol p*I cDNA clones as heterologous probes to isolate corresponding genes in other related grasses. Changes in the DNA or amino acid sequences might then be used as molecular indicators of evolutionary change in the grasses. The extraordinary similarity in antigenic and allergenic determinants between *Lol p*I of rye-grass and the corresponding protein in related grasses suggests that some regions of the molecules may be highly conserved, whereas others are not. Differences in homology may be useful to indicate evolutionary relationships.

Acknowledgments

We thank the Australian Research Council, the Australian National Health and Medical Research Council, and the Asthma Foundation of Victoria for financial support.

Literature Cited

Albani, D., L.S. Robert, P.E. Donaldson, I. Altosaar, P.G. Arnison, and S.F. Fabijanski. 1990. Characterization of a pollen-specific gene family from *Brassica napus* which is activated during early microspore development. Plant Mol. Biol. **15**:605–622.

Ansari, A.A., T.K. Kihara, and D.G. Marsh. 1987. Immunochemical studies of *Lolium perenne* (rye grass) pollen allergens, *Lol p*I, II, and III. J. Immunol. **139**:4034–4041.

Ansari, A.A., P. Shenbagamurthi, and D.G. Marsh. 1989a. Complete amino acid sequence of a *Lolium perenne* (perennial rye-grass) pollen allergen, *Lol p*II. J. Biol. Chem. **264**:11181–11185.

Ansari, A.A., P. Shenbagamurthi, and D.G. Marsh. 1989b. Complete primary structure of a *Lolium perenne* (perennial rye-grass) pollen allergen, *Lol p*III: Comparison with known *Lol p*I and II sequences. Biochemistry **28**:8665–8670.

Ashida, E.R., and V.L. Scofield. 1987l Lymphocyte major histocompatibility complex-encoded class II structures may act as sperm receptors. Proc. Natl. Acad. Sci. U.S.A.: **84**:3395–3399.

Beattie, A.J., C. Turnbull, T. Hough, S. Jobson, and R.B. Knox. 1985. The vulnerability of pollen and fungal spores to ant secretions: Evidence and some evolutionary implications. Am. J. Bot. **72**:606–614.

Bino, R.J. 1985. Histological effects of microsporogenesis in fertile and cytoplasmic male sterile and restored fertile *Petunia hybrida*. Theor. Appl. Genet. **69**:425–428.

Brown, S.M., and M.L. Crouch. 1990. Characterization of a gene family abundantly expressed in *Oenothera organensis* pollen that shows sequence similarity to polygalacturonase. Plant Cell **2**:263–274.

Castresana, C., F.D. Carvalho, G. Gheysen, M. Habets, D. Inze, and M. Van Montagu. 1990. Tissue-specific pathogen-induced regulation of a *Nicotiana plumbaginifolia* β-1,3 glucanase gene. Plant Cell **2**:1131–1143.

Cottam, G.P., D.M. Moran, and R. Standring. 1986. Physicochemical and immunochemical characterization of allergenic proteins from rye-grass (*Lolium perenne*) pollen prepared by a rapid and efficient and immunochemical characterization of allergenic proteins from rye-grass (*Lolium perenne*) pollen prepared by a rapid and efficient purification method. Biochem. J. **234**:305–310.

Dickinson, H.G. 1987. The physiology and biochemistry of meiosis in the anther. Int. Rev. Cytol. **107**:79–110.

Domec, C., B. Garbay, M. Fournier, and J. Bennet. 1990. cDNA library construction from small amounts of unfractionated RNA: Association of cDNA synthesis with polymerase chain reaction amplification. Anal. Biochem. **188**:422–426.

Dumas, C., J. Duplan, C. Said, and J. Soulier. 1983. ^{1}H Nuclear magnetic resonance to correlate water content and pollen viability. *In* D. Mulcahy and E. Ottaviano, eds., Pollen: Biology and Implications for Plant Breeding. Elsevier, Amsterdam, pp. 15–25.

Dumas, C., R.B. Knox, C.A. McConchie, and S.D. Russell. 1984. Emerging physiological concepts in fertilization. What's New Plant Physiol. **15**:177–220.

Dupuis, I., P. Roeckel, E. Matthys-Rochon, and C. Dumas. 1987. Procedure to isolate viable sperm cells from corn (*Zea mays* L.) pollen grains. Plant Physiol. **85**:876–878.

Esch, R.E., and D.G. Klapper. 1989. Identification and localisation of allergenic determinants on grass group I antigens using monoclonal antibodies. J. Immunol. **142**:179–184.

Evans, D.E., M.B. Singh, and R.B. Knox. 1990. Pollen development: Applications in biotechnology. *In* S. Blackmore and R.B. Knox, eds., Microspores: Evolution and Ontogeny. Academic Press, London, pp. 309–338.

Ford, S.A., and B.A. Baldo. 1987. Identification of bermuda grass (*Cynodon dactylon*)-pollen allergens by electroblotting. J. Allergy Clin. Immunol. **79**:711–720.

Frankel, R., S. Izhar, and J. Nitsan. 1969. Timing of callase activity and cytoplasmic male sterility in *Petunia*. Biochem. Genet. **3**:451–459.

Gavel, Y., and G. von. Heijne. 1990. Sequence comparisons of transit peptides targeting proteins to chloroplasts. FEBS Lett. **261**:455–458.

Griffith, I.J., P.M. Smith, J. Pollock, P. Theerakulpisut, A., Avjioglu, S. Davies, T. Hough, M.B. Singh. R.J. Simpson, L.D. Ward, and R. B. Knox. 1991. Cloning and sequencing of *Lol p*I, the major allergenic protein of rye-grass pollen. FEBS Lett. **279**:210–215.

Hanson, D.D., D.A. Hamilton, J.L. Travis, D.M. Bashe, and J.P. Mascarenhas. 1989. Characterization of a pollen-specific cDNA clone from *Zea mays* and its expression. Plant Cell **1**:173–179.

Heslop-Harrison, J., Y. Heslop-Harrison, and K.R. Shivanna. 1984. The evaluation of pollen quality, and a further appraisal of the fluorochromatic (FCR) test procedure. Theor. Appl. Genet. **67**:367–375.

Hough, T., M.B. Singh, I. Smart, and R.B. Knox. 1986. Immunofluorescent screening of monoclonal antibodies to surface antigens of animal and plant cells bound to polycarbonate membranes. J. Immunol. Methods **92**:103–107.

Howlett B.J., and R.B. Knox. 1984. Allergic interactions. Encycl. Plant Physiol. **17**:655–674.

Johnson, P., and D.G. Marsh. 1965. "Isoallergens" from rye-grass pollen. Nature (London) **206**:935–937.

Johnson, P., and D.G. Marsh. 1966. Allergens from common rye-grass pollen (*Lolium perenne*). Immunochemistry **3**:91–100.

Knox, R.B. 1984. Pollen-pistil interactions. Encycl. Plant Physiol. **17**:508–608.

Knox, R.B., and J. Heslop-Harrison. 1966. Control of pollen fertility through the agency of the light regime in the grass *Dichanthium aristatum*. Phyton **11**:256–267.

Knox, R.B., and M.B. Singh. 1987. New Perspectives in pollen biology and fertilization. Ann. Bot. **60**:15–37.

Knox, R.B., and M.B. Singh. 1990. Reproduction and recognition phenomena in the Poaceae. *In* G.P. Chapman, ed., Reproductive Versatility in the Grasses. Cambridge University Press, Cambridge, pp. 220–239.

Knox, R.B., M. Gaget, and C. Dumas. 1987. Mentor pollen techniques. Int. Rev. Cytol. **107**:315–332.

Knox, R.B., D. Southworth, and M.B. Singh. 1988. Sperm cell determinants and control of fertilization in plants. *In* G.C. Chapman, ed., Eukaryote Cell Recognition. Cambridge University Press, Cambridge, pp. 175–193.

Knox, R.B., M.B. Singh, T. Hough, and P. Theerakulpisut. 1989. The rye-grass pollen allergen, *Lol p*I. *In* T. Merrett, ed., Allergy and Molecular Biology. Pergamon Press, Oxford, pp. 161–171.

Koltunow, A.M., J. Truettner, K.H. Cox, M. Wallroth and R.B. Goldberg. 1990. Different temporal and spatial gene expression patterns occur during anther development. Plant Cell **2**:1201–1224.

Kranz, E., J. Bautor, and H. Lorz. 1991. *In vitro* fertilization of single isolated gametes of maize by electro-fusion. Sex. Plant Reprod. **4**:12–16.

Ladyman, J.A.R., and R.E. Taylor. 1988. ^{31}P and ^{1}H NMR as a non-destructive method for measuring pollen viability. *In* Cresti, M., P. Gori, and E. Pacini, eds. Sexual Reproduction in Higher Plants. Springer Verlag, Berlin, pp. 69–74.

Leiferman, K.M., and G.J. Gleich. 1976. The cross-reactivity of IgE antibodies with pollen allergens. I. Analyses of various species of grass pollens. J. Allergy Clin. Immunol. **58**:129–139.

Loose, M. de, T. Alliote, G. Gheysen, C. Genetello, J. Gielen, P. Soetaert, M. Van Montagu, and D. Inze. 1988. Primary structure of a hormonally-regulated β-glucanase of *Nicotiana plumbaginifolia*. Gene **70**:13–23.

Lowenstein, H. 1978. Quantitative immunoelectrophoretic methods as a tool for the analysis and isolation of allergens. Prog. Allergy **25**:1–62.

Mariani, C., M.De. Beuckeleer, J. Truettner, J. Leemans, and R.B. Goldberg. 1990. Induction of male sterility in plants by chimaeric ribonuclease gene. Nature (London) **347**:737–741.

Marsh, D.G., Z.H. Haddad, and D.H. Campbell. 1970. A new method for determining the distribution of allergenic fractions in biological materials. J. Allergy 46:107–112.

Mascarenhas, J.P. 1975. The biochemistry of angiosperm pollen development. Bot. Rev. 41:259–314.

Mascarenhas, J.P. 1989. The male gametophyte of flowering plants. Plant Cell 1:657–664.

Mascarenhas, J.P. 1990a. Gene activity during pollen development. Annu. Rev. Plant Physiol. Plant Mol. Biol. 41:317–338.

Mascarenhas, J.P. 1990b. Gene expression in the angiosperm male gametophyte. In S. Blackmore and R.B. Knox, eds., Microspores, Ontogeny and Evolution. Academic Press, London, pp. 265–280.

McConchie, C.A., S.D. Russell, C. Dumas, M. Tuohy, and R.B. Knox. 1987. Quantitative cytology of the sperm cells of Brassica campestris and B. oleracea. Planta 170:446–452.

McCormick, S., A. Smith, C. Gasser, K. Sachs, M. Hinchee, R. Horsch, and R. Fraley. 1987. The identification of genes specifically expressed in reproductive organs of tomato. In D. Nevins and R. Jones, eds., Tomato Biotechnology. Allan R. Liss, New York. pp. 255–265.

Mulcahy, G.B., and D.L. Mulcahy. 1987. The effect of supplemented media on the growth in vitro of bi- and tri-nucleate pollen. Plant Sci. 55:213–216.

Murphy, L.D., C.E. Herzog, J.B. Rudick, A.T. Fojo, and S.E. Bates. 1990. Use of the polymerase chain reaction in the quantitation of mdr-1 gene expression. Biochemistry 29:10351–10356.

Perez, M., G.Y. Ishioka, L.E. Walker, and R.W. Chesnut. 1990. cDNA cloning and immunological characterization of the rye-grass allergen Lol pI. J. Biol. Chem. 265:16210–16215.

Roeckel, P., A. Chaboud, E. Matthys-Rochon, S. Russell, and C. Dumas. 1990. Sperm cell structure, development, and organization. In S. Blackmore and R.B. Knox, eds., Microspores: Evolution and Ontogeny. Academic Press, London, pp. 281–308.

Russell, S.D. 1985. Preferential fertilization in Plumbago: Ultrastructural evidence for gamete level recognition in an angiosperm. Proc. Natl. Acad. Sci. U.S.A. 82:6129–6132.

Russell, S.D. 1986. Isolation of sperm cells from the pollen of Plumbago zeylanica. Plant Physiol. 81:317–319.

Russell, S.D., and D.D. Cass. 1983. Unequal distribution of plastids and mitochondria during sperm cell formation in Plumbago zeylanica. In D.L. Mulcahy and E. Ottaviano, eds., Pollen: Biology and Implications for Plant Breeding. Elsevier, Amsterdam, pp. 135–140.

Sasaki, Y., H. Yasuda, Y. Ohoba, and H. Harada. 1990. Isolation and characterization of a novel nuclear protein from pollen mother cells of lily. Plant Physiol. 94:1467–1471.

Silvanovich, A., J. Astwood, L. Zhang, E. Olsen, F. Kisil, A. Sehon, S. Mohapatra, and R. Hill. 1991. Nucleotide sequence analysis of three cDNAs coding for *Poa pIX* isoallergens of kentucky bluegrass pollen. J. Biol. Chem. **266**:1204–1210.

Shivanna, K.R., H. Xu, P. Taylor, and R.B. Knox. 1988. Isolation of sperms from the pollen tubes of flowering plants during fertilization. Plant Physiol. **87**:647–650.

Singh, M.B., and R.B. Knox. 1984. Quantitative cytochemistry of β-galactosidase in normal and enzyme-deficient pollen of *Brassica campestris*: Application of the indigogenic method. Histochem. J. **16**:1273–1296.

Singh, M.B., and R.B. Knox. 1985a. A gene controlling beta-galactosidase deficiency in pollen of oilseed rape *Brassica campestris*. J. Hered. **76**:199–201.

Singh, M.B., and R.B. Knox. 1985b. Grass pollen allergens: Antigenic relationships detected using monoclonal antibody and dot-blotting immunoassay. Int. Arch. Allergy Appl. Immunol. **78**:300–304.

Singh, M.B., P. O'Neill, and R.B. Knox. 1985. Initiation of post-meiotic beta-galactosidase synthesis during microsporogenesis in oilseed rape. Plant Physiol. **77**:225–228.

Singh, M.B., P.M. Smith, and R.B. Knox. 1990. Molecular biology of rye-grass pollen allergens. *In* B. Baldo, ed., Molecular Approaches to the Study of Allergens. Karger, Basel, pp. 101–120.

Singh, M.B., T. Hough, P. Theerakulpisut, A. Avjioglu, S. Davies, P. Smith, P. Taylor, R.J. Simpson, L. Ward, J. McCluskey, R. Puy, and R.B. Knox. 1991. Isolation of complementary DNA encoding newly identified major allergenic protein of rye-grass pollen: Intracellular targeting to the amyloplast. Proc. Natl. Acad. Sci. U.S.A. **88**:1384–1388.

Singh, M.B., P. Taylor, and R.B. Knox. 1992. Special preparation methods for immunocytochemistry of plant material. *In* J. Beesley, ed., Immunocytochemistry—A Practical Approach. IRL Press, Oxford, in press.

Skriver, K., and J. Mundy. 1990. Gene expression in response to abscisic acid and osmotic stress. Plant Cell **2**:503–512.

Stewart, G.A., and P.G. Holt. 1985. Submicronic airborne allergens. Med. J. Aust. **143**:426–427.

Stieglitz, H. 1977. Role of β-1,3 glucanase in post-meiotic microspore release. Dev. Biol. **57**:87–97.

Stinson, J., and J.P. Mascarenhas. 1985. Onset of alcohol dehydrogenase synthesis during microsporogenesis in maize. Plant Physiol. **77**:222–224.

Stinson, J.R., A.J. Eisenberg, R.P. Willing, M. Enrico Pe, D.D. Hansen, and J.P. Mascarenhas. 1987. Genes expressed in the male gametophyte of flowering plants and their isolation. Plant Physiol. **83**:442–447.

Suphioglu, C., M.B. Singh, P.E. Taylor, R. Bellomo, R. Puy, and R.B. Knox. 1992. Mechanism of grass-pollen-induced asthma. Lancet 339, in press.

Taylor P., J. Kenrick, Y. Li, V. Kaul, B.E.S. Gunning, and R.B. Knox. 1989. The male germ unit of *Rhododendron*: Quantitative cytology, three-dimensional reconstruction, isolation and detection using fluorescent probes. Sex. Plant Reprod. **2**:254–264.

Theerakulpisut, P., M.B. Singh, S. Strother, and R.B. Knox. 1989. Isolation of cDNA clones specifically expressed in the pollen grains of *Brassica campestris*. *In* R.B. Knox, M.B. Singh, and L.F. Troiani, eds., Pollination '88. School of Botany, University of Melbourne, pp. 110–114.

Theerakulpisut, P., M.B. Singh, H. Xu, J.M. Pettitt, and R.B. Knox. 1991. Isolation and developmental expression of *Bcp*1, an anther specific cDNA clone in *Brassica campestris*. Plant Cell **3**:1073–1084.

Tunen, A.J. van, S.A. Hartmann, L.A. Mur, and J.N.M. Mol. 1989. Regulation of chalcone flavanone isomerase gene expression in *Petunia hybrida*: The use of alternative promoters in corolla, anthers and pollen. Plant Mol. Biol. **12**:539–551.

Twell, D., R. Wing, J. Yamaguchi, and S. McCormick. 1989. Isolation and expression of an anther-specific gene from tomato. Mol. Gen. Genetics **247**:240–245.

Ursin, V.M., J. Yamaguchi, and S. McCormick. 1989. Gametophytic and sporophytic expression of anther-specific genes in developing tomato anthers. Plant Cell **1**:727–736.

Walsh, D.J., J.A. Matthews, R. Denmeade, P. Maxwell, M. Davidson, and M.R. Walker. 1990. Monoclonal antibodies to proteins from cocksfoot grass (*Dactylis glomerata*) pollen: Isolation and N-terminal sequence of a major allergen. Int. Arch. Allergy Appl. Immunol. **91**:419–425.

Willing, R.P., and J.P. Mascarenhas. 1984. Analysis of the complexity and diversity of mRNAs from pollen and shoots of *Tradescantia*. Plant Physiol. **75**:865–868.

Willing, R.P., D. Bashe, and J.P. Mascarenhas. 1988. An analysis of the quantity and diversity of messenger RNAs from pollen and shoots of *Zea mays*. Theoret. Appl. Genet. **75**:751–753.

Wing, R.A., J. Yamaguchi, S.K. Larabell, V.M. Ursin, and S. McCormick. 1989. Molecular and genetic characterization of two pollen-expressed genes that have sequence similarity to pectate lyases of the plant pathogen *Erwinia*. Plant Mol. Biol. **14**:17–28.

11

Components of Reproductive Success in the Herbaceous Perennial *Amianthium muscaetoxicum*

Joseph Travis
Florida State University

Introduction

One of the goals of evolutionary biology is to explain how current pressures of natural selection maintain the distributions of phenotypic variation that are observed in plant populations. An accurate characterization of the form and strength of phenotypic selection requires an estimator of fitness whose accuracy and precision are within tolerable margins of error. With such an estimator one can explore where in the life cycle the critical fitness differences among individuals are generated, whether the critical phase for generating fitness differences is consistent in space and time, and whether spatial and temporal variations in absolute fitness lend hierarchical structure to the overall process of selection.

This process of inference is not easily accomplished for many organisms because of the difficulty of obtaining a satisfactory estimator of fitness. Individual fitness, the contribution of the individual to future population growth, is usually estimated by a surrogate, such as the number of propagules produced over the course of a lifetime. In many contexts, such as studies of annual plants, this surrogate is either an accurate estimator of fitness or sufficiently close that no dramatic distortions are introduced. In some contexts, such as studies of long-lived perennial plants, data on lifetime production of progeny cannot be obtained within a practical time span. For these contexts biologists have suggested a surrogate, such as the mean seed production per year (Cohen and Dukas, 1990). Temporal variation in selective pressures within a generation is absorbed into the final level of variation among individuals in average annual seed production and the covariance between average seed production and the values of phenotypic characters hypothesized to be adaptive.

Here I explore the extent to which patterns of annual seed production in a perennial herb can offer insights into the nature of fitness variation in a natural population. This exploration is based on 11 years of study of *Amianthium muscaetoxicum* (Walter) Gray (Liliaceae). I review published results and offer new data that address (1) how annual seed production varies spatially and temporally, (2) which phase of the reproductive process is the most crucial for generating variation in seed production, and (3) whether predation on fruits and seeds might change the relationship between the number of seeds produced and the actual level of absolute fitness that the number represents. I apply the conclusions from this study to three general problems: (1) the extent to which observed fitness variation is environmentally induced, (2) the evolution of flower number and the production of "excess" flowers, and (3) the evolution of iteroparity.

Background

Natural History

Amianthium muscaetoxicum is a perennial lily that is widespread in open woods throughout the eastern United States. It is extremely common under the 50- to 60-year-old second-growth oak-maple canopy at Mountain Lake Biological Station, Giles County, Virginia. The plants occur in patches of high abundance throughout the area and in the surrounding Jefferson National Forest (see Stephenson, 1982, for a description of vegetation of the area). Herbarium records at the Biological Station include collections made before chestnut blight swept the area in the 1920s and include notations that *Amianthium* was very common under the chestnut (*Castanea dentata*) canopy. Thus, it is unlikely that the modern abundance and distribution of the plant are artifacts of the altered conditions induced by the loss of chestnuts from the forest. The density of plants within a large patch can be high: the average density on 200 1-m² plots ranges from 2.5 to 5.1 plants/ m², but the highly aggregated nature of the plant's distribution on a small scale can generate local crowding up to 10 mature plants/m² and > 30 juveniles/m² (Fig. 11.1). Average densities of plants in flower can range from 0.37/m² to 1.2 plants/m².

Amianthium has a modular structure imposed by the perennating bulb below ground. The bulb has one or more separate lobes that are bound together by a fibrous coating; each lobe is topped by a cluster of linear leaves. The distribution of the number of leaf clusters or lobes varies from location to location but is always highly skewed, and < 5% of plants have more than seven lobes (maximum observed in mapped plots was 18 lobes). An individual plant may add a lobe between years (rarely two lobes and never more than two) but will not always do so. Each lobe is capable of sending out an inflorescence, but usually only a minority of lobes do so. Plants at or above the minimum size observed in flowering plants (based

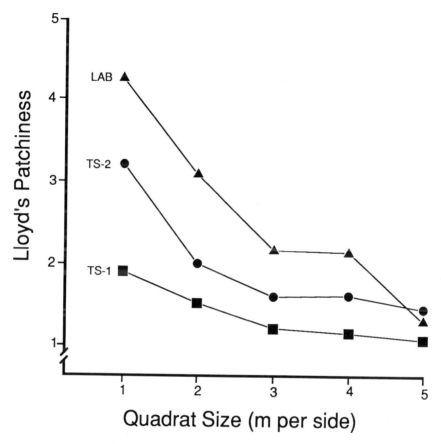

Figure 11.1. Lloyd's patchiness index for plants in reproductive size classes as a function of quadrat size for three populations of *Amianthium muscaetoxicum*: Laboratory (Lab), Twin Springs 2 (TS-2), and Twin Springs 1 (TS-1). Values above 1 indicate a clumped distribution of plants. Plants in the Lab population are the most clumped at all scales of measurement; plants approach a random pattern at the largest quadrat sizes.

on total leaf area) average 5.5 lobes (standard error = 0.3, n = 222), but the average number of inflorescences produced does not exceed two (Robbins, 1986). The clumped pattern of the plants and their ability to produce two or more inflorescences produce local inflorescence densities that range from $0.32/m^2$ to $6.4/m^2$. Individuals do not reproduce asexually, and although they are often quite crowded, individual plants can always be clearly distinguished.

The inflorescence is a compact, bracteate receme with small, white, perfect flowers. Inflorescence size varies widely but has an approximately normal distribution around 60 flowers (Fig. 11.2), and lobes with greater total leaf areas produce more flowers per inflorescence (Robbins, 1986). The number of flower buds is determined by the time at which the flowering stalk begins to elongate. Individual flowers open sequentially from bottom to top on the inflorescence at a rate of about 4–5 flowers per day. Several of the most distal buds do not open. Stigmas are receptive for 6 days, beginning on the first day the flower is open. Anther dehiscence starts on the day the flower opens and lasts 2 days; pollen is viable for 4 days (Palmer et al., 1989). Duration of flowering for an inflorescence is proportional to its number of flowers (Robbins, 1986; Palmer et al., 1988): a small inflorescence with 20 flowers lasts about 5 days, and larger ones with ≥ 60 flowers last from 11 to 21 days. The number of ovules per flower is 14 ± 0.5 (mean ± standard error: Robbins, 1986). Pollination is accomplished by insect vectors, primarily beetles and butterflies (Travis, 1984). The plants are largely self-incompatible (Travis, 1984), and self-pollen can occlude the stigma and block the effectiveness of subsequently arriving outcross pollen (Palmer et al., 1989). Studies of fluorescent-dust dispersal

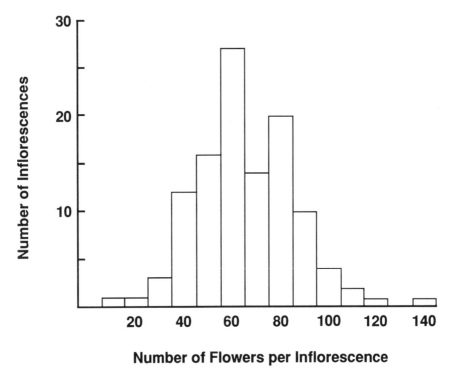

Figure 11.2. Histogram of the number of flowers per inflorescence of *Amianthium muscaetoxicum* in 1984 (from Robbins, 1986).

suggest that pollen dispersal rates decrease linearly with increasing distance from the donor up to a distance of 30 m, beyond which some dispersal occurs independently of distance from the donor (Palmer et al., 1988). Pollen dispersal distances vary with phase of the flowering season and yield neighborhood areas of 262 m² in the middle of the season and over 1700 m² at the end of the season, when fewer inflorescences are available.

The perianth turns green when the stigma is no longer receptive and persists through seed dispersal. The fruit is a trilocular capsule that dehisces 3–4 weeks after the cessation of flowering. The brownish, ellipsoid seeds average 3 mg air-dried mass and disperse either by falling out of the capsule or by collapsing of the scape supporting the infructescence. Germination occurs early in the following spring.

Allozyme variation at 10 loci is at a level typical for outcrossers: observed heterozygosity values average 0.27 (Palmer et al., 1988; J. Travis, C. Werth, and A. Redmond, unpublished data). There is no genetic structuring of the Salt Pond Mountain population at the scale of the large patches in which the plants are found: estimated values of Wright's (1969) F_{ST} (the standardized variance of allele frequencies among large patches) do not exceed 0.006 for any locus (J. Travis, C. Werth, and A. Redmond, unpublished data). This result is not surprising in light of the potential for long pollen-dispersal distances. Crosses between nearest-neighbor plants either produce no seeds or produce seeds of extremely small size. This result suggests a small-scale genetic structure in which near-neighbor plants are closely related (Redmond et al., 1989). Subsequent allozyme analyses have confirmed that neighboring individuals share more alleles than would be expected on the basis of a random pattern (J. Travis, C. Werth, and A. Redmond, unpublished data).

Seed Production as a Measure of Fitness

The use of some function of a hermaphroditic individual's seed production as a surrogate for individual fitness is arguably incomplete. This measure is technically only a surrogate for the individual's fitness as a female and takes no account of its fitness as a male through pollen donation. It ignores any effect the genotypic identity of the pollen donors may have on the quality of the zygote; such effects require weighting seed numbers by paternal identity to obtain an accurate estimation of even the female component of fitness (unless all male effects are distributed in the same manner in all seed parents). It also ignores any effects of seed size or other environmental maternal influences on progeny quality (Roach and Wulff, 1987).

The consequences of what is not embraced by this surrogate for fitness are not so obviously drastic that they automatically impeach its usefulness. Variation among individuals in fitness through female function can be far greater than variation through male function (Schlicting and Devlin, 1989; Devlin and Ellstrand, 1990), although the generality of this result is unknown. The evidence that paternal identity must be incorporated into calculations of seed production for *Amianthium* is not compelling either. Statistically significant effects of the distance separating pollen donors form recipients on seed production paled in importance when compared to differences among pollen recipients in their abilities to produce seeds (Redmond et al., 1989). Experiments indicate that seed production in *Amianthium* is often pollinator limited (Travis, 1984; Robbins, 1986), which could be used to argue that abortion that is selective with respect to genotype of pollen is an unimportant source of variation in seed quality (cf. Marshall and Ellstrand, 1988). The importance of interactive effects of male and female genetic identities (aside from those involved in consanguineous crosses) that would compromise the use of seed number is still unclear. Although such effects are known (Robbins and Travis, 1986; Travis, 1988), they are not general (Schlichting and Devlin, 1989). The effects of variation in seed size on subsequent viability are unknown in *Amianthium*, but wide variation in seed size suggests that such effects might be large.

These considerations suggest that the use of seed production as a surrogate for fitness is likely to account for the majority of fitness variation. The levels of variation in seed numbers are such that the factors that are ignored are unlikely to alter the conclusions to be drawn. Those factors may still have evolutionary significance: even tiny effects on fitness will prevail in a large enough population over a long enough time. It does imply that the role they do play must be considered within the context of the variation in seed production.

Patterns of Annual Seed Production

Variation among Locations and Years

Although there is wide, statistically significant variation among locations and years (Table 11.1), those factors explain only a small fraction of the total variance (Table 11.2). The data are derived from a stratified sampling plan executed in each of three populations over three successive years; stratification is by number of lobes, such that each sample of plants reflects the size structure of the population from which it was drawn. If there are effects of plant size on seed production and if plant size distributions vary among locations, then the average effects noted here may be due solely to different plant size distributions. Seed production data were log-transformed for analysis.

Variation in seed production can arise from variation in any of four components: number of inflorescences, number of flowers per inflores-

Table 11.1. Average Values of Individual Seed Production and Its Components in *Amianthium muscaetoxicum* in Three Locations in 3 Years

Year	Variable	Twin Springs 1	Twin Springs 2	White Pine Woods
1981	Seeds per plant	82	82	41
	Number of inflorescences	2.1	2.1	1.4
	Flowers per inflorescence	55	55	58
	Fruit-set	0.34	0.34	0.26
	Seeds per fruit	2.1	2.1	1.9
1982	Seeds per plant	70	51	24
	Number of inflorescences	1.4	1.3	1.0
	Flowers per inflorescence	46	50	48
	Fruit-set	0.45	0.36	0.30
	Seeds per fruit	2.4	2.2	1.7
1983	Seeds per plant	60	55	9
	Number of inflorescences	1.4	1.7	1.0
	Flowers per inflorescence	55	53	43
	Fruit-set	0.34	0.26	0.16
	Seeds per fruit	2.3	2.4	1.3

Table 11.2. Analysis of Variance on Log-Transformed Seed Production in *Amianthium muscaetoxicum*

Source	df	MS	F
Year	2	1.12	2.12
Location	2	2.95	5.61**
Year × location	4	2.79	5.30***
Residual	186	0.53	—

$**p < 0.01$
$***p < 0.001.$

cence, fruit-set, and seeds per fruit. These components operate sequentially in time, combine multiplicatively, and may be correlated with one another. Variation from site to site may be due to components different from those that influence variation from year to year. Multivariate analysis of variance provides a useful mechanism for pinpointing the components that determine variation in seed production when it is used on the logarithms of each component (see Timm, 1975, for statistical details of the following explanation). In a general multivariate analysis of variance, the test statistic operates on the sum of several variables. The absolute values of the correlations of the variables with the scores of the observations along the first discriminant function serve as reliable indicators of which variable contributed to the statistical significance. Because the components of seed production combine multiplicatively, a multivariate analysis of their logarithms will combine them additively in the test statistic. This analysis is thus ideally suited to the problem at hand. The multivariate analysis cannot be used as the sole test of statistical significance of the differences in fecundity because variables, such as the number of inflorescences, are not normally distributed. This problem does not affect the estimates of the correlations and should not affect a conservative interpretation of the results.

A multivariate analysis of variance on the data from Table 11.1 indicated that the major contributions to variation among years in annual seed production came from variation in number of inflorescences per plant (correlation of -0.77 with the discriminant function) and the number of seeds per fruit ($r = 0.66$). The importance of seeds per fruit is paradoxical unless one recalls that the multiplicative nature of seed production amplifies small changes in the number of seeds per fruit into large changes in total seed production. Variation among sites is based primarily on variation in the number of inflorescences per plant ($r = 0.76$) and variation in fruit-set ($r = 0.59$). The fact that the magnitude of location differences was far greater in 1983 than in 1981, which is the source of the statistical interaction between the effects of location and year, stems from a change in the number of seeds per fruit, as suggested by the raw data and confirmed by the discriminant function correlation for this variable ($r = 0.91$). The average value of seeds per fruit in White Pine Woods dropped 33% across years, whereas those of the other populations rose 10–14% during the same period.

Spatial and temporal variation in average annual seed production appears to be based largely on the number of inflorescences a plant produces and, secondarily, on fruit-set and seeds per fruit. Differences in plant size structure play the dominant role in explaining the small amount of total variation in seed output that can be accounted for by knowledge of location and year. The three populations included in this discussion displayed the greatest range of annual seed production and were chosen to illustrate the "best case" scenario for maximal spatial and temporal variation. Inclusion of more populations would lower the proportion of variance explained. Obviously, there is little repeatable hierarchical structure to variation in seed production: most variation exists among individuals.

A lack of hierarchical structure in seed production does not necessarily imply a lack of hierarchical structure in fitness. If seed size is important for survivorship (and it usually is: see Winn, 1988) and if locations produce seeds of different sizes, then the differences in propagule quality that result from weighting seed numbers by their likely survival will introduce hierarchical variation in fitness. There is considerable potential for this effect in *Amianthium*. Seed sizes from a stratified sample of plants with 1, 2, or ≥ 3 lobes vary among populations (Table 11.3). Average seed size for the White Pine Woods population, weighted by proportional representation of plant size classes, is 30% larger than the comparable average from the Twin Springs 2 population. It is not clear that the organization of fitness will be hierarchical in *Amianthium* because the population that makes the largest seeds (White Pine Woods) makes the fewest of them (Table 11.1). This pattern may result from a response to a strong hierarchical selection regime in the past that has induced a compromise between numbers and sizes of seeds so as to produce equivalent absolute fitness in the present through opposing routes.

There is also an interesting effect of plant size on seed size. Larger plants in the Twin Springs 2 population make smaller seeds (Table 11.3), despite considerable heterogeneity among individual plants within a size class. There is no comparable effect in White Pine Woods, where there is also much less heterogeneity among individuals in seed size. As a result, seed-size variation is less likely to contribute to fitness variation among individuals in White Pine Woods than in Twin Springs 2.

Variation among Individuals

Fully 85% of the overall variance in seed production is among individuals. In this section I examine this variation more fully.

A Prospective Study of Individual Production Rates

In 1981, 72 plants were selected in one population (Twin Springs 1) for a detailed study of individual variation in seed production. Plants were selected by a stratified sampling plan from the roster of plants in a mapped plot that were large enough to flower. They were chosen at random within strata of lobe number class, whether actually in flower that year or not. The goal of this plan was to follow their seed production patterns for the next 3 years to assay the cumulative variation among individuals and the repeatability of differences among individuals in annual production.

Of the 72 plants, 13 never flowered in the 3-year period; 42 flowered in a single year; 15 in two of the three years; and only two in all three years. The range in cumulative seed production among the 59 plants that flowered at least once was 3–420 seeds (standard error of 11 seeds). The two plants that flowered in all 3 years produced 45 and 214 seeds. Plants that flowered in any 2 years produced from 32 to 420 seeds cumulatively (Table 11.4). Despite the wide range of variation in the data as a whole, there is a surprising level of repeatability in average annual seed production (one-way analysis of variance among individuals on square root-transformed data: $F_{14,15} = 2.54$; $p < 0.05$; intraclass correlation = 0.44). Seed pro-

Table 11.3. Seed-Size Variation in *Amianthium muscaetoxicum* as a Function of Population and Plant Size

Population	Plant number	Number of lobes	Number of seeds measured	Average (mg)	Standard error
Twin Springs 2	1	1	26	3.01	0.18
	4	1	15	3.49	0.08
	7	1	17	1.55	0.12
	8	1	13	3.29	0.12
	9	1	23	2.95	0.14
	15	1	13	2.20	0.16
			Average 2.75		
	3	2	14	2.60	0.16
	5	2	21	2.00	0.06
	6	2	14	2.99	0.17
	14	2	14	1.83	0.10
			Average 2.35		
	2	≥3	12	2.61	0.75
	10	≥3	16	0.93	0.05
	11	≥3	20	1.67	0.16
	12	≥3	19	2.45	0.16
	13	≥3	17	2.09	0.07
			Average 1.95		
White Pine Woods	2	1	15	2.64	0.01
	4	1	14	2.82	0.15
	5	1	13	2.75	0.33
	6	1	13	2.77	0.11
	7	1	12	2.63	0.26
	11	1	18	4.11	0.25
	15	1	11	3.22	0.19
			Average 2.99		
	1	2	15	2.99	0.18
	3	2	15	2.89	0.16
	8	2	18	2.61	0.13
	9	2	20	2.76	0.11
	12	2	14	5.01	0.10
	13	2	12	2.62	0.47
			Average 3.15		
	10	≥3	15	4.01	0.12
	14	≥3	11	2.52	0.32
			Average 3.27		

Table 11.4. Annual Seed Production in *Amianthium muscaetoxicum* for Each of 2 Years for the 15 Plants That Flowered at Least Twice over a 3-Year Period (out of 72 Examined)

Plant	Year 1	Year 2
1	12	148
2	0	32
3	40	128
4	8	17
5	114	306
6	170	166
7	54	57
8	118	60
9	72	109
10	29	7
11	46	80
12	149	72
13	96	56
14	73	88
15	54	9

duction per year appears continuously distributed, and the repeatability is not produced by the undue influence of one or two plants. This repeatability is even more surprising because in no single year was there a correlation between a plant's seed production and its total leaf area (correlations in 1981, 1982, and 1983 of 0.05, 0.16, and −0.01, respectively; sample sizes of 42, 22, and 37, respectively). Plants with greater leaf areas tended to produce more seeds per fruit (correlations of 0.48, 0.15, 0.11 in the 3 successive years), but the correlation was significant only in 1981.

The Sources of Individual Variation in Annual Seed Production

Individual variation can be examined in more detail if the variation is partitioned among its components (number of inflorescences, number of flowers per inflorescence, fruit-set, and seeds per fruit). This approach is analogous to the partitioning used to examine sources of spatial and temporal variation in average seed production, but a different technique is necessary. The method of partitioning standardized variance among individuals into multiplicative components, as derived by Arnold and Wade (1984) from Crow's technique (1958), can be used. This method partitions the total variance among individuals in seed production into levels attributable to each component by computing sequentially weighted means and weighted variances at each temporal stage of the reproductive process. The variance component at each stage is adjusted for the effects of the prior stage such that each can be interpreted as the extent of standardized variance in seed production at that stage, given what effects have preceded that stage. The substantive criticisms of the use of this method (e.g., Endler, 1986; Downhower et al., 1987; Koenig and Albano, 1987) revolve around the difficulty of interpreting its meaning when studies on different organisms are compared. The difficulties inherent in those comparisons do not arise in this case [see Campbell (1989) for another use of this method in plant reproductive biology].

The samples of plants from the Twin Springs populations used for the analysis of spatial and temporal differences were of sufficient size for this analysis (Table 11.5). The most consistent source of variance among individuals in all 3 years was fruit-set. In the Twin Springs 1 population, fruit-set was overwhelmingly important in 1981, the most important source of variation in 1982, and as important as the number of inflorescences in 1983. In the Twin Springs 2 population, fruit-set was the dominant factor every year. Seeds per fruit and number of flowers per inflorescence played distinctly subordinate roles.

Fruit-set is the critical phase in the genesis of individual variation. Experimental manipulations of inflorescence size and comparisons of natural and hand pollinations reveal that fruit-set is limited by pollinator availability, especially in smaller inflorescences (Travis, 1984; Robbins, 1986). Pollinator limitation suggests, in turn, that part of the variation among individuals in fruit-set may be due to variation in the local density of inflorescences, especially during the height of the flowering season when pollinators move short distances (Palmer et al., 1988). Indeed fruit-set is often positively density dependent (Robbins, 1986). But density dependence in fruit-set is not a consistent phenomenon (Fig. 11.3). At the Twin Springs 1 population in 1991, fruit-set increased as the density of neigh-

Table 11.5. The Total Opportunity for Selection in *Amianthium muscaetoxicum* and Its Partitioning into Sequential Components for Three Populations in Each of 3 Years[a]

Trait	Twin Springs 1			Twin Springs 2		
	1981	1982	1983	1981	1982	1983
Total	0.59	0.62	0.86	0.79	1.40	1.56
Number of inflorescences	0.17	0.12	0.22	0.16	0.38	0.37
Number of flowers per inflorescence	0.05	0.14	0.09	0.14	0.05	0.13
Fruit-set	0.43	0.18	0.20	0.29	0.63	0.58
Seeds per fruit	0.06	0.07	0.10	0.05	0.24	0.15

[a]Sample sizes range from 35 to 45 in each column.

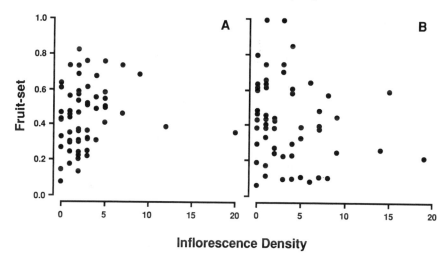

Figure 11.3. Fruit-set as a function of inflorescence density in two populations of *Amianthium muscaetoxicum*, Twin Springs 1 (A) and Moonshine Dell (B), in 1991. Fruit-set is density dependent in Twin Springs 1 but not in Moonshine Dell.

boring inflorescences increased (Fig. 11.3A; $F_{1,58} = 5.05$; $p < 0.03$; $r^2 = 0.08$). The two points with the highest inflorescence densities included three or four extraordinarily large individuals that may have been close relatives; omitting these points, the trend is stronger ($F_{1,56} = 10.34$; $p < 0.002$; $r^2 = 0.16$). In the same year there was no relationship between local inflorescence density and fruit-set in the Moonshine Dell population (Fig. 11.3B: $F_{1,58} = 10.34$; $p < 0.002$; $r^2 = 0.16$). Thus, variation among individuals in the density of neighbors is only a partial explanation of the source of their variation in fruit-set.

Hand pollination does not increase fruit-set above natural levels in larger inflorescences (i.e., those with ≥ 40 flowers: Robbins, 1986). Larger inflorescences are open for a longer period, and although they are not more likely to be visited by pollinators on any single day, once visited, they are likely to have more of their flowers pollinated (Robbins, 1986). This latter effect occurs only late in the flowering season, however, and is not repeatable during the period of peak flowering (Palmer et al., 1988). Hand pollinations also highlight an effect of flower position: proximal flowers are more likely to set seed than distal flowers, and the seeds they yield are heavier, despite equal pollination intensity (Robbins, 1986; Redmond et al., 1989). These observations suggest that the upper limit to seed production in larger plants is set by resources. Larger plants produce more flowers than they could possibly fill with seeds. Thus another fraction of the variation among individuals in fruit-set is due to variation in what sets the limit of fruit-set.

It is possible that the last flowers to open on the inflorescences of larger plants function primarily as pollen donors [see Sutherland and Delph (1984) for a review of the hypotheses for excess flower production]. This extra source of variation among individuals in fitness would not be reflected in total seed production. This hypothesis can be evaluated by examining whether flower position has an effect on natural patterns of fruit-set in larger inflorescences that are not pollinator-limited.

Two data sets address this issue. The first set examines fruit-set patterns in the 10 most proximal flowers, the 10 most distal flowers, and the 10 central flowers in larger inflorescences (≥ 40 flowers) in four locations in a single year (Fig. 11.4). The most distal flowers have higher fruit-set rates than the most proximal flowers in two locations, and the most proximal have highest fruit-set in only one location. These suggestions are confirmed by a repeated-measure univariate analysis of variance using the log-odds transform (Trexler et al., 1988) of fruit-set rates. Locations varied significantly ($F_{3,58} = 4.64$; $p < 0.001$), and there was a significant main effect of flower position ($F_{2,112} = 8.06$; $p < 0.001$), but the effect of flower position on fruit-set varied with location, as revealed by the significant interaction term ($F_{6,112} = 4.60$; $p < 0.001$). The second data set examines whether these position effects vary across 3 years in a single population (Table 11.6). An analogous repeated-measure ANOVA revealed significant variation in fruit-set among years ($F_{2,42} = 3.69$; $p < 0.05$) and among positions ($F_{2,84} = 15.14$; $p < 0.001$). The position effect varied significantly from one year to another ($F_{4,84} = 2.80$; $p < 0.05$). Both data sets indicate inconsistency in the relative importance of the specific positions for fruit-set. If the production of extra flowers allows enhanced male function without concomitant female function, it does so only in some locations in some years, suggesting that it is not a consistent source of individual variation in fitness.

The Effect of Predation

Overview

Seed number will translate into a suitable surrogate for fitness if the two variables have a rank-order correlation over the entire range of seed-production values. Two effects can scuttle this correlation. First, if seed number increases at the expense of seed size and seed size plays a critical role, then there is a seed number beyond which increases in number lead to increased net fitness. I cannot address this issue. Second, if there are

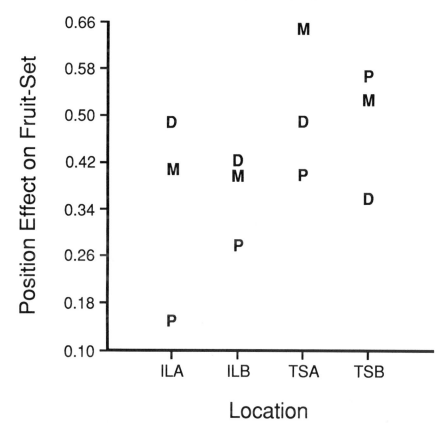

Figure 11.4. Fruit-set as a function of flower position in the inflorescence (*P* = average of the 10 most proximal flowers from each of 15 plants; *M* = average of the 10 central flowers; *D* = average of the 10 most distal flowers) and location (ILA, Illegal Trail, site A; ILB, Illegal Trail, site B; TSA, Twin Springs 1, site A; TSB, Twin Springs 1, site B) in *Amianthium muscaetoxicum*.

predators that respond to increased seed numbers in patches or on a plant by aggregating and thereby increasing predation rate as a function of fruit or seed number in a patch, then this density-dependent predation will also set a seed number beyond which further increases lead to decreased fitness through increased attractiveness to predators [see Taylor (1988) for a review]. To visualize this result, allow seed survival rate to decrease in linear fashion with an increase in seed number. Net seed production will equal the original number multiplied by the survival rate. In this case net seed

Table 11.6.   Average Fruit-Set (Standard Error in Parentheses) for 15
Inflorescences in Each of 3 Years from the Twin Springs 1 Population of
Amianthium muscaetoxicum as a Function of Position

	Position		
Year	Proximal	Middle	Distal
1981	0.40 (0.06)	0.65 (0.06)	0.49 (0.07)
1982	0.39 (0.07)	0.49 (0.08)	0.13 (0.05)
1983	0.55 (0.09)	0.57 (0.07)	0.24 (0.08)

production will be a quadratic function of the original number of seeds,
and the point of maximum fitness for a plant will be at an intermediate
seed number.

The aggregated distribution of *Amianthium*, its limited dispersal of seeds
as a group, and its presentation of swollen, splitting capsules just before
seed dispersal make the possibility of aggregated predator attack and density-
dependent mortality a realistic one. In the next two sections, I examine
the patterns of fruit and seed predation on *Amianthium* in this light.

Fruit Predation

Ripening capsules are attacked by chewing insects that rip the locules
in characteristic fashion and consume all of the seeds inside. The incidence
of this predation has varied dramatically across years from 0 of 130 inflorescences examined in 1982 to 91 of 120 in 1991. Even in years when the
overall incidence was low, such as 1981, there was variation among localities. In the Twin Springs 1 population in that year, predation affected only
3% of infructescences, but it affected 25% of the infructescences at the
Illegal Trail population. Density dependence can occur as a function of
the density of local infructescences, as a function of the number of flowers
on an infructescence (if the animals used the overall size as a cue), or as
a function of the number of ripening capsules. I illustrate analysis of each
possibility: density dependence is indicated by an increase in predation
rate with increases in density, and statistical tests for that increase were
made according to the logistic regression methods outlined by Trexler et
al. (1988).

In 1989 local infructescence density varied extraordinarily relative to
other years (compare Figs. 11.3 and 11.5), which makes this year the most
likely one in which this form of density-dependent predation might arise.
Nonetheless, no evidence indicated that rate of predation had any relationship to local infructescence density (Fig. 11.5). In this case there is no
relationship between number of fruits destroyed and local density, so there

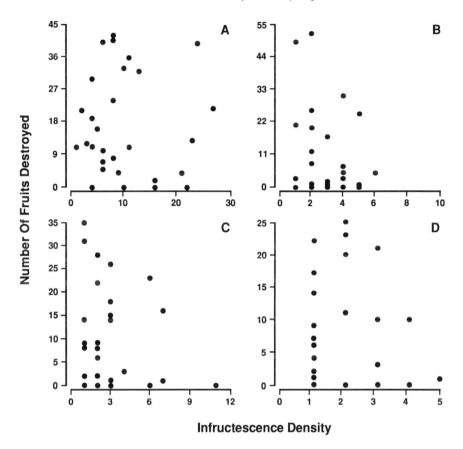

Figure 11.5. Number of fruits of *Amianthium muscaetoxicum* destroyed by insect seed predators on an infructescence as a function of infructescence density (number of infructescences within a circle of 1-m radius centered on the focal infructescence) for plants in each of two habitat types in each of two populations in 1989. (A) Plants in light gaps in Twin Springs 1; (B) plants underneath the canopy in Twin Springs 1; (C) plants in light gaps in Moonshine Dell; (D) plants underneath the canopy in Moonshine Dell.

will be no chance for an increasing rate of predation. In no other year was there any indication of this type of density dependence.

Density-dependent predation at the level of flower number occurred in 1989 at Moonshine Dell (Fig. 11.6, gap and canopy pooled: $F_{1,57} = 4.58$; $p < 0.05$) but not at Twin Springs ($F_{1,57} = 0.19$; $p > 0.05$). No density dependence was detected in any other year.

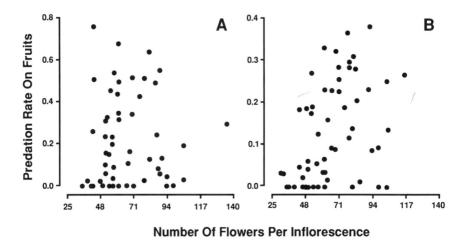

Figure 11.6. Rate of predation on fruits of *Amianthium muscaetoxicum* by insect seed predators as a function of original number of flowers on the inflorescence for plants in two populations in 1989. (A) Twin Springs 1; (B) Moonshine Dell.

Predation rates, as functions of the actual number of fruits available on an infructescence, vary widely among years and locations (Fig. 11.7). In 1981 at Twin Springs 1 (Fig. 11.7A), predation rate showed a nonsignificant tendency to decrease as the number of fruits increased from 2 to 19. In the same location 10 years later (Fig. 11.7B), fruit production was dramatically higher, but predation rates were much lower and showed no particular relationship to the number of fruits. At Moonshine Dell in the productive year of 1991 (Fig. 11.7C), predation rates were very high and displayed a tendency to decrease with increasing numbers of fruits. Despite this wide variation in numbers of fruits and predation rates, there is no indication in these or any other data that there is any density-dependent predation as a function of numbers of available fruits.

Fruit predation rates vary widely among locations, as shown in Figures 11.5–11.7. An analysis of variance on the fruit predation rates depicted in Figure 11.5 reveals that the major difference is between the gap location at Twin Springs and the other three locations ($F_{3,116} = 4.20$; $p < 0.05$). As the figure also implies, the location effect explains only a small fraction (7%) of the total variance in fruit predation rates. Predation rates in the same locations 2 years later varied from an average of 13% in the canopy of Twin Springs to 84% in the canopy at Moonshine Dell. Thus, locations do not differ consistently in fruit predation rates from one year to another, even when those differences are statistically significant in any single year.

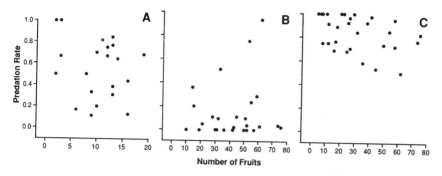

Figure 11.7. Rate of predation on fruits of *Amianthium muscaetoxicum* by insect seed predators as a function of the original number of fruits available for plants underneath the canopy. (A) Plants in Twin Springs 1 in 1981; (B) plants in Twin Springs 1 in 1989; (C) plants in Moonshine Dell in 1989.

Seed Predation

Small, nocturnally active rodents will readily consume the seeds of *Amianthium*. In the laboratory, individual *Peromyscus maniculatus* and *P. leucopus* consumed >100 seeds in 3 hr. When mice ate the seeds, they first stripped the seed coats. No seeds emerged in the feces of any of the six animals to which we fed *Amianthium* seeds, and no animals suffered any ill effects from the seeds after 72 hr.

The two species of *Peromyscus* appear to be the only significant seed predators. Seed baits were placed in *Amianthium* patches eight times during the period of capsule dehiscence in August 1982 (when seeds were falling naturally) and consisted of one of five densities of seeds (5, 10, 20, 40, 80) placed either outside of a wire mesh (6.4-mm) cage or inside the cage (to preclude access by vertebrates), either at night or during the day, at one of five locations. Seed baits consisted of the seeds placed inside a plastic plate of approximately 20 cm diameter; baits were set out after dusk and checked just before dawn (night exposure) or set out just after dawn and checked before dusk (day exposure). Seeds were taken from only the exposed baits set out at night (the other treatments lost an average of less than one seed per exposure: 0.6 ± 0.3 seeds, $n = 120$), and nearly all missing seeds were represented by their seed coats.

The combined density of the two species of mice has varied enormously in space and time during the last 12 years [J. O. Wolff, personal communication; see Baccus and Wolff (1989) for an overview of the dynamics of these populations]. Average total mouse densities have ranged from 3 to over 100 mice per hectare during this period, and average total densities during the months of August from 1980 though 1991 have ranged from 6 to 46 mice per hectare (average of 24 mice per hectare). The range of variation of density from one location to another in any single year can be large (Wolff, 1985). The mice are interspecifically territorial (Wolff et al., 1983), and average home-range sizes vary from 500 to 700 m^2 (Wolff, 1985). The mice do not compress their home ranges below 500 m^2 per individual; at very high densities individual ranges overlap frequently, and little of the available area is not occupied, whereas at low densities considerable area can be unoccupied. Mice will not alter the size or location of their home ranges in response to food addition (Wolff, 1985), so the bait treatments were not likely to attract mice that were not resident in those spots. These considerations of the ecology of the mice indicate that the predation pressure on seeds will vary enormously in space and time as mouse densities vary.

Seed predation rates declined with increases in density of seed baits (Fig. 11.8), as would be expected if the mice foraged until they were sated. Predation rates also differed among the localities; the lowest predation rate in a locality averaged 0.02 ± 0.01 across eight nights, whereas the highest averaged 0.62 ± 0.12. The best statistical model for the data (Table 11.7) indicated that the variation among localities was five times more important than the variation among density treatments. Localities accounted for 41% of the variation in seed predation rate. This effect, the result of where mice are and where they are not, imposes a spatial structure on seed production rates, which will produce random variation in seed production and random variation in absolute fitness among localities.

The density of *Amianthium* infructescences and their average seed production per infructescence are likely to combine to generate an enormous rain of seeds, and mice may not make an appreciable dent in this number. However, the patchy distribution of the plants and the patchy distribution of the mice can mean that, in some localities in some years, some plants will suffer enormous seed-predation rates. Like fruit predation, this factor will function as an effect that is stochastic in time and space with respect to the plants.

Implications

Average seed-production rates show little large-scale spatial or temporal organization. Most of the variation is among individuals and has its major genesis in differences in fruit-set. The vagaries of pollinator service, the effects of local density that facilitate pollination, the timing of flowering, and the size of the floral display affect fruit-set, but each factor contributes a small amount to the variance among individuals and each factor varies

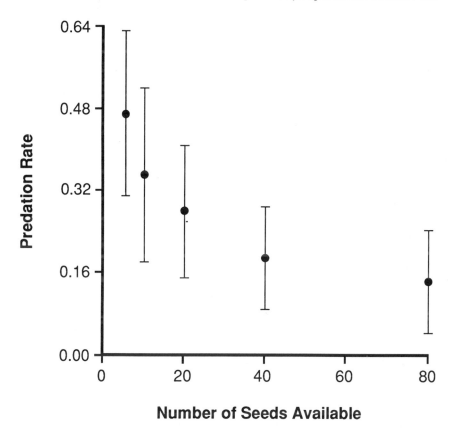

Figure 11.8. Predation rate by *Peromyscus* spp. on experimental patches of seeds of *Amianthium muscaetoxicum* as a function of the number of seeds available in a patch in 1982. Data are shown as average values from eight replicates per experimental density surrounded by error bars that are one standard error in either direction from the average.

in its strength from year to year. Individual differences are repeatable over 2–3 years, but repeatable differences among individuals need not imply intrinsic differences. A major fraction of the variance among individuals in seed-production rates can be environmentally induced by microhabitat factors such as soil or light. Indeed, appreciable observational and experimental evidence indicates that such factors do play the major role in this system (A. Redmond, unpublished data). Nevertheless, these differences are likely to be filtered through the largely stochastic effects of fruit and seed predation. The conclusion to be drawn is that, if the average annual

Table 11.7. Best Model for the Analysis of Variance on Predation Rate of Mice on *Amianthium* Seeds

Source	df	MS	F
Location	4	26.5	7.51**
Density given location	1	23.0	6.51*
Time period given location, density	7	5.1	1.46
Residual	27	3.5	—

*$p < 0.05$
**$p < 0.01$.

net seed production is a reasonable measure of fitness, most of its variance is due to environmental causes.

This conclusion is not surprising (Crow, 1958). It does imply that other sources of fitness variation, such as paternal function, must be considered in the face of an enormous stochastic component to maternal function (stochastic components to paternal function have been described: Devlin and Ellstrand, 1990). It also implies that the actual intensity of selection that occurs through breeding values will be more consonant with the traditional assumptions of weak selection in population genetics and will not compel a rethinking of population-genetic theory (Endler, 1986). Stochastic variation in maternal function can introduce biases into functional gender calculations (Robbins and Travis, 1986). It also indicates that experiments to detect the effect of paternal identity on offspring fitness, or some factor involved with paternal identity such as distance from the maternal plant (Waser and Price, 1989; Redmond et al., 1989), will maximize power by using split-plot designs or other variants in which the maternal plant is treated as a whole-plot factor [see Palmer et al. (1989) for another example or Woodward et al. (1988) for an analogue in an animal system].

Several hypotheses have been advanced to explain the production of excess flowers in hermaphroditic plants (Sutherland and Delph, 1984). No single hypothesis may be sufficient to explain the patterns in *Amianthium*. It is hard to give credence to the hypothesis that excess flowering effort in *Amianthium* evolved slowly for the purpose of enhancing male function. There is no temporal separation of gender function in the inflorescence (Palmer et al., 1989) and no consistency in the lack of female function in distal flowers. Enhanced male function may be an ancillary benefit in some locations in some years, when pollinator service is not limiting. The gradual opening of the entire inflorescence, the abortion of terminal buds on larger inflorescences, the inconsistency of just where (and by implication when) pollination will occur on the inflorescence, the consistency of where seeds are produced when all flowers are hand pollinated, and the natural limitation of seed production by the level of fruit-set seem more consonant with the "bet-hedging" hypothesis for excess flower production. Cohen and Dukas (1990) have verified the logic of the appealing notion that, as the variance in pollination success increases, the optimal number of flowers increases and the apparent excess of flowers will increase. In many ways *Amianthium* is an excellent example.

Amianthium may also be the supreme example of the evolution of iteroparity in response to uncertainty. As the coefficient of variation in the age- and time-indexed vital rates rises, selection favors a more iteroparous life history (Orzack and Tuljapurkar, 1989). With greater iteroparity, the individual is more able to "ride out" stochastic variations in vital rates. The patterns of net seed production in *Amianthium* support this interpretation for its life history. Average annual seed production varies temporally and spatially, but most of the variation is among individuals. Risk of loss of seeds to predation is best described as a stochastic process that operates on a small spatial scale and induces temporal variation in an individual's net annual seed production and spatial variation for any genotype. Variation in seedling survival rates will only enhance the stochastic effect on net annual seed production.

Although the study of annual seed production and the factors that limit it can offer a number of insights into the structure of fitness variation, it is still an incomplete surrogate. Most *Amianthium* seedlings die, and, if data on other perennial plants are any indication, successful recruitment into the reproductive size classes may itself be a process that is stochastic in space and time. Mortality rates of reproductively sized plants are negligible on an annual basis at Mountain Lake; after 11 years nearly 75% of the original marked individuals are still thriving. In such a situation the best surrogate for an individual's real fitness (contribution to population growth) may not be how many seeds it has produced over its lifetime (or how high its average annual output has been) but simply how often it has put a minimum number of seeds into the pool. The more often it has reproduced, the more likely it will circumvent the stochastic hazards and have an offspring in the right place at the right time. I suggest that such a surrogate for fitness will be a proper guide to the outcome of selection only when temporal stochastic variation exceeds some threshold level, when the repeatability of individual variation in annual seed production is low (unlike *Amianthium*), and when there is a considerable cost to overproduction of seeds in any one year. If overproduction in one year precludes effective reproduction for the next several years, then the individual is effectively less iteroparous and may be less able to "ride out" the stochastic effects than an individual that reproduces only moderately in any given year but does so more regularly. The severity of the phenotypic cost of reproduction becomes a key empirical parameter in this case, and experiments that estimate that cost (e.g., Primack and Hall, 1990) offer important information. This last suggestion is at odds with some views on the cost of reproduction (Reznick, 1985; Reznick et al., 1986), but perhaps the special problems posed by the study of iteroparous perennials illuminate the points at which some of our "general" notions of evolutionary processes may be distortions.

Acknowledgments

Many people have participated in the *Amianthium* project over the years, and I take this opportunity to thank them all. Special thanks are due to two exemplary assistants, Mildred Pinnell and Anne Hoover, who executed complex experimental and sampling designs flawlessly and accomplished incredibly tedious tasks cheerfully. I have had the benefit of the knowledge, intuition, and companionship of excellent colleagues on this project over the years, and I thank M. Palmer, A. Redmond, L. Robbins, and C. Werth for their efforts and patience. This project has outlasted several directors at the Mountain Lake Biological Station (B. J. Cole, J. L. Riopel, J. J. Murray, and J. O. Wolff), and I thank them for access to the facilities of the Station and for creating an atmosphere conducive to the conduct of good science. I owe a great personal and intellectual debt to Dr. J. Antonovics for his encouragement of this work in its early years; this paper is dedicated to him in humble gratitude. This work has been supported by a number of sources, including the Pratt Fund of the University of Virginia and the President's Fund of Florida State University. Some analysis of data and the writing of this paper occurred while I was supported by the National Science Foundation (BSR 88-18001).

Literature Cited

Arnold, S.J., and M.J. Wade. 1984. On the measurement of natural and sexual selection: Theory. Evolution **38**:709–719.

Baccus, R., and J.O. Wolff. 1989. Genetic composition of fluctuating populations of *Peromyscus leucopus* and *Peromyscus maniculatus*. J. Mammal. **70**:592–602.

Campbell, D.R. 1989. Measurement of selection in a hermaphroditic plant: Variation in male and female pollination success. Evolution **43**:318–334.

Cohen, D., and R. Dukas. 1990. The optimal number of female flowers and fruits-to-flowers ratio in plants under pollination and resource limitation. Am. Nat. **135**:218–241.

Crow, J.F. 1958. Some possibilities for measuring selection intensities in man. Human Biol. **30**:1–13.

Devlin, B., and N.C. Ellstrand. 1990. Male and female fertility variation in wild radish, a hermaphrodite. Am. Nat. **136**:87–107.

Components of Reproductive Success / 279

Downhower, J.F., L.S. Blumer, and L. Brown. 1987. Opportunity for selection: An appropriate measure for evaluating variation in the potential for selection? Evolution **41**:1395–1400.

Endler, J.A. 1986. Natural Selection in the Wild. Princeton University Press, Princeton, NJ.

Koenig, W.D., and S.S. Albano. 1987. Lifetime reproductive success, selection, and the opportunity for selection in the white-tailed skimmer *Plathemis lydia* (Odonata: Libellulidae). Evolution **41**:22–36.

Marshall, D.L., and N.C. Ellstrand. 1988. Effective mate choice in wild radish: Evidence for selective seed abortion and its mechanism. Am. Nat. **131**:739–756.

Orzack, S.H., and S. Tuljapurkar. 1989. Population dynamics of variable environments. VII. The demography and evolution of iteroparity. Am. Nat. **133**:901–923.

Palmer, M., J. Travis, and J. Antonovics. 1988. Seasonal pollen flow and progeny diversity in *Amianthium muscaetoxicum*: Ecological potential for multiple mating in a self-incompatible, hermaphroditic perennial. Oecologia **76**:19–24.

Palmer, M., J. Travis, and J. Antonovics. 1989. Temporal mechanisms influencing gender expression and pollen flow within a self-incompatible perennial, *Amianthium muscaetoxicum* (Liliaceae). Oecologia **78**:231–236.

Primack, R.B., and H. Hall, 1990. Costs of reproduction in the pink lady's slipper orchid: A four-year experimental study. Am. Nat. **16**:638–656.

Redmond, A.M., L. Robbins, and J. Travis. 1989. The effects of pollination distance on seed production in three populations of *Amianthium muscaetoxicum* (Liliaceae). Oecologia **79**:260–264.

Reznick, D. 1985. Cost of reproduction: An evaluation of the empirical evidence. Oikos **44**:257–267.

Reznick, D., E. Perry, and J. Travis. 1986. Measuring the cost of reproduction: A comment on papers by Bell (1984a,b). Evolution **40**:1338–1344.

Roach, D.A., and R.D. Wulff. 1987. Maternal effects in plants. Annu. Rev. Ecol. Syst. **18**:209–235.

Robbins, L. 1986. Inflorescence size and reproductive success in *Amianthium muscaetoxicum*, a perennial lily. M.S. Thesis, Florida State University, Tallahassee, FL.

Robbins, L., and J. Travis. 1986. Examining the relationship between functional gender and gender specialization in hermaphroditic plants. Am. Nat. **128**:409–415.

Schlichting, C.D., and B. Devlin. 1989. Male and female reproductive success in the hermaphroditic plant *Phlox drummondii*. Am. Nat. **133**:212–227.

Stephenson, S.L. 1982. A gradient analysis of slope forest communities of the Salt Pond Mountain area in southwestern Virginia. Castanea **47**:201–215.

Sutherlund, S., and L.F. Delph. 1984. On the importance of male fitness in plants: Patterns of fruit-set. Ecology **65**:1093–1104.

Taylor, A.D. 1988. Large-scale spatial structure and population dynamics in arthropod predator-prey systems. Ann. Zool. Fenn. **24**:63–74.

Timm, N.H. 1975. Multivariate Analysis with Applications in Education and Psychology. Brooks/Cole, Monterey, CA.

Travis, J. 1984. Breeding system, pollination, and pollinator limitation in a perennial herb, *Amianthium muscaetoxicum* (Liliaceae). Am. J. Bot. **71**:941–947.

Travis, J. 1988. Differential fertility as a major mode of selection. Trends Ecol. Evol. **3**:227–230.

Trexler, J.C., C. McCulloch, and J. Travis, 1988. How can the functional response best be determined? Oecologia **76**:206–214.

Winn, A.A. 1988. Ecological and evolutionary consequences of seed size in *Prunella vulgaris*. Ecology **69**:1537–1544.

Waser, N.M., and M.V. Price. 1989. Optimal outcrossing in *Ipomopsis aggregata*: Seed set and offspring fitness. Evolution **43**:1097–1109.

Woodward, B.D., J. Travis, and S. Mitchell. 1988. The effects of the mating system on progeny performance in *Hyla crucifer* (Anura, Hylidae). Evolution **42**:784–794.

Wolff, J.O. 1985. The effects of density, food, and interspecific interference on home range size in *Peromyscus leucopus* and *Peromyscus maniculatus*. Can. J. Zool. **63**:2657–2662.

Wolff, J.O., M.H. Freeberg, and R.D. Dueser. 1983. Interspecific territorality in two sympatric species of *Peromyscus* (Rodentia: Cricetidae). Behav. Ecol. Sociobiol. **12**:237–242.

Wright, S. 1969. Evolution and the Genetics of Populations. Vol. 2: The Theory of Gene Frequencies. University of Chicago Press, Chicago, IL.

12

Environmental and Genetic Sources of Variation in Floral Traits and Phenotypic Gender in Wild Radish: Consequences for Natural Selection

Susan J. Mazer

University of California

Introduction

Evolutionists have long been interested in the causes and evolutionary implications of labile sex expression in wild plant species. The causes and evolutionary significance of variation in sex expression within and among species have been explored in a variety of empirical and theoretical contexts. In the simplest case, variation in levels of flower or fruit abortion creates variation among inflorescences, individuals, and species in the proportion of functionally female flowers (Lloyd, 1979, 1980a,b,c; Stephenson, 1981; Sutherland and Delph, 1984; Garnock-Jones, 1986; Sutherland, 1986; Pellmyr, 1987; Mazer et al., 1989). Temporal changes in gender following a change in plant size, age, or floral position (and, presumably, a change in the availability of resources) has been predicted by theory and observed in many field and greenhouse situations (Ghiselin, 1969; Charnov and Bull, 1977; Policansky, 1981, 1982; Bertin, 1982; Charnov, 1982; Lovett Doust and Cavers, 1982a; Bierzychudek, 1984; Solomon, 1985; Cid-Benevento, 1987; Thompson, 1987; Condon and Gilbert, 1988; Thomson et al., 1989; but see Charnov, 1986 for a discussion of exceptions). Several studies have illustrated that phenology of pollen and ovule production among population members may influence the functional gender of individuals, resulting in temporal variation in the correlation between phenotypic and functional gender (Thomson and Barrett, 1981; Devlin and Stephenson, 1987; Nakamura et al., 1989; Mazer et al., 1989). In addition, insect herbivory has been found to induce changes in host plant sex expression (Hendrix and Trapp, 1981). Among species, the relative investment in male and female gametes has been related to mating system. As predicted by theory, a number of studies have shown that highly outcrossed species allocate relatively more energy or nutrients to male reproductive organs than do highly selfed species (Cruden, 1976, 1977; Charlesworth and Charlesworth, 1981; Lovett Doust and Cavers, 1982c; Schoen, 1982; Cruden and Lyon, 1985; McKone, 1987).

Finally, adaptive environmental sex determination (ESD), the ability of nonmobile individuals to assess their local environment prior to sexual reproduction and then to produce flowers of the gender with the highest relative fitness in that environment, has been observed or implied in numerous monoecious and dioecious species (Schaffner, 1922; Gregg, 1975; McArthur, 1977; Onyekwelu and Harper, 1979; Freeman et al., 1980; McArthur and Freeman, 1982; Lovett Doust and Cavers, 1982b; Freeman and McArthur, 1984; Freeman et al., 1984; Freeman and Vitale, 1985). Spatial segregation of the sexes (one expected result of ESD) seems to be a common feature of dioecious species (Cole, 1979; Cox, 1981; Fox and Harrison, 1981; Freeman et al., 1976, 1981; Waser, 1984; Lovett Doust et al., 1987; Bierzychudek and Eckhart, 1988; Dawson and Bliss, 1989; Cameron and Wyatt, 1990; but see Bawa and Opler, 1977; Iglesias and Bell, 1989 for discussions of exceptions and caveats).

Models of the evolution of sex allocation, particularly models of the evolution of gender modification in response to local environmental conditions, share a number of assumptions, some of which have more empirical support than others (Goldman and Willson, 1986). Two assumptions are relatively noncontroversial, even if not fully empirically explored: first, plants cannot choose the environment in which they grow; and second, there exists in natural populations heritable variation in sex expression and in the plastic response of sex expression to environmental variation. Other assumptions are more tenuous and require rigorous testing. For example, it is not clear that individual plant genotypes can assess the condition of their local environment and respond by producing a more fit phenotype. More critically, a universal assumption in sex allocation theory is that an increased level of investment in one gametic type necessarily incurs a cost with respect to the allocation of resources to the alternate gametic type. This assumption has been seriously questioned (Cruden and Lyon, 1985; Devlin, 1989), although no detailed measures exist of the genetic correlation between pollen and ovule production among genotypes or entire plants. Because phenotypic correlations at the level of individual flowers cannot be used to infer the presence of evolutionary constraints, it is critical that we obtain estimates of the true genetic correlations between these traits. Finally, models of the evolution of ESD assume that local environments consistently have differential effects on male and female fitness (i.e., the relationship between sex expression and expected individual fitness is environment-specific). One shared prediction of these resource-based models is that, under stressful conditions, natural selection will favor those genotypes that invest preferentially in the least costly gamete (generally considered to be pollen).

In view of their rarity relative to hermaphroditic species, dioecious, monoecious, and subdioecious plants have received disproportionate attention from empiricists. This asymmetry is probably due to the fact that sex expression in such species can be measured simply by recording the gender of individual ramets or genets in a qualitative way (male or female), or by counting the proportions of male, female, and bisexual flowers. In contrast, accurate measures of phenotypic gender (the relative investment in male and female reproductive parts) in hermaphrodites require precise counts of pollen and ovules per flower (or seeds per fruit, fruit weight, etc.). Until electronic particle counters recently became available, this was an extremely tedious task. Consequently, there exist few studies of gender variation in hermaphroditic species (but see Bawa and Webb, 1983; Devlin and Stephenson, 1987; Thomson, 1989).

The purpose of the current study is 3-fold. First, I provide data concerning the limits of and sources of variation in phenotypic gender and the floral traits that comprise it in a hermaphroditic species, *Raphanus sativus* L. (wild radish). Second, I evaluate three of the above assumptions for genotypes sampled from a feral population and grown under a natural range of conditions: (1) that there is statistically detectable, heritable variation in gender, (2) that there is heritable variation in the plastic response of sex expression to one environmental variable (planting density), and (3) that there exists a strong phenotypic or genetic trade-off between allocation of resources to pollen vs. ovules. Finally, I propose a hypothesis to explain male-biased gender in stressful environments that is an alternative to the more common evolutionary explanation: that it is primarily the high cost of seed production that causes male-biased phenotypes to have the highest expected individual fitness in low-resource environments. To achieve these goals, I address the following sets of questions.

1. How does planting density influence individual survivorship and the mean phenotype of a population with respect to primary sex characters (pollen grain number, pollen size, ovule number, individual seed mass, and phenotypic gender) and secondary sex characters (floral advertisement: petal area, petal area/pollen grain, and petal area/ovule)? Is the effect of density trait-specific? If so, do correlations among traits change with density?

2. Does density influence the magnitude of detectable, expressed heritable variation in floral traits (i.e., do parental influences on progeny phenotype depend on the environmental conditions in which progeny are grown)? If not, do parental effects on progeny

phenotype detected within density treatments predict parental effects across densities?

3. Does density influence the phenotypic rank of different genotypes with respect to these floral traits? For example, does the genotype that produces the most ovules per flower at one density also produce the most ovules at other densities?

4. Is there genetic variation in the phenotypic response to increasing density? Do some traits exhibit greater genetic variation in phenotypic plasticity than others?

The first set of questions allows me to resolve the two processes that may cause phenotypic change across an environmental gradient: selection and direct environmental modification of phenotype. First, density-specific survivorship of distinct phenotypes may create differences among densities in the mean phenotype of surviving adult plants. Second, density may invoke a common phenotypic response in most or all genotypes, resulting in a population-level change in mean phenotype independently of selection.

These particular floral traits are of interest to evolutionists for several reasons. The study of pollen production, pollen size, and ovule number is necessary to generate a relatively simple measure of phenotypic gender and to explore the possibility of trade-offs among these traits. Corolla size and pollen production influence pollinator visitation rates and pollen removal rates in *R. sativus* (Young and Stanton, 1990b), so one might expect natural selection to have purged wild populations of genetic variation in these traits. Finally, estimates of the allocation of biomass to petals vs. gametophytes (petal area/pollen grain or ovule) allow one to ask whether individuals or genotypes at the "male" end of the phenotypic gender spectrum invest disproportionately in floral display, as might be expected if genotypes that have specialized in pollen production have also been under strong selection to increase pollinator attraction.

The last three sets of questions examine the effects of density on the potential for natural selection to effect the evolution of floral traits and gender. If genotypes differ significantly in phenotype at some but not all densities, then heritability estimates and the potential rate of evolutionary change by natural selection will also be density-specific. In addition, even if the magnitude of phenotypic differences among genotypes (intergenotypic variance) is consistent across densities, the phenotype associated with each genotype may change such that when densities are pooled, phenotypic differences among genotypes are obscured. That is, traits that exhibit genetic variation within densities may not do so when expressed in heterogeneous environments. In addition, if the phenotypic ranks of different genotypes alternate across densities, then the identity of the genotype favored by natural selection may also vary, assuming a consistent relationship between phenotype and fitness. Such Genotype × Density interactions may constrain the rate of evolution by natural selection in a heterogeneous environment (Clare and Luckinbill, 1985; Luckinbill and Clare, 1985; Shaw, 1986; Via and Lande, 1985, 1987; Via, 1987). Finally, if there exists genetic variation in the magnitude or direction of phenotypic plasticity in gender or floral traits in response to density, this suggests that phenotypic plasticity itself may be open to natural selection, a condition for environmental sex determination to evolve.

Study Organism

Raphanus sativus (Brassicaceae) is an annual, weedy crucifer that occurs in disturbed habitats, fallow old fields, and coastal dune communities throughout California. It has escaped cultivation since colonial times. *Raphanus sativus* and *R. raphanistrum* have been the focus of many demographic, pollination, and population genetic studies examining the role of natural selection in molding phenotypic variation in quantitative and discrete characters related to individual fitness (Stanton, 1984a,b, 1987a,b; Marshall and Ellstrand, 1986, 1988; Mazer et al., 1986; Stanton et al., 1986; Mazer 1987a,b, 1989; Stanton and Preston, 1988; Snow and Mazer, 1988; Stanton et al., 1989; Karron and Marshall, 1990; Young and Stanton, 1990a,b).

Wild radish has several features that make it a convenient study organism. It grows readily under controlled conditions in the greenhouse, in experimental gardens, and as seed introductions in natural populations. It has a short life cycle (3–4 months), large flowers in which pollen dehiscence is spatially separated from the stigma, a high proportion of fruit-set following hand-pollinations, and large seeds in comparison to other rapidly growing species. These features make it possible to measure reproductive components of fitness relatively easily. Because individual plants can produce hundreds of seeds in the greenhouse (depending on pot size) and field, it is possible to replicate lineages of seeds across experimental treatments.

The floral biology of wild radish permits controlled pollinations among plants to produce genetically distinct lineages. Flowers do not readily self-pollinate, and the bending of anthers away from the exserted receptive stigma prevents stigma clogging by self-pollen. Inbreeding in the field and in controlled breeding programs is restricted, because wild radish exhibits a combination of gametophytic and sporophytic self-incompatibility. Alleles expressed by the haploid genome (G-locus) and by the diploid pollen wall (S-locus) have been identified, and both may contribute to incompatibility reactions (Putrament, 1960; Lewis et al., 1988).

Feral populations of *R. sativus* are probably derived from cultivars of the domesticated radish. Nevertheless, populations of wild radish behave as a wild species. Ellstrand and Marshall (1985a,b) observed a high level of allozyme variation and Hardy–Weinberg ratios in many populations. This high degree of genetic polymorphism, in addition to the high degree of phenotypic variation readily observed in all wild and greenhouse populations, probably partly results from radish cultivars having been maintained as genetically variable, outcrossing populations. Some California populations of wild radish show evidence of introgression from *R. raphanistrum* (Panetsos and Baker, 1967), as shown by the appearance of yellow floral pigment, typical of *R. raphanistrum*, in the normally purple or white flowers of *R. sativus*. In the population of *R. sativus* used in this study, however, yellow pigment is completely absent.

The source population for the genotypes used in this study inhabits an old field on University of California property south of the Goleta Slough State Ecological Reserve. This field is also the site of my experimental garden. This close proximity of the source population to the experimental garden helps to ensure that the phenotypic variation observed among genotypes is typical for the species, and not strongly influenced by observing the genotypes in a novel environment (Service and Rose, 1985).

Material and Methods

In August 1988, seeds from 100 maternal plants growing in an abandoned field at the margin of the East Storke Campus Wetlands of the University of California (Santa Barbara) were collected and stored in the dark at room temperature until December 1988. At that time, they were planted in the greenhouse in 10-cm plastic pots in a sand-loam-peat mixture (1 seed per pot). Seventy-five synchronously flowering individuals, each a progeny of a distinct field-collected plant, were used in the following breeding design. Sixty plants were divided into 15 groups of four individuals, which were used as seed-bearing or maternal plants (pollen recipients). Each group of maternal plants was assigned at random to one of 15 pollen donors. This generation is the parental generation in a standard breeding design in which females are nested within males (Fig. 12.1).

To ensure that each maternal plant was genetically compatible with the donor assigned to it, compatibility tests were performed. If an ovary did not expand within 48 hr after hand-pollination with pollen of the assigned donor, then the cross was considered incompatible and the maternal plant was replaced with a genetically distinct one until a compatible cross was achieved. Approximately 20 flowers on each maternal plant were pollinated, and mature, dry fruits were removed from the plants in March 1989.

Field-Collected Generation: Seeds from 75 individuals

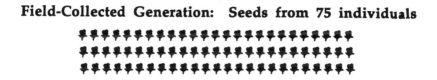

Greenhouse Parental Generation: 75 distinct genotypes

15 Pollen Donors

15 Groups of 4 Maternal Plants

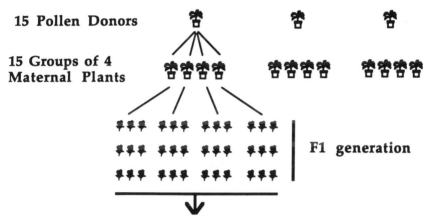

F1 generation

Each paternal "family" includes all progeny produced by the four maternal plants pollinated by the donor

F1 Garden-Grown Generation: 1800 individuals representing 60 maternal and 15 paternal families

Figure 12.1. Nested Breeding Design. Garden-grown individuals were derived from matings among a total of 75 greenhouse-grown and hand-pollinated individuals. Each of the 75 parental plants represented a distinct maternal plant whose seeds were collected from a single natural population of wild radish. Each paternal family is represented by four maternal families nested within it.

From the 60 maternal families, a total of 1800 seeds were assigned at random to three planting densities (600 seeds/density). Because the families were sampled at random (with replacement), the number of seeds per family sown in each density ranged from 5 to 17; however, the number of seeds used per maternal family was statistically independent of density (log-likelihood chi-square test: $\chi^2 = 104.89$, $p > 0.8004$, model df $= 118$, total df $= 1798$). In June, the 1800 F_1 seeds were sown in a fenced, irrigated experimental garden situated directly adjacent to the population that provided the source seeds for the parental generation.

The garden consisted of a total of nine plots representing three blocks. Each block included a high-, medium-, and low-density plot. A plot was an 8×25 grid of 200 seeds positioned randomly with respect to maternal family membership. In the high-density plot, seeds were sown 5 cm apart ($\sim$400 plants/m^2); in the medium-density plot, seeds were sown 10 cm apart ($\sim$100 plants/m^2); in the low-density plot, seeds were sown 20 cm apart ($\sim$30 plants/m^2). These densities represent a range commonly found in natural populations of wild radish. To avoid edge effects, I planted around the perimeter of each plot a border of genetically variable wild radish seeds collected from the source population. Naturally occurring competitors (*Avena*, *Bromus*, and *Convolvulus*) were not removed from the garden plots; the primary difference between density treatments was the level of intraspecific competition.

Measurement of Life History and Reproductive Characters

Germination date and flowering date were recorded for each member of the F_1 generation; these dates were then categorized as "cohorts." The date on which the first seed in the garden germinated or began to flower was assigned a value of 1, and each subsequent day was represented by a successive integer. The effects of density on life history traits are treated in detail elsewhere (Mazer and Schick, 1991a,b) and will not be discussed here except as necessary. As each surviving plant flowered, one flower per plant was removed to measure the visible petal area (estimated as visible petal length $\times$ petal width for one petal/flower) and ovules/flower. A second flower was used to measure pollen grains/flower and the modal volume of the individual pollen grains produced by each flower. For a subsample of adults in each density, 30 seeds per plant were chosen at random and weighed to provide an estimate of mean individual seed mass.

In addition to these primary sexual characters, a number of "composite" traits were recorded to derive measures of gender and of floral display per unit investment in male and female gametophytes. These traits included pollen:ovule ratio, petal area per pollen grain, petal area per ovule, and standardized phenotypic gender (sensu Lloyd and Bawa, 1984). Phenotypic gender measures were based on the investment in pollen and ovules per flower relative to the population at each density or to the entire garden-wide population. "Density-specific" standardized phenotypic gender was defined for each individual in each density as

$$G_i = \text{phenotypic gender (female) of individual } i = \frac{d_i}{d_i + l_i E}$$

where d_i = the number of ovules/flower produced by individual i, l_i = the number of pollen grains/flower produced by individual i, and E = an equivalence factor that expresses pollen production in units equivalent to ovules. Phenotypic gender may vary between zero and one, with higher values indicating greater relative investment in female gamete production. Each density had its own value of E, which is defined as the total number of ovules per flower produced by the population within the density treatment divided by the total number of pollen grains per flower produced by the population in that density. This measure of E is a reliable equivalence factor if the number of flowers open per individual is independent of individual flower pollen:ovule ratios. E for the low-density treatment (E_{low}) equalled 6.0×10^{-5}; $E_{medium} = 5.7 \times 10^{-5}$; $E_{high} = 5.8 \times 10^{-5}$. "Garden-wide" phenotypic gender was defined in the same way as the density-specific measures, except that the equivalence factor was estimated using pollen and ovule production per flower for the entire garden ($E_{garden} = 5.9 \times 10^{-5}$). Because density-specific and garden-wide estimates of phenotypic gender are extremely similar, I focus on garden-wide phenotypic gender, which reflects an individual's relative investment in pollen and ovules in comparison to the garden mean.

Pollen grain number per flower and modal pollen grain volume were measured with an Elzone 180PC Particle Counter (Devlin, 1989). The modal value of pollen grain volume was used instead of the arithmetic mean because the frequency distribution of pollen grain size was sometimes skewed due to either an abundance of inviable pollen grains (very small grains one-quarter or less the volume of the normal pollen grains in a frequency distribution) or a large number of pollen grain "doublets" (cases where pollen grains adhered to one another, creating a second peak in the distribution). When inviable pollen grains were apparent, these values were deleted from pollen counts. The area under the "doublet" peak was doubled when pollen counts were evaluated in order to count each doublet as two grains.

Flowers were collected from the garden prior to anther dehiscence. The six anthers from each flower were then removed and stored in an open plastic microcentrifuge tube for approximately 1 week to dry, facilitating pollen dehiscence. Anthers and pollen were then washed into a glass vial containing a known quantity of a 2% NaCl solution (about 25 ml) and sonicated for 60 sec to promote further pollen release and separation. These samples were analyzed with the particle counter 18 hr later, following a second sonication to homogenize the solution. The number of pollen grains and the modal pollen grain size (measured in μm^3) in 0.5 ml samples of NaCl solution was measured. The total number of pollen grains per flower was estimated by using the mean pollen count from three 0.5-ml aliquots of each sample (the mean number of pollen grains in 0.5 ml was multiplied by twice the number of milliliters of solution in the vial). We measured pollen volume using the frequency distribution of the pollen provided by the same three aliquots.

Effects of Density on Mean Phenotype

To detect significant differences among densities with respect to progeny phenotype, we used PROC GLM (SAS Institute Inc., 1987) to partition phenotypic variance into components due to block, density, and the interaction between them. In these and the following analyses, to approach normality, petal area and ovule number were ln-transformed prior to analysis and the reciprocal of petal area/pollen grain was used. The other traits were not transformed because their frequency distributions were unimodal and approximately normal.

Parental Effects on Progeny Phenotype within and across Densities

The analysis of variance of the nested design described above detects significant maternal and paternal influences on progeny phenotype (Comstock and Robinson, 1948; Hallauer and Miranda, 1981; Becker, 1984). For a single population density within the experimental garden, the model for the analysis of variance is

$$Y_{ijk} = \mu + m_i + f_{ij} + p_k + e_{ijk},$$

where Y is the phenotype of the kth offspring of the ith pollen donor and the jth female, μ is the population mean, m_i is the effect of the ith pollen donor (or male), f_{ij} is the effect of the jth maternal plant nested within the ith donor, p_k is the block effect and e_{ijk} is the experimental error. Given a few assumptions (e.g., no dominance or epistasis for these quantitative traits and that differences among paternal families in the garden were not due to environmental or extranuclear genetic differences among pollen donors in the greenhouse; see Young and Stanton, 1990a), significant paternal effects on progeny phenotype for a given trait indicate significant additive genetic variance among the paternal (parental) genotypes for the observed trait. Significant maternal effects on progeny phenotype may be due to additive genetic, nonadditive genetic, or nongenetic maternal environmental effects. Block × Paternal Family and Block × Maternal Family (nested within Paternal Family) interactions were also estimated.

For each density treatment, PROC GLM was used to conduct three-way mixed model ANOVAs to partition variance into components due to block (fixed effect), paternal family (random effect), and maternal family (random). Type III sums of squares were used to detect effects of each class variable independently of the others. F-tests for paternal effects used the mean square of maternal family (nested within paternal family) as the error term; F-tests for block effects used the mean square of the Block ×

Paternal Family interaction term in the denominator. F-tests to detect significant Block $\times$ Paternal Family interactions used the Block $\times$ Maternal Family (Paternal Family) interaction term in the denominator. Analogous ANOVAs were also conducted on the entire garden population (all densities pooled) in order to determine whether parental effects on progeny phenotype within densities predicted the strength of parental effects on progeny phenotype across densities.

In general, inconsistent paternal effects among densities indicate statistically detectable levels of additive genetic variance relative to total phenotypic variance (a measure of narrow-sense heritability; Falconer, 1989), which also differ among densities. The data I had available in this study, however, presented several problems. First, because mortality rates differed greatly among densities, I could not conduct identical statistical tests (with respect to the number of degrees of freedom) for each density because final sample sizes differed markedly among densities. As a result, the comparison of ANOVAs conducted on each density's adult population is subject to the criticism that differences in the statistical results of these ANOVAs could be due either to the phenotypic consequences of growing plants in different densities (i.e., density may influence the magnitude of intergenotypic relative to total phenotypic variance) or to the effects of conducting F-tests with different numbers of degrees of freedom. The simple comparison of the statistical significance of paternal or maternal effects on progeny phenotype at different densities cannot identify which factor is causal. It is critical to note, however, that differences between densities in sample size are not statistical "artifacts" without biological meaning; they are due to a density-related biological process (differential mortality). In other words, although the effects of changes in intergenotypic variation and sample size are confounded among densities, both of these features are due to biological factors that may create differences among densities in the ability of ANOVAs to detect significant parental effects on progeny phenotype.

A second difficulty in the analysis of these data is the problem of significance testing when multiple ANOVAs are conducted. In this study, I conducted 44 ANOVAs [11 traits $\times$ (3 densities + pooled densities)] to detect significant parental influences on progeny phenotype and many additional ANOVAs to detect the effects of density and block on phenotype within and across families, and to detect significant Family $\times$ Density interactions. Due to repeated tests of significance, some of the main effects would be expected to be statistically significant simply due to chance. This is a perennial problem in the analysis of environmental and genetic influences on the phenotype of quantitative traits. Although I use a p-value of 0.05 as a criterion to report statistically significant results (except for the effects of density within families, for which I use $p < 0.01$), p-values should be closely observed when attempting to judge the biological or statistical significance of any individual ANOVA.

Family × Density Interactions for Floral Traits

Family × Density interactions were sought to determine whether the effect of parental family membership on progeny phenotype depends strongly on density and whether the effect of density on mean phenotype depends strongly on parental family for floral traits in wild radish (seed mass is treated by Mazer and Wolfe, unpublished manuscript). To detect Family × Density interactions, two-way ANOVAs were conducted for each trait, using the entire garden-wide data set. These ANOVAs partitioned phenotypic variance in each trait into components due to the effects of density, paternal or maternal family, and the interaction between them. Type III sums of squares (SS) were used to detect significant paternal or maternal (random), density (fixed), and interaction effects independently of the others. F-tests conducted to detect density effects used the mean square of the interaction as the error term. To provide an additional and less conservative test for significant Family × Density interactions, I performed an order-dependent ANOVA for each trait to extract the Type I SS for the interaction term, which was placed first in the model. Because the degrees of freedom in the numerator of the F-value associated with the Type I SS is much higher than for the Type III SS, the F-test associated with the Type I SS is more likely to be statistically significant.

Genetic Variation in Phenotypic Plasticity

To detect differences among paternal families in the plastic response to density, two-way ANOVAs were conducted on each paternal family to detect the effects of density on phenotype independently of the block effect.

Phenotypic and Genetic Integration of Floral Traits

The integration of floral traits among individuals and genotypes was examined at each density. Pearson correlation coefficients were estimated for the population of surviving adults within each density to establish phenotypic correlations among floral traits. Even though the garden plots provided relatively uniform conditions within densities, the variation in germination time that occurred within treatments created the potential for neighbor effects to generate heterogeneous environments within each treatment. Individuals that germinated early grew faster than, and had the opportunity to suppress the growth of, late-germinating individuals. Consequently, phenotypic correlations between traits were likely to be inflated relative to genetic correlations, with early-emerging individuals (or individuals in resource-rich microsites) tending to have large phenotypic values of most traits and late-emerging individuals (or those in resource-poor microsites) tending to have low phenotypic values. Such tendencies would result in positive phenotypic correlations that could mask any underlying negative genetic correlations. To provide qualitative estimates of the genetic correlations between traits expressed in each density, correlation coefficients were estimated using the mean value of each character for each paternal family. In contrast to the phenotypic correlations estimated in each density, the degrees of freedom for these genetic correlation estimates were constant across densities (all 15 paternal families were present in each density treatment). Again, due to sample size differences between these tests for phenotypic and genetically based correlations, the tests are not equivalent in power. Nevertheless, pairs of traits that do not exhibit significant correlation coefficients among individuals ("phenotypic" correlations), but that do exhibit significant correlation coefficients among family means ("genetic" correlations), can be interpreted as exhibiting genetically based correlations that are not due to microenvironmental variation.

Results

Density Effects on Survivorship

Survivorship was strongly dependent on density. Although nearly 90% of sown seeds germinated at each density, seedling mortality increased rapidly with increasing density. The low-density plots had 45.5% mortality (327 of the original 600 seeds survived); the medium-density plots suffered 68.3% mortality (190 survived); and the high-density plots suffered 79% mortality (125 survived). Time constraints prevented the sampling of many of the high-density individuals for floral traits because many of them flowered for only a few days. Although some families had higher survivorship than others, mortality differences among families were independent of density. A log-likelihood chi-square test indicated that differential mortality among families was not density-specific for paternal families ($\chi^2 = 27.93$; $p > 0.4681$; model df $= 28$) or for maternal families ($\chi^2 = 136.67$; $p > 0.1153$; model df $= 118$). Thus, phenotypic differences among density treatments were not due to density-specific selection against certain genotypes. Rather, environmental modification of phenotype was responsible for the observed differences among treatments in mean phenotype.

Range of Phenotypic Variation

All floral traits, composite traits, and phenotypic gender exhibited marked phenotypic variation within the experimental treatments of this study. In general, though the standard errors were largest at high density due to low sample size, the phenotypic range of individuals was less under high-density than under low-density conditions (Figs. 12.2 and 12.3). Ovule number varied by a factor of 6 among individuals grown at low density and by a factor of 2.7 among individuals grown at high density; pollen production per flower varied by a factor of 24.8 among low-density individuals and by a factor of 5.5 among high-density individuals; pollen grain volume varied by a factor of 2.3 and 1.5 in the low- and high-density plots, respectively. Petal area spanned a 4.8-fold range of variation in the low-density plots and a 3.5-fold range at high density. Pollen:ovule ratios varied by a factor of 52.7 in the low-density and by a factor of 6 in the high-density plots, whereas garden-wide phenotypic gender spanned a 3.4-fold range at low density and a 2.3-fold range at high density. In contrast, the range in the mean mass of individual seeds produced by garden-grown individuals was similar at all densities (6.6 at high density vs. 7.2 at low density).

Density Effects on Phenotype

Due to a higher potential for intraspecific competition for water and nutrients with increasing density, I predicted that increasing density would result in a decline in mean phenotype for floral traits reflecting investment in gametes and floral display. Furthermore, if environmental modification of gender in wild radish were the adaptive outcome of natural selection as predicted by sex allocation theory, I expected that phenotypic gender should become more male biased as stress increased.

In addition to higher rates of mortality, evidence of stress in the high-density treatment appeared as a significant delay in flowering time. This delay occurred even though germination in these plots occurred significantly earlier than in the medium- and low-density plots (see Mazer and Schick, 1991b). Moreover, the individuals in the high-density treatment that survived to flower were derived ultimately from seeds that germinated significantly earlier than the flowering survivors in the other treatments. Development time was simply increased in the high-density plots.

Phenotypic changes in the mean values of individual floral traits did appear to reflect a decline in resource availability as density increased, but not all traits responded equally (Figs. 12.2 and 12.3). There was a statistically significant effect of density on ovule number per flower; high-density plots contained plants with fewer ovules per flower than medium- or low-density plots. Mean seed mass also declined from low density to medium density; the standard error of the mean for mean seed mass in the high-density plots was extremely high due to the small sample size. Petal area and pollen production per flower also declined as density increased, although these changes were not statistically significant. Pollen grain volume was the most phenotypically constant trait. In general, ovule production

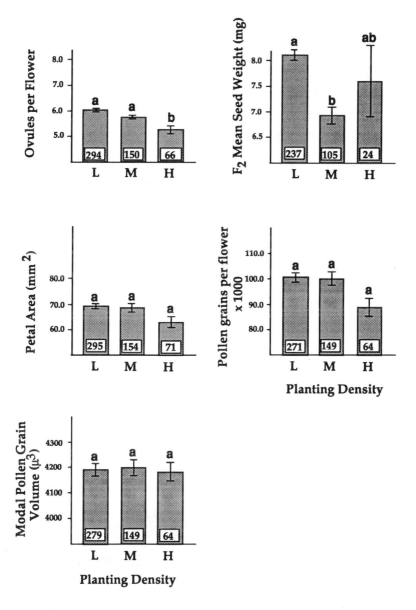

Figure 12.2. Effects of density on mean phenotype of floral traits in each of three planting densities. Bars indicate the mean phenotypic values; lines indicate the standard error of the mean for each trait. Sample sizes are given at the base of each bar. L, low density; M, medium density; H, high density. Within each graph, shared letters indicate that mean phenotypic values do not differ significantly between densities, as determined by Tukey's test following a two-way ANOVA (detecting the effects of block and density on mean phenotype).

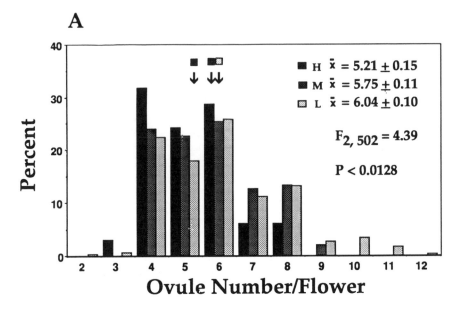

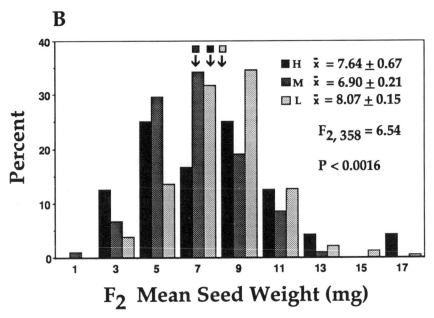

Figure 12.3. Frequency distributions by density for floral traits and gender. In each graph, the arrows situated below each small box indicate the mean phenotypic value associated with high-, medium-, or low-density treatments (mean values ± the standard error of the mean are indicated to the right). L, low density; M, medium density; H, high density. *F*-values summarize the results of the two-way ANOVAs conducted to detect significant effects of

C

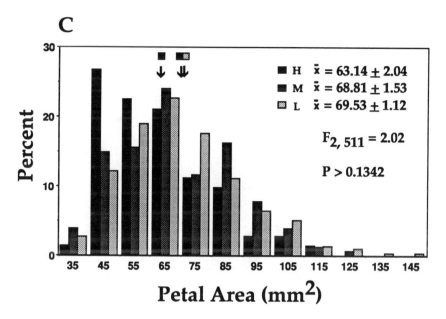

H x̄ = 63.14 ± 2.04
M x̄ = 68.81 ± 1.53
L x̄ = 69.53 ± 1.12

$F_{2, 511} = 2.02$

$P > 0.1342$

Petal Area (mm^2)

D

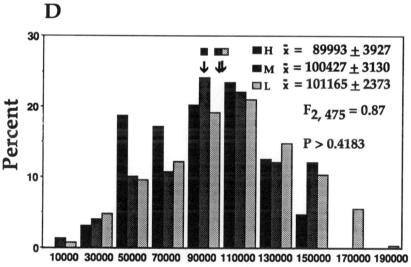

H x̄ = 89993 ± 3927
M x̄ = 100427 ± 3130
L x̄ = 101165 ± 2373

$F_{2, 475} = 0.87$

$P > 0.4183$

Pollen Grains/Flower

(*continued*) density on phenotype independent of the block effect. *p*-values refer to the level of significance associated with the density effect. (A) Ovule number/ flower. (B) Mean individual seed mass produced by adults in the experimental garden. (C) Petal area (estimated as the visible length × width of a petal emerging from a corolla). (D) Pollen grains produced per flower. (E) Modal volume of a single pollen grain. (F) Garden-wide standardized phenotypic gender (0 = pollen production only; 1 = ovule production only).

E

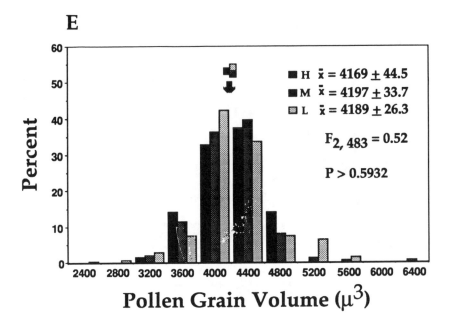

F

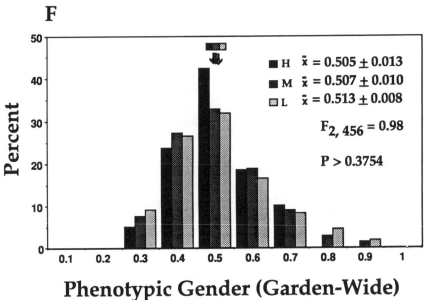

Figure 12.3. (*Continued*).

and seed mass were much more sensitive to planting density than was pollen production.

In spite of the response of ovule number to planting density, the composite traits that included ovule number did not show any significant phenotypic change with density. Pollen:ovule ratio, garden-wide phenotypic gender, and petal area/ovule showed no significant change with density [*pollen:ovule ratio:* $F_{2,456}$ (density effect, independent of block effect) = 0.50, $p > 0.6048$; high-density $\overline{X}$ = 18,006, SE = 941; medium-density $\overline{X}$ = 18,480, SE = 716; low-density $\overline{X}$ = 18,343, SE = 554; *petal area/ ovule:* $F_{2,507}$ = 0.66, $p > 0.7054$; high-density $\overline{X}$ = 12.81, SE = 0.54; medium-density $\overline{X}$ = 12.64, SE = 0.37; low-density $\overline{X}$ = 12.40, SE = 0.26]. It is noteworthy, however, that the highest values of phenotypic gender (the most female phenotypes: $G > 0.75$) were represented only by the medium- and low-density plots (Fig. 12.3). Finally, although there was a slight tendency for the low-density plots to exhibit more petal area/pollen grain than the higher density plots, this difference in "advertisement" was not significant [$F_{2,460}$ = 0.41, $p > 0.6637$; high-density $\overline{X}$ (mm²/10,000 grains) = 8.08, SE = 0.56; medium-density $\overline{X}$ = 8.40, SE = 0.46; low-density $\overline{X}$ = 8.92, SE = 0.56].

Parental Influences on Progeny Phenotype within and across Densities

The ability of the ANOVAs to detect significant variation among paternal families differed among densities and traits. Density-specific ANOVAs for individual life history and floral traits are reported by Mazer and Schick (1991b), but the significant results of these traits are summarized here. The ANOVAs reported here include the two-way interaction terms, which were omitted from the analyses of Mazer and Schick (1991b) because they were almost never statistically significant in the single-density ANOVAs. Across densities, however, these interactions were commonly statistically significant, so for consistency I provide ANOVAs with these two-way interaction terms included. In one case, a maternal effect reported as significant by Mazer and Schick (1991b) was not detected in the ANOVA that included the interaction terms (germination date at low density). In another case, Mazer and Schick (1991b) did not detect a maternal effect that was significant when the interaction terms were included (flowering date at medium density).

In low-density plots, there were significant paternal effects on flowering date and modal pollen grain volume, independent of the block effect and of maternal effects nested within sires (Appendix). In medium-density plots, there were significant paternal effects on germination date, flowering date, and number of ovules per flower. In high-density plots, there were significant paternal effects only on pollen grain volume. Phenotypic gender exhibited no strong paternal effects at any density. Maternal effects also differed among density treatments, but in a pattern distinct from that of the paternal effects (Appendix). Maternal effects on progeny phenotype were significant for petal area and pollen:ovule ratios at low density. At medium and high density, no maternal effects on progeny phenotype were statistically significant for any of the floral traits, but flowering date was subject to maternal effects at medium and high density.

The discrepancy between paternal and maternal effects may in part be explained by the fact that the individuals in this breeding design (unlike diallel designs commonly used by evolutionary ecologists) did not serve both as maternal plants and as pollen donors. That is, paternal effects reflect additive genetic variance among the 15 pollen donors, whereas maternal effects nested within paternal families reflect additive genetic, nonadditive genetic, and maternal environmental effects on progeny phenotype. In addition, the degrees of freedom associated with the F-value used to test for significant paternal effects were fewer than those used to detect maternal effects.

Parental effects detected by ANOVAs performed on the pooled populations from all densities were not always consistent with density-specific parental effects (Appendix and Table 12.1). Significant paternal effects were detected for flowering time and pollen grains/flower; significant ma-

Table 12.1. Summary of Parental Influences on Progeny Phenotype within and across Densities[a]

	Garden-wide		Density-specific	
	Paternal effects	Maternal effects	Paternal effects	Maternal effects
Germination date	−	+	+	+
Flowering date	+	+	+	+
Petal area	−	+	−	+
Pollen grain number	+	−	−	−
Ovule number	−	+	+	−
Pollen volume	−	−	+	−
Petal area/ovule	−	+	−	−
Pollen:ovule ratio	−	+	−	+
Garden-wide phenotypic gender	−	+	−	−

[a]Plus signs indicate that paternal or maternal effects were significant in a three-way nested ANOVA (Tables 12.A and 12.B, Appendix). A plus sign in the "density-specific" columns indicates that paternal or maternal effects on progeny phenotype for the specified trait were significant in at least one density. See Appendix for details concerning significance levels and the particular densities in which density-specific parental effects were detected.

ternal effects were detected for germination date, flowering date, petal area, ovules/flower, petal area/ovule, pollen:ovule ratio, density-specific phenotypic gender, and garden-wide phenotypic gender. Of the two floral traits that exhibited detectable levels of additive genetic variance within at least one density (ovule number and pollen volume), neither of them exhibited significant V_A across densities. In contrast, although paternal influences on pollen production per flower were not detectable within densities, this effect was significant across densities.

The comparison of the ANOVAs conducted on the combined data (all densities pooled) with those conducted on each density separately presents a problem that is similar to that of comparing the ANOVAs of different densities. It should generally be easier to detect significant parental effects on progeny phenotype using the pooled data because the degrees of freedom are so much higher than in any of the component data sets. Consequently, cases in which the density-specific ANOVAs detect significant parental effects, but for which the pooled data ANOVAs do not detect them, suggest that the genotypes are truly behaving differently within vs. across densities. For example, there are significant paternal effects on progeny ovule number at medium density ($F = 3.24$, $p < 0.0025$), but not for the pooled data ($F = 1.33$, $p > 0.2200$). A difference in sample size is not the cause of the difference between these ANOVAs in the significance of this paternal effect; rather, the magnitude of intergenotypic relative to total phenotypic variance is diminished in the pooled densities due to the high level of variation associated with density (which inflates the denominator of the F-value).

Family × Density Interactions for Floral Traits

Using the Type III sum of squares, a significant Paternal Family × Density interaction was detected only for petal area and petal area/ovule (Table 12.2). It is surprising that no significant family × density interaction was detected for germination cohort, flowering cohort, ovule number, or pollen grain volume, given that the statistical effect of paternal family on these characters varied among density treatments. No significant Maternal Family × Density interaction term appeared for any of the floral traits. The absence of significant family × density interactions in these ANOVAs suggests that the effect of density on phenotype does not truly differ among paternal or maternal families for most traits. This result implies that the differences between ANOVAs performed on different data sets are due primarily to sample size differences and not to the effects of density on intergenotypic variation.

Using the Type I sum of squares, however, significant Paternal Family × Density interactions were tested for germination cohort ($F_{44,1610} = 3.26$,

Table 12.2. Summary of Two-Way ANOVAs That Detected Significant Interactions between Paternal Family and Density Treatment[a]

Source	df	MS	F	p	r^2
Ln (Petal Area)					
Paternal family (P)	14	0.1509	2.16	0.0062	
Density (D)	2	0.3592	2.92	0.0710	
D × P	27	0.1229	1.76	0.0115	
Model	43	0.1453	2.08	0.0001	0.16
Error	476	0.0699			
Total	519				
Petal Area/Ovule					
Paternal family (P)	14	45.77	2.2	0.0027	
Density (D)	2	0.13	0	0.9961	
D × P	27	33.41	1.77	0.0107	
Model	43	32.34	1.71	0.0043	0.14
Error	464	18.88			
Total	507				

[a]Type III mean squares are reported for each of the main effects and interaction term. Paternal family is considered to be a random effect, whereas density is considered to be a fixed effect. F-tests to determine statistical significance of the density effect used the mean square for the interaction term in the denominator. The coefficient of determination, r^2, is the proportion of the total sum of squares accounted for by the model sum of squares. For all other traits, there were no significant Paternal Famiy × Density or Maternal Family × Density interaction terms.

$p < 0.0001$), flowering cohort ($F_{44,597} = 3.04, p < 0.0001$), ovule number ($F_{43,466} = 2.06, p < 0.0002$), petal area ($F_{43,476} = 2.08, p < 0.0001$), petal area/ovule ($F_{43,464} = 1.71, p < 0.0043$), pollen grain number ($F_{43,440} = 1.52, p < 0.0214$), and pollen grain volume ($F_{43,448} = 1.56, p < 0.0151$). Significant Maternal Family × Density interactions were detected for germination cohort ($F_{179,1475} = 1.75, p < 0.0001$), flowering cohort ($F_{160,641} = 1.94, p < 0.0001$), ovule number ($F_{145,364} = 1.45, p < 0.0028$), petal area ($F_{148,371} = 1.71, p < 0.0001$), and petal/area ovule ($F_{145,362} = 1.64, p < 0.0001$). Given the density-specific strength of parental effects on petal area, petal area/ovule, pollen grain volume, pollen:ovule ratio, and ovule number, these results suggest that the phenotypic response to density is not the same for all maternal or paternal families.

Genetic Variation in Phenotypic Plasticity

The family-specific effects of density on phenotype also suggest that there exists a low level of genetic variation in phenotypic plasticity. Differences among paternal families in their plastic response to density were detected in germination cohort, flowering cohort, petal area, ovules/flower, and petal area/ovule (Table 12.3 and Fig. 12.4). Using a significance level of $p < 0.01$, nine paternal families exhibited significant differences among densities for germination cohort, whereas six families showed no such differences. Two families exhibited significant delays in flowering time as density increased, but 13 families did not show a statistically significant delay (Fig. 12.4A). Two families showed significant changes in petal area with density, and the responses were in different directions (Fig. 12.4B and C). Only one family showed a significant phenotypic response to density for ovule number and petal area/ovule (a different family for each trait). Pollen production, pollen:ovule ratio, pollen size, and garden-wide phenotypic gender did not differ among densities within any of the paternal families (Fig. 12.4D). Norms of reaction for all of these traits did frequently cross, but the absence of significant Type III Paternal Family × Density interaction terms (except for petal area and petal area/ovule) suggests that these changes in phenotypic rank do not account for much of the total phenotypic variance in most floral traits.

Table 12.3.   Density Effects on Phenotype within Paternal Families[a]

Paternal family	Germi-nation cohort	Flower-ing cohort	Petal area	Ovules per flower	Petal area/ ovule	Pollen grains/ flower	Pollen volume	Garden-wide phenotypic gender
1	*	ns	ns	ns	ns	ns	ns	ns
2	*	ns	ns	ns	ns	ns	ns	ns
3	****	*	ns	ns	ns	ns	ns	ns
4	***	ns	**	*	ns	ns	ns	ns
5	*	ns	ns	ns	ns	ns	ns	ns
6	ns	ns	ns	***	*	ns	ns	ns
7	ns	ns	ns	ns	ns	ns	ns	ns
8	ns	**	ns	ns	ns	ns	ns	ns
9	**	ns	ns	ns	ns	ns	ns	ns
10	****	ns	ns	ns	ns	ns	ns	ns
11	**	ns	ns	ns	ns	ns	ns	ns
12	***	****	ns	ns	ns	ns	ns	ns
13	**	ns	**	ns	**	ns	ns	ns
14	**	ns	ns	ns	ns	ns	ns	ns
15	**	ns	ns	ns	ns	ns	ns	ns

[a]This table summarizes cases in which a two-way ANOVA conducted on each paternal sibship detected significant effects of density on phenotype independent of the block effect. *$p < 0.05$; **$p < 0.01$; ***$p < 0.001$; ****$p < 0.0001$.

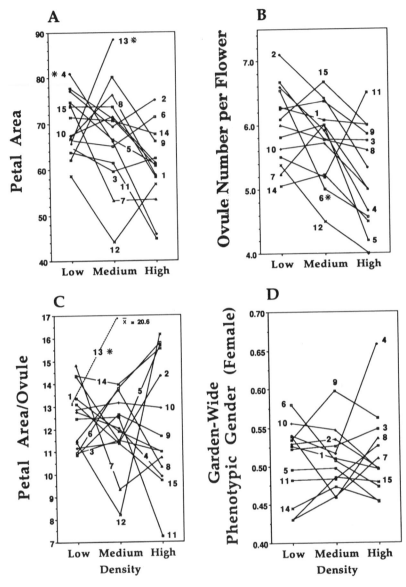

Figure 12.4. Norms of reaction for each of the 15 paternal families represented at low, medium, and high density. Asterisks indicate those families in which a significant effect of density on phenotype (independent of the block effect) was detected by a two-way ANOVA. (A) Petal area. (B) Ovule number/flower. (C) Petal area/ovule (the dotted line indicated for paternal family #13 should extend to a value of 20.6 at medium density). (D) Garden-wide standardized phenotypic gender.

Density Effects on Phenotypic and Genetic Integration of Floral Traits

The magnitude and pattern of phenotypic correlations differed among densities (Table 12.4). For example, the phenotypic relationships between allocation to petals and allocation to pollen and ovule production varied among densities. Petal area was significantly positively correlated with ovule number per flower only in the low-density treatment, but was positively correlated with pollen production in low- and medium-density plots. Petal area was independent of male and female gamete production at high density, with correlation coefficients very close to zero. A positive relationship between petal area and pollen volume was significant only at medium density.

Only the low-density plot exhibited any strong evidence for a significant inverse relationship between male and female function; individuals that

Table 12.4. Pearson Product–Moment Correlation Coefficients for Floral and Gender Traits in Garden-Grown Wild Radish Representing Each of Three Density Treatments[a]

Density	Ovules/ flower	Pollen/ flower	Pollen volume	Pollen: ovule ratio	Density- specific gender (female)
Low					
Petal area	0.12*	0.22***	0.06	0.09	−0.10
Ovules/flower		−0.08	−0.17**	−0.64****	0.60****
Pollen grains/flower			0.09	0.78****	−0.82****
Pollen grain volume				0.14*	−0.18**
Pollen:ovule ratio					−0.95***
Medium					
Petal area	0.08	0.17*	0.18*	0.13	−0.15
Ovules/flower		0.01	0.02	−0.54****	0.60****
Pollen grains/flower			−0.02	0.80****	−0.83****
Pollen grain volume				−0.02	−0.02
Pollen:ovule ratio					−0.95****
High					
Petal area	0.04	0.06	0.09	0.06	−0.04
Ovules/flower		0.18	−0.22	−0.49****	0.43***
Pollen grains/flower			−0.06	0.76****	−0.79****
Pollen grain volume				−0.03	−0.04
Pollen:ovule ratio					−0.97****

[a]*$p < 0.05$; **$p < 0.01$; ***$p < 0.001$; ****$p < 0.0001$. Sample sizes are approximately 260–300 for low density, 140–150 for medium density, and 60–70 for high density.

produced large pollen grains produced relatively few ovules per flower (though the negative correlation coefficient at high density is also suggestive). Under none of the conditions observed here was there evidence for a trade-off between pollen production and ovule production. There was also no evidence for a phenotypic trade-off between pollen size and pollen production/flower in any of the treatments.

Among families, however, there was evidence for resource limitation reflected in a trade-off between pollen production and pollen size. At medium-density, pollen grain volume and pollen production per flower were negatively correlated among paternal families [$r(14) = -0.53$; $p < 0.05$]; the phenotypic correlation between these traits, however, was not significant. In the high-density and low-density plots, no significant correlations among family means appeared between any floral traits.

Discussion

Does Planting Density Influence Individual Survivorship and the Mean Phenotype of a Population with Respect to Primary and Secondary Sex Characters?

One of the most striking results of this study is the fact that some reproductive traits are much more buffered against the effects of density than others. The most drastically affected trait was the probability of survival to reproductive age; the level of mortality at high density was much higher than at low density. The relative abundance of different paternal families, however, remained constant across densities, so densities did not provide family-specific selection regimes. Another clear result was that traits associated with female reproductive success were much more sensitive to density than those associated with male reproductive success. Ovule number and seed mass declined significantly with density, whereas pollen grain size and pollen production changed little with increasing density. The observation that phenotypic gender remains virtually unchanged across densities is particularly noteworthy and contrasts strongly with the response to density observed in many monoecious and gynodioecious species. This result raises the question: might hermaphroditic species in general exhibit much greater constancy in phenotypic gender than monoecious and subdioecious species? The high degree of canalization of phenotypic gender at the level of individual flowers suggests that this trait may experience strong natural selection for phenotypic constancy in heterogeneous environments. The ecological and evolutionary significance of phenotypic or genetic variation in gender have yet to be explored in hermaphroditic species.

Do Parental Influences on Progeny Phenotype Depend on the Environmental Conditions in Which Progeny Are Grown?

To determine whether the ability to detect parental influences on progeny phenotype depends upon initial planting density, a legitimate but tricky first step is to compare the results of the ANOVAs conducted on data provided by an initially identical number of progeny raised under different densities. This is a precarious step because differences between ANOVAs associated with distinct densities (and mortality rates) may be due to differences between densities in the magnitude of intergenotypic relative to total phenotypic variance (which influence the F-value), or differences in the number of degrees of freedom available for significance testing. This type of comparison in this study, however, does illustrate one disturbing result: the initial density that an experimental ecologist uses to raise plants will have a strong influence on the ability to detect maternal or paternal influences or quantitative life history and floral traits when using the most common statistical tool: analysis of variance. The issue of whether the observed differences among densities in the strength of parental effects are due to phenotypic plasticity or to a reduction in available degrees of freedom is an important but separate question.

In some instances it is clear that factors other than a change in sample size cause differences among densities in the strength of parental effects on progeny phenotype. In this study, given similar F-values, the likelihood that an ANOVA will detect a statistically significant parental effect on progeny phenotype declines with increasing density. This is due to the fact that the degrees of freedom available in the denominator of all F-tests are much greater for the low-density data than for the high-density data. So, if the low-density ANOVA detects a statistically significant parental effect on progeny phenotype, but the high-density ANOVA does not (given similar F-values), this difference in significance would not necessarily imply *biologically* meaningful differences among densities in the magnitude of intergenotypic relative to total phenotypic variance. The difference could simply reflect an F-value that was statistically significant when associated with a high number of degrees of freedom (as in the low-density data) but not with a low number (as in the medium- or high-density data). On the other hand, if the high-density or medium-density ANOVA detects a statistically significant parental effect on progeny phenotype where the low-density treatment does not, this difference probably is not due to sample size, but rather to the effects of density on intergenotypic variance. In this study, ovule number per flower shows this pattern.

Can Gender or Environmental Modification of Gender Evolve in Wild Radish?

All floral and gender traits exhibited marked phenotypic variation within the relatively uniform plots of my experimental treatments, but most of this variation could not be attributed to genetic lineage consistently. Consequently, much of the phenotypic variation does not represent variation on which natural selection can operate to effect evolutionary change in this population. Nevertheless, certain traits expressed at certain densities did exhibit strong parental influences on progeny phenotypes. In the case of highly significant paternal effects (e.g., pollen volume at low and high density and ovule number per flower at medium density), these results predict that natural selection on these traits could result in an evolutionary response on a local geographical scale. If the density-specific parental effects on progeny phenotype observed in this study truly reflect changes in intergenotypic variance with density, then the potential for natural selection to act in field populations of wild radish will depend on the degree of local population structure and gene flow. It is important to note that the range of densities used in this study does not represent the full range of environments encountered by wild radish in nature, nor does the range of genotypes observed span the probable range of genetic variation available in field populations. Accordingly, the results of this study provide minimum estimates of heritable variation in floral traits, the degree to which local environmental conditions influence the magnitude of expressed phenotypic differences among paternal or maternal families, genetic variation in the phenotypic response to density, and the importance of genotype by environment interactions in wild populations of *R. sativus*.

Even given these limitations, ovule number per flower did exhibit significant heritable variation (at medium density), and there were differences among paternal families with respect to the plastic response of ovule number to density. Because ovule production is a major component of phenotypic gender in wild radish, it is possible that the evolution of phenotypic gender can be affected indirectly by natural selection on ovule number and by selection on the phenotypic response to density exhibited by ovule number. Note, however, that because the detectable genetic component to variation in floral traits and the level of phenotypic integration between them (sensu Schlichting, 1989a,b) varied among densities, the degree to which selection on one trait will result in correlated phenotypic and evolutionary responses by other traits may also depend on local environmental conditions.

Several results suggest that natural selection on gender in wild radish may be constrained in natural populations. Most importantly, the inability to detect paternal effects on phenotypic gender either within or across densities in this study indicates that gender is not a highly heritable trait, at least among the 15 pollen donors represented in the greenhouse generation. On the other hand, the strong maternal effect on phenotypic gender across densities suggests that there may be additive genetic, nonadditive genetic, or maternal environmental influences that were not detectable among the paternal families. The evolution of gender *modification* in response to local environmental conditions requires also that there be genetic variation in phenotypic plasticity associated with sex expression; the current study suggests that this type of genetic variation may be relatively rare in wild radish. The absence of genetic variation in a trait, of course, is not evidence that natural selection is irrelevant to its evolution; it may simply mean that natural selection in the past has been so strong as to purge the population of all available genetic variation in the plastic response of sex expression to density.

Application of Sex Allocation Theory to Wild Radish

Because individual fitness was not estimated in this wild radish population, this study does not permit one to conclude whether the observed changes in reproductive allocation in response to density are an adaptive response to a decline in resources as predicted by sex allocation theory or, alternatively, the proximate result of a decline in some resource critical to ovule, but not to pollen, production. Also, because gender allocation was measured at the individual flower level, it is not possible to say with certainty that genets as a whole changed their phenotypic gender with density. Stanton and Preston (1986, 1988) reported, however, that there is greater variation among, rather than within, individuals of *R. sativus* with respect to flower size and pollen production. This suggests that the phenotypic gender of individual flowers is a reasonable predictor of phenotypic gender at the level of entire plants. Finally, this study reveals little about changes in *functional* gender (actual contributions to the next generation through pollen and through seeds) with density, which is the trait that is the real focus of sex allocation theory.

A convincing demonstration that the phenotypic changes in this radish population reflect an adaptive response to density would require evidence that the expected fitness of a male-biased phenotype in high-density conditions is actually higher than that of a more female-biased phenotype. Still, if the resource-based predictions of sex allocation theory apply here, then one might infer that wild radish is responding in an adaptive way and that the relative cost of male and female gamete *production* is the major evolutionary factor that favors the maintenance of pollen production at the expense of ovule production in low-resource environments. Fundamental to this explanation, however, is the assumption that there is a necessary trade-off between pollen and ovule production, a pattern not found in this study (nor in *Lobelia cardinalis*: Devlin, 1989). The observation of no negative correlation in this study suggests that a perspective concerning "optimal" allocation of resources to male and female function should not be purely resource based.

Adaptive Environmental Gender Modification: An Alternative Hypothesis

The environmental modification of floral traits seen in wild radish is consistent with (although weaker than) empirical results in many monoecious and dioecious species: increased stress is associated with an increase in the proportion of biomass representing male relative to female reproductive structures. There are some exceptions to this pattern in other species, which suggests that the resource-based hypothesis is incomplete. For example, Aizen and Kenigsten (1990) report a counterintuitive result: floral sex ratios of tall stems (number of male inflorescences:number of female flowers) in *Quercus ilicifolia*, a monoecious scrub oak, became less male biased as stress increased along a topographic gradient (lower elevations are associated with a relatively stressful short growing season subject to a high frequency of fatal frosts). In *Ambrosia artemisiifolia* (a monoecious annual herb), larger plants in the greenhouse and taller plants in the field and greenhouse were relatively more male than smaller plants, an unexpected result if larger plants command more resources than smaller plants (McKone and Tonkyn, 1986; see also Abul-Fatih et al., 1979 for analogous data on *Ambrosia trifida* and Burd and Allen, 1988 for data concerning teosinte). In *Acalypha rhomboidea* and *Pilea pumila*, highly female-biased floral ratios were maintained under a variety of experimental conditions, although low soil moisture did result in a small increase in functional maleness in *Acalypha* due to a reduction in seed production, and increased plant size was associated with a greater increase in female than male reproductive function (Cid-Benevento, 1987). Cid-Benevento (1987) observed that in annual species for which transmission of gametes to the next generation through pollen depends strongly on wind velocity and the distribution of conspecifics, the fitness advantages associated with increased investment in pollen may be highly unpredictable. In such cases, it may not be advantageous to maintain or to increase male reproductive investment in low-resource environments, even if seed production is highly costly.

I propose that, in addition to the absolute resource cost of male vs. female gamete production in terms of water or nutrients, an additional factor that will have a strong effect on the fitness gain associated with increased investment in pollen or ovules is the dispersibility of gametes. This hypothesis may be stated as follows: In stressful environments (i.e., those in which there is a high risk of seed abortion or seedling mortality for any reason, including low resource availability), flowering genotypes that preferentially invest in the gametic type with the higher probability of dispersing to a favorable environment will be favored by natural selection over those that do not. In this context, the degree to which an environment is "favorable" depends on its effect on the pollen or seed that disperses to it, or on the fertilized ovule and seed that result following pollen dispersal and germination. For natural selection to operate in a deterministic way, there must be some predictable relationship between dispersal distance (or the probability of dispersal) and the quality of the environment experienced by the dispersed seed or pollen grain (and the ovule it fertilizes).

Under the simplest conditions, this hypothesis suggests several testable predictions that do not appear in the resource-based considerations of "optimal" sex expression in resource-poor environments. First, the gamete dispersal hypothesis would predict that (all other things being equal) species with seeds of low dispersibility should show a stronger phenotypic response to increased stress, preferentially investing in pollen production, than species with seeds of high dispersibility. In other words, if the ability of individuals to disperse progeny as seeds to potentially better environments is severely limited, then fitness gains through the dispersal of pollen may favor an increase in pollen production, regardless of the relative absolute costs of producing pollen and seeds ("If you can't thrive where you are, then run"). Second, species with reliable, long-distance pollen dispersal (defined as distances greater than the average radius of a high-stress "patch") should show increased allocation to pollen in high-stress environments relative to species that typically disperse pollen to shorter distances ("If you have a means to escape easily, then do it and do it often"). Extending this prediction to the extreme generates a third prediction: relative to obligately outcrossing species, obligately autogamous species should not preferentially invest in pollen production in high-stress environments, because their male gametes simply cannot disperse ("If you can't escape, then don't bother to change").

These predictions provide some clear alternatives to the resource-based hypothesis concerning the allocation of resources to male and female gametes. Under some conditions (e.g., the third prediction), even if male gametes cost very little to produce, one would not always predict that a reduction in resources would favor genotypes that invest preferentially in pollen. Similarly, differences between species in the dispersibility of their seeds or pollen may result in the evolution of different phenotypic response to a decline in resources, even if the relative and absolute costs of gamete production are identical between the species. On a practical level, it may be very difficult to assign species to alternative classes and to assess the evolutionary benefits of pollen dispersal relative to seed dispersal. The magnitude of seed and pollen dispersal distances relative to patch size will be a critical determinant of the possibility of escape to a better environment. Moreover, the probability of environmental deterioration between generations will determine the evolutionary gain of dispersal. Dispersibility in time as well as space must also be considered as an important escape mechanism available to seeds but not to pollen. Consequently, the evolution of seed dormancy might be a more reliable mechanism of coping with local environmental stress than the evolution of resource reallocation to maximize dispersal to a better environment. At the very least, a test of each of these predictions will require observations of characteristic norms of reaction for pollen, ovule, and seed production and gender across an array of species representing each class.

Appendix

Table 12.A. Three-Way ANOVAs for Garden-Grown Progeny Representing Each of Three Planting Densities[a]

Source	df	MS	F	p	r^2
Low Density: Germination Cohort					
Block (B)	2	8.28	5.12	0.0128	
Paternal parent (P)	14	4.63	1.86	0.0577	
Maternal parent (Paternal parent) (M[P])	46	2.49	1.27	0.1179	
B × P	28	1.62	0.89	0.6191	
B × (M[P])	81	1.82	0.92	0.6596	
Model	171	2.41	1.23	0.0494	0.35
Error	392	1.96			
Total	563				
Low Density: Flowering Cohort					
Block (B)	2	1094.95	23.02	0.0001	
Paternal parent (P)	14	208.56	2.09	0.0321	
Maternal parent (Paternal parent) (M[P])	44	99.92	1.05	0.3993	
B × P	26	47.56	0.41	0.9913	
B × (M[P])	50	114.73	1.21	0.1873	
Model	136	130.81	1.37	0.0216	0.50
Error	190	95.18			
Total	326				
Low Density: Ln (Petal Area)					
Block (B)	2	0.1133	1.97	0.1605	
Paternal parent (P)	14	0.0982	1.12	0.3653	
Maternal parent (Paternal parent) (M[P])	44	0.0873	1.51	0.0338	
B × P	25	0.0575	0.82	0.6943	
B × (M[P])	43	0.0699	1.21	0.1994	
Model	128	0.0950	1.64	0.0013	0.56
Error	166	0.0578			
Total	294				

Table 12.A.   (Continued)

Source	df	MS	F	p	r^2
Low Density: Pollen Grain Volume					
Block (B)	2	312,205	1.70	0.2032	
Paternal parent (P)	14	483,434	2.42	0.0130	
Maternal parent (Paternal parent) (M[P])	44	199,682	1.12	0.3087	
B × P	25	183,668	1.06	0.4307	
B × (M[P])	39	174,019	0.97	0.5242	
Model	124	209,476	1.17	0.1769	0.49
Error	154	179,028			
Total	278				
Low Density: Petal Area/Ovule					
Block (B)	2	0.40	0.02	0.9765	
Paternal parent (P)	14	16.36	0.64	0.8154	
Maternal parent (Paternal parent) (M[P])	44	25.51	1.44	0.0536	
B × P	25	16.65	1.05	0.4320	
B × (M[P])	41	15.82	0.89	0.6580	
Model	126	23.45	1.32	0.0460	0.50
Error	166	17.73			
Total	292				
Low Density: Pollen:Ovule					
Block (B)	2	1180×10^5	1.36	0.2754	
Paternal parent (P)	14	1129×10^5	1.19	0.3169	
Maternal parent (Paternal parent) (M[P])	44	949×10^5	1.56	0.0280	
B × P	25	869×10^5	0.90	0.5986	
B × (M[P])	36	961×10^5	1.58	0.0328	
Model	121	1028×10^5	1.69	0.0015	0.50
Error	140	610×10^5			
Total	261				
Medium Density: Germination Cohort					
Block (B)	2	385.67	57.82	0.0001	
Paternal parent (P)	14	15.74	2.33	0.0160	
Maternal parent (Paternal parent) (M[P])	45	6.74	1.29	0.1073	
B × P	28	6.67	1.32	0.1721	
B × (M[P])	76	5.06	0.97	0.5544	
Model	165	11.95	2.29	0.0001	0.50
Error	371	5.22			
Total	536				

Table 12.A. (Continued)

Source	df	MS	F	p	r^2
Medium Density: Flowering Cohort					
Block (B)	2	812.80	6.89	0.0043	
Paternal parent (P)	14	286.11	2.13	0.0307	
Maternal parent (Paternal parent) (M[P])	40	134.02	1.75	0.0162	
B × P	24	118.00	1.04	0.4558	
B × (M[P])	27	113.22	1.48	0.0901	
Model	107	173.95	2.28	0.0001	0.75
Error	82	76.38			
Total	189				
Medium Density: Ln (Ovules/Flower)					
Block (B)	2	0.0737	1.13	0.3402	
Paternal parent (P)	14	0.1088	3.24	0.0024	
Maternal parent (Paternal parent (M[P])	36	0.0335	0.60	0.9495	
B × P	22	0.0651	1.40	0.2702	
B × (M[P])	13	0.0465	0.83	0.6282	
Model	87	0.0618	1.80	0.3490	0.35
Error	62	0.0562			
Total	149				
Medium Density: Petal Area/Ovule					
Block (B)	2	12.70	1.32	0.3085	
Paternal parent (P)	14	36.41	1.95	0.0546	
Maternal parent (Paternal parent) (M[P])	35	18.67	1.08	0.3919	
B × P	22	17.10	1.32	0.3085	
B × (M[P])	13	12.97	0.75	0.7089	
Model	87	22.22	1.28	0.1526	0.65
Error	61	17.33			
Total	148				
High Density: Germination Cohort					
Block (B)	2	0.39	0.29	0.7500	
Paternal parent (P)	14	2.10	1.39	0.1964	
Maternal parent (Paternal parent) (M[P])	45	1.51	0.71	0.9212	
B × P	28	1.33	0.76	0.7919	
B × (M[P])	83	1.75	0.83	0.8541	
Model	172	1.85	0.87	0.8523	0.28
Error	381	2.12			
Total	553				

Table 12.A. (*Continued*)

Source	df	MS	F	p	r²
High Density: Flowering Cohort					
Block (B)	2	207.94	2.79	0.1008	
Paternal parent (P)	14	231.54	0.94	0.5301	
Maternal parent (Paternal parent) (M[P])	33	246.52	1.95	0.0145	
B × P	12	74.40	1.23	0.3780	
B × (M[P])	10	60.62	0.48	0.8954	
Model	71	198.59	1.57	0.0428	0.68
Error	53	126.23			
Total	124				
High Density: Pollen Grain Volume					
Block (B)	2	10,666	29.73	0.0001	
Paternal parent (P)	13	191,295	2.67	0.0255	
Maternal parent (Paternal parent) (M[P])	20	71,665	0.46	0.9520	
B × P	5	177,895	1.42	0.1107	
B × (M[P])	3	48,013	0.31	0.8194	
Model	43	113,282	0.73	0.8130	0.61
Error	20	155,916			
Total	63				

[a]ANOVAs were conducted to detect effects of paternal parent, maternal parent nested within paternal parent, block, block × paternal parent interactions, and block × maternal parent (paternal parent) interactions on progeny phenotype for life history and floral traits. Only those ANOVAs that detected significant paternal or maternal effects on progeny phenotype within densities or across densities (as in Table 12.B) are reported. Type III mean squares are reported for each main effect and interaction term; these mean squares are determined for each effect when it is placed last into the model. The coefficient of determination, r^2, indicates the proportion of the total sum of squares represented by the model sum of squares. F-tests for the paternal effect use the Type III Mean Square for the maternal parent nested within paternal parent in the denominator; F-tests for the block effect use the Type III Mean Square for the Block × Paternal parent interaction in the denominator; F-tests for the Block × Paternal parent interaction use the Type III Mean Square for the Block × Maternal parent (Paternal parent) in the denominator.

Table 12.B. Three-Way ANOVAs to Detect Effects of Paternal Parent, Maternal Parent Nested within Paternal Parent, Block, Block × Paternal Parent Interactions, and Block × Maternal Parent (Paternal Parent) Interactions on Progeny Phenotype for Life History and Floral Traits[a]

Source	df	MS	F	p	r^2
Germination Cohort					
Block (B)	2	143.19	29.73	0.0001	
Paternal parent (P)	14	8.89	1.40	0.1902	
Maternal parent (Paternal parent) (M[P])	46	6.33	1.74	0.0017	
B × P	28	4.82	1.42	0.1107	
B × (M[P])	90	3.40	0.93	0.6549	
Model	180	6.56	1.80	0.0001	0.18
Error	1474	3.64			
Total	1654				
Flowering Cohort					
Block (B)	2	484.86	3.78	0.0351	
Paternal parent (P)	14	520.28	2.82	0.0041	
Maternal parent (Paternal parent) (M[P])	46	184.45	1.68	0.0048	
B × P	28	128.17	0.96	0.5295	
B × (M[P])	83	133.27	1.21	0.1158	
Model	173	183.96	1.67	0.0001	0.38
Error	468	110.11			
Total	641				
Ln (Petal Area)					
Block (B)	2	0.2700	3.77	0.0354	
Paternal parent (P)	14	0.0914	0.64	0.8151	
Maternal parent (Paternal parent) (M[P])	46	0.1423	2.34	0.0001	
B × P	28	0.0716	0.98	0.5041	
B × (M[P])	75	0.0729	1.20	0.1421	
Model	165	0.1093	1.80	0.0001	0.46
Error	354	0.0608			
Total	519				
Pollen Grains/Flower					
Block (B)	2	3262×10^6	3.09	0.0612	
Paternal parent (P)	14	3404×10^6	2.18	0.0239	
Maternal parent (Paternal parent) (M[P])	46	1560×10^6	1.27	0.1245	
B × P	28	1055×10^6	0.62	0.9170	
B × (M[P])	71	1299×10^6	1.38	0.0350	

Table 12.B. (*Continued*)

Source	df	MS	F	p	r²
Model	161	1870 × 10⁶	1.52	0.0008	0.43
Error	322	1229 × 10⁶			
Total	483				

Ln (Ovules/Flower)

Source	df	MS	F	p	r²
Block (B)	2	0.1125	1.63	0.2147	
Paternal parent (P)	14	0.1392	1.33	0.2272	
Maternal parent (Paternal parent) (M[P])	46	0.1047	1.62	0.0094	
B × P	28	0.0692	0.87	0.6448	
B × (M[P])	73	0.0790	1.22	0.1238	
Model	163	0.0956	1.48	0.0015	0.41
Error	346	0.0647			
Total	509				

Pollen Grain Volume

Source	df	MS	F	p	r²
Block (B)	2	543,585	3.81	0.0345	
Paternal parent (P)	14	384,603	1.84	0.0616	
Maternal parent (Paternal parent) (M[P])	46	209,471	1.26	0.1306	
B × P	28	142,834	1.02	0.4606	
B × (M[P])	71	140,454	0.85	0.8030	
Model	161	196,834	1.18	0.1021	0.37
Error	330	166,213			
Total	491				

Petal Area/Pollen Grain

Source	df	MS	F	p	r²
Block (B)	2	146,065	0.25	0.7811	
Paternal parent (P)	14	689,811	1.54	0.1350	
Maternal parent (Paternal parent) (M[P])	46	448,106	1.06	0.3717	
B × P	28	585,990	1.34	0.1635	
B × (M[P])	69	437,433	1.04	0.4077	
Model	159	557,551	1.32	0.0197	0.41
Error	308	421,796			
Total	467				

Petal Area/Ovule

Source	df	MS	F	p	r²
Block (B)	2	7.88	0.36	0.7020	
Paternal parent (P)	14	23.36	0.76	0.7054	
Maternal parent (Paternal parent) (M[P])	46	30.78	1.86	0.0010	
B × P	28	21.98	1.11	0.3484	
B × (M[P])	73	19.74	1.19	0.1512	

Table 12.B.   (Continued)

Source	df	MS	F	p	r^2
Model	163	27.39	1.66	0.0001	0.44
Error	344	16.53			
Total	507				
Pollen:Ovule					
Block (B)	2	577×10^5	0.89	0.4237	
Paternal parent (P)	14	1254×10^5	1.20	0.3050	
Maternal parent (Paternal parent) (M[P])	46	1041×10^5	1.78	0.0024	
B × P	28	651×10^5	0.64	0.9026	
B × (M[P])	69	1013×10^5	1.74	0.0009	
Model	159	1049×10^5	1.80	0.0001	0.48
Error	304	583×10^5			
Total	463				
Density-Specific Phenotypic Gender					
Block (B)	2	0.0196	1.73	0.1962	
Paternal parent (P)	14	0.0251	1.08	0.4030	
Maternal parent (Paternal parent) (M[P])	46	0.0233	1.72	0.0041	
B × P	28	0.0113	0.54	0.9635	
B × (M[P])	69	0.0209	1.54	0.0072	
Model	159	0.0216	1.59	0.0003	0.45
Error	304	0.0136			
Total	463				
Garden-Wide Phenotypic Gender					
Block (B)	2	0.0111	0.97	0.3909	
Paternal parent (P)	14	0.0247	1.06	0.4202	
Maternal parent (Paternal parent) (M[P])	46	0.0234	1.73	0.0039	
B × P	28	0.0114	0.54	0.9644	
B × (M[P])	69	0.0211	1.56	0.0063	
Model	159	0.0215	1.59	0.0003	0.45
Error	304	0.0136			
Total	463				

[a]Type III mean squares are reported for each main effect and interaction term. The coefficient of determination, r^2, indicates the proportion of the total sum of squares represented by the model sum of squares. F-tests for the paternal effect use the Type III Mean Square for the maternal parent nested within paternal parent in the denominator; F-tests for the block effect use the Type III Mean Square for the Block × Paternal parent interaction in the denominator; F-tests for the Block × Paternal parent interaction use the Type III Mean Square for the Block × Maternal parent (Paternal parent) in the denominator.

Acknowledgments

This work was supported by a University of California General Research Grant. I would like to thank Johanne Brunet, John Damuth, Veronique Delesalle, Lynda Delph, John Endler, David Lloyd, Bruce Mahall, Craig Osenberg, Terry Schick, Colette St. Mary, Nancy Vivrette, Robert Warner, and Lorne Wolfe, with whom I had many stimulating discussions concerning this work. Christine Noé and Lisa Meuller helped with pollinations in the greenhouse and population monitoring in the garden.

Literature Cited

Abul-Fatih, H.A., F.A. Bazzaz, and R. Hunt. 1979. The biology of *Ambrosia trifida*. III. Growth and biomass allocation. New Phytol. **83**:829–838.

Aizen, M.A., and A. Kenigsten. 1990. Floral sex ratios in scrub oak (*Quercus ilicifolia*) vary with microtopography and stem height. Can J. Bot. **68**:1364–1368.

Bawa, K.W., and P.A. Opler. 1977. Spatial relationships between staminate and pistillate plants of dioecious tropical forest trees. Evolution **31**:64–68.

Bawa, K.S., and Webb, C.J. 1983. Floral variation and sexual differentiation in *Muntingia calabura* (Elaeocarpaceae), a species with hermaphrodite flowers. Evolution **37**:1271–1282.

Becker, W.A. 1984. Manual of Quantitative Genetics, 4th ed. Academic Enterprises, Pullman, WA.

Bertin, R.I. 1982. The ecology of sex expression in red buckeye. Ecology **63**:445–456.

Bierzychudek, P. 1984. Determinants of gender in Jack-in-the-pulpit: The influence of plant size and reproductive history. Oecologia **65**:14–18.

Bierzychudek, P., and V. Eckhart. 1988. Spatial segregation of the sexes of dioecious plants. Am. Nat. **132**:34–43.

Burd, M., and T.F.H. Allen. 1988. Sexual allocation strategy in wind-pollinated plants. Evolution **42**:403–407.

Cameron, R.G., and R. Wyatt. 1990. Spatial patterns and sex ratios in dioecious and monoecious mosses of the genus *Splachnum*. The Bryologist **93**:161–166.

Charlesworth, D., and B. Charlesworth. 1981. Allocation of resources to male and female functions in hermaphrodites. Biol. J. Linn. Soc. **15**:57–74.

Charnov, E.L. 1982. The Theory of Sex Allocation. Princeton University Press, Princeton, NJ.

Charnov, E.L. 1986. Size advantage may not always favor sex change. J. Theor. Biol. **119**:283–285.

Charnov, E.L., and J. Bull. 1977. When is sex environmentally determined? Nature (London) **266**:228–230.

Cid-Benevento, C.R. 1987. Relative effects of light, soil moisture availability and vegetative size on sex ratio of two monoecious woodland annual herbs: *Acalypha rhomboidea* (Euphorbiaceae) and *Pilea pumila* (Urticaceae). Bull. Torrey Bot. Club 114:293–306.

Clare, M.J., and L.S. Luckinbill. 1985. The effects of gene-environment interaction on the expression of longevity. Heredity 55:10–29.

Cole, S.L. 1979. Aberrant sex ratios in Joyoba associated with environmental factors. Desert Plants 1:8–11.

Comstock, R.E., and H.F. Robinson. 1948. The components of genetic variance in populations of biparental progenies and their use in estimating the average degree of dominance. Biometrics 4:254–266.

Condon, M.A., and L.E. Gilbert. 1988. Sex expression of *Gurania* and *Psiguria* (Cucurbitaceae): Neotropical vines that change sex. Am. J. Bot. 75:875–884.

Cox, P.A. 1981. Niche partitioning between sexes of dioecious plants. Am. Nat. 117:295–307.

Cruden, R.W. 1976. Intra-specific variation in pollen-ovule ratios and nectar secretion—preliminary evidence of ecotypic adaptation. Ann. Missouri Bot. Gard. 63:277–289.

Cruden, R.W. 1977. Pollen:ovule ratios: A conservative indicator of breeding systems in flowering plants. Evolution 31:32–46.

Cruden, R.W., and D.L. Lyon. 1985. Patterns of biomass allocation to male and female functions in plants with different mating systems. Oecologia 66:299–306.

Dawson, T.E., and L.C. Bliss. 1989. Patterns of water use and the tissue water relations in the dioecious shrub, *Salix arctica*: The physiological basis for habitat partitioning between the sexes. Oecologia 79:332–343.

Devlin, B. 1989. Components of seed and pollen yield of *Lobelia cardinalis*: Variation and correlations. Am. J. Bot. 76:204–214.

Devlin, B., and A.G. Stephenson. 1987. Sexual variations among plants of a perfect-flowered species. Am. Nat. 130:199–218.

Ellstrand, N.C., and D.L. Marshall. 1985a. Interpopulation gene flow by pollen in wild radish, *Raphanus sativus*. Am. Nat. 126:606–616.

Ellstrand, N.C., and D.L. Marshall. 1985b. The impact of domestication on the distribution of allozyme variations within and among cultivars of radish, *Raphanus sativus* L. Theor. Appl. Genet. 69:393–398.

Falconer, D.S. 1989. Introduction to Quantitative Genetics, 3rd ed. Longman, London.

Fox, J.F., and A.T. Harrison. 1981. Habitat assortment of sexes and water balance in a dioecious grass. Oecologia 49:233–235.

Freeman, D.C., and E.D. McArthur. 1984. The relative influences of mortality, nonflowering, and sex change on the sex ratios of six *Atriplex* species. Bot. Gaz. 145:385–394.

Freeman, D.C., and J.J. Vitale. 1985. The influence of environment on the sex ratio and fitness of spinach. Bot. Gaz. 146:137–142.

Freeman, D.C., L.G. Klikoff, and K.T. Harper. 1976. Differential resource utilization by the sexes of dioecious plants. Science **193**:597–599.

Freeman, D.C., K.T. Harper, and E.L. Charnov. 1980. Sex change in plants: Old and new observations and new hypotheses. Oecologia **47**:222–232.

Freeman, D.C., E.D. McArthur, K.T. Harper, and A.C. Blauer. 1981. Influence of environment on the floral sex ratio of monoecious plants. Evolution **35**:194–197.

Freeman, D.C., E.D. McArthur, and K.T. Harper. 1984. The adaptive significance of sexual lability in plants using *Atriplex canescens* as a principal example. Ann. Missouri Bot. Gard. **71**:265–277.

Garnock-Jones, P.J. 1986. Floret specialization, seed production and gender in *Artemisia vulgaris* L. (Asteraceae, Anthemideae). Bot. J. Linn. Soc. **92**:285–302.

Ghiselin, M.T. 1969. The evolution of hermaphroditism among animals. Quart. Rev. Biol. **44**:189–209.

Goldman, D.A., and M.F. Willson. 1986. Sex allocation in functionally hermaphroditic plants: A review and critique. Bot. Rev. **2**:157–194.

Gregg, K.B. 1975. The effect of light intensity on sex expression in species of *Cycnoches* and *Catasetum* (Orchidaceae). Selbyana **1**:101–113.

Hallauer, A.R., and J.B. Miranda. 1981. Quantitative Genetics in Maize Breeding. Iowa State University Press, Ames, IA.

Hendrix, S.D., and E.J. Trapp. 1981. Plant-herbivore interactions: Insect induced changes in host plant sex expression and fecundity. Oecologia **49**:119–122.

Iglesias, M.C., and G. Bell. 1989. The small-scale spatial distribution of male and female plants. Oecologia **80**:229–235.

Karron, J.D., and D.L. Marshall. 1990. Fitness consequences of multiple paternity in wild radish, *Raphanus sativus*. Evolution **44**:260–268.

Lewis, D., S.C. Verma, and M.I. Zuberi. 1988. Gametophytic-sporophytic incompatibility in the Cruciferae—*Raphanus sativus*. Heredity **61**:355–366.

Lloyd, D.G. 1979. Parental strategies of angiosperms. N.Z. J. Bot. **17**:595–606.

Lloyd, D.G. 1980a. Sexual strategies in plants. I. An hypothesis of serial adjustment of maternal investment during one reproductive session. New Phytol. **86**:69–79.

Lloyd, D.G. 1980b. Sexual strategies in plants. II. Data on the temporal regulation of maternal investment. New Phytol. **86**:81–92.

Lloyd, D.G. 1980c. Sexual strategies in plants. III. A quantitative method for describing the gender of plants. N.Z. J. Bot. **18**:103–108.

Lloyd, D.G., and K.S. Bawa. 1984. Modification of the gender of seed plants in varying conditions. Evol. Biol. **17**:255–336.

Lovett Doust, J., and P.B. Cavers. 1982a. Resource allocation and gender in the green dragon *Arisaema dracontium* (Araceae). Am. Midl. Nat. **108**:144–148.

Lovett Doust, J., and P.B. Cavers. 1982b. Sex and gender dynamics in jack-in-the-pulpit, *Arisaema triphyllum* (Araceae). Ecology **63**:797–808.

Lovett Doust, J., and P.B. Cavers. 1982c. Biomass allocation in hermaphrodite flowers. Can. J. Bot. **60**:2530–2534.

Lovett Doust, J., G. O'Brien, and L. Lovett Doust. 1987. Effect of density on secondary sex characteristics and sex ratio in *Silene alba* (Caryophyllaceae). Am. J. Bot. **74**:40–46.

Luckinbill, L.S., and M.J. Clare. 1985. Selection for life span in *Drosophila melanogaster*. Heredity **55**:9–18.

Marshall, D.L., and N.C. Ellstrand. 1986. Sexual selection in *Raphanus sativus*: Experimental data on nonrandom fertilization, maternal choice, and consequences of multiple paternity. Am. Nat. **127**:446–461.

Marshall, D.L., and N.C. Ellstrand. 1988. Effective mate choice in wild radish: Evidence for selective seed abortion and its mechanism. Am. Nat. **131**:739–756.

Mazer, S.J. 1987a. The quantitative genetics of life history and fitness components in *Raphanus raphanistrum* L. (Brassicaceae): Ecological and evolutionary consequences of seed-weight variation. Am. Nat. **130**:891–914.

Mazer, S.J. 1987b. Parental effects on seed development and seed yield in *Raphanus raphanistrum*: Implications for natural and sexual selection. Evolution **41**:355–371.

Mazer, S.J. 1989. Family mean correlations among fitness components in wild radish: Controlling for maternal effects on seed weight. Can. J. Bot. **67**:1890–1897.

Mazer, S.J., and C.T. Schick. 1991a. Constancy of population parameters for life history and floral traits in *Raphanus sativus* L. I. Norms of reaction and the nature of genotype by environment interactions. Heredity **67**:143–156.

Mazer, S.J., and C.T. Schick. 1991b. Constancy of population parameters for life history and floral traits in *Raphanus sativus* L. II. Effects of planting density on phenotype and heritability estimates. Evolution **45**:1888–1907.

Mazer, S.J., A.A. Snow, and M.L. Stanton. 1986. Fertilization dynamics and parental effects upon fruit development in *Raphanus raphanistrum*: Consequences for seed size variation. Am. J. Bot. **73**:500–511.

Mazer, S.J., R.R. Nakamura, and M.L. Stanton. 1989. Seasonal changes in components of male and female reproductive success in *Raphanus sativus* L. (Brassicaceae). Oecologia **81**:345–353.

McArthur, E.D. 1977. Environmentally induced changes of sex expression in *Atriplex canescens*. Heredity **38**:97–103.

McArthur, E.D., and D.C. Freeman. 1982. Sex expression in *Atriplex canescens*: Genetics and environment. Bot. Gaz. **143**:476–482.

McKone, M. 1987. Sexual allocation and outcrossing rate: A test of theoretical predictions using bromegrass (*Bromus*). Evolution **41**:591–598.

McKone, M.J., and D.W. Tonkyn. 1986. Intrapopulation gender variation in common ragweed (Asteraceae: *Ambrosia artemisiifolia* L.), a monoecious, annual herb. Oecologia **70**:63–67.

Nakamura, R.R., M.L. Stanton, and S.J. Mazer. 1989. Effects of mate size and mate number on male reproductive success in plants. Ecology **70**:71–76.

Onyekwelu, S., and J. Harper. 1979. Sex ratio and niche differentiation of spinach (*Spinacia oleracea* L.). Nature (London) **282**:609–611.

Panetsos, C.A., and H.G. Baker. 1967. The origin of variation in "wild" *Raphanus sativus* (Cruciferae) in California. Genetica **38**:243–274.

Pellmyr, O. 1987. Temporal patterns of ovule allocation, fruit set, and seed predation in *Anemopsis macrophylla* (Ranunculaceae). Bot. Mag. Tokyo **100**:175–183.

Policansky, D. 1981. Sex choice and the size-advantage model in jack-in-the-pulpit (*Arisaema triphyllum*). Proc. Natl. Acad. Sci. U.S.A. **78**:1306–1308.

Policansky, D. 1982. Sex change in plants and animals. Annu. Rev. Ecol. Syst. **13**:471–495.

Putrament, A. 1960. Studies in self-sterility in *Raphanus sativus* (L.) var. *radicula* (DC). Act. Soc. Bot. Pol. XXIV:289–313.

SAS Institute Inc. 1987. SAS/STAT Guide for Personal Computers, Version 6 Edition. SAS Institute Inc., Cary, NC.

Schaffner, J.H. 1922. Control of the sexual state in *Arisaema triphyllum* and *Arisaema dracontium*. Am. J. Bot. **9**:72–78.

Schlichting, C.D. 1989a. Phenotypic plasticity in *Phlox*. II. Plasticity of character correlations. Oecologia **78**:496–501.

Schlichting, C.D. 1989b. Phenotypic integration and environmental change. BioScience **39**:460–464.

Schoen, D.J. 1982. Male reproductive effort and breeding system in an hermaphrodite plant. Oecologia **53**:255–257.

Service, P.M., and M.R. Rose. 1985. Genetic covariation among life-history components: The effect of novel environments. Evolution **39**:943–945.

Shaw, R.G. 1986. Response to density in a wild population of the perennial herb *Salvia lyrata*: Variation among families. Evolution **40**:492–505.

Snow, A.A., and S.J. Mazer. 1988. Gametophytic selection in *Raphanus raphanistrum*: A test for heritable variation in pollen competitive ability. Evolution **42**:1065–1075.

Solomon, B.P. 1985. Environmentally influenced changes in sex expression in an andromonoecious plant. Ecology **66**:1321–1332.

Stanton, M.L. 1984a. Developmental and genetic sources of seed weight variation in *Raphanus raphanistrum* L. (Brassicaceae). Am. J. Bot. **71**:1090–1098.

Stanton, M.L. 1984b. Seed variation in wild radish: Effect of seed size on components of seedling and adult fitness. Ecology **65**:1105–1112.

Stanton, M.L. 1985. Seed size and emergence time within a stand of wild radish (*Raphanus raphanistrum* L.): The establishment of a fitness hierarchy. Oecologia **67**:524–531.

Stanton, M.L. 1987a. The reproductive biology of petal color variants in wild populations of *Raphanus sativus* L.: I. Pollinator response to color morphs. Am. J. Bot. **74**:178–187.

Stanton, M.L. 1987b. The reproductive biology of petal color variants in wild populations of *Raphanus sativus* L.: II. Factors limiting seed production. Am. J. Bot. **74**:188–196.

Stanton, M.L., and R.E. Preston. 1986. Pollen allocation in wild radish: Variation in pollen grain size and number. *In* D. L. Mulcahy, G. B. Mulcahy, and E. Ottaviano, eds., Biotechnology and Ecology of Pollen. Springer-Verlag, New York, pp. 461–466.

Stanton, M.L., and R.E. Preston. 1988. Ecological consequences and phenotypic correlates of petal size variation in wild radish, *Raphanus sativus* (Brassicaceae). Am. J. Bot. **75**:528–539.

Stanton, M.L., A.A. Snow, and S.N. Handel. 1986. Floral evolution: Attractiveness to pollinators increases male fitness. Science **232**:1625–1627.

Stanton, M.L., A.A. Snow, S.N. Handel, and J. Bereczky. 1989. The impact of a flower-color polymorphism on mating patterns in experimental populations of wild radish (*Raphanus raphanistrum* L.). Evolution **43**:335–346.

Stephenson, A.G. 1981. Flower and fruit abortion: Proximate causes and ultimate functions. Annu. Rev. Ecol. Syst. **12**:157–165.

Sutherland, S. 1986. Patterns of fruit-set: What controls fruit-flower ratios in plants? Evolution **40**:117–128.

Sutherland, S., and L. Delph. 1984. On the importance of male fitness in plants: Patterns of fruit-set. Ecology **65**:1093–1104.

Thompson, J.N. 1987. The ontogeny of flowering and sex expression in divergent populations of *Lomatium grayi*. Oecologia **72**:605–611.

Thomson, J.D. 1989. Deployment of ovules and pollen among flowers within inflorescences. Evol. Trends Plants **3**:65–68.

Thomson, J.D., and S.C.H. Barrett. 1981. Temporal variation of gender in *Aralia hispida* Vent. (Araliaceae). Evolution **35**:1094–1107.

Thomson, J.D., M.A. McKenna, and M.B. Cruzan. 1989. Temporal patterns of nectar and pollen production in *Aralia hispida*: Implications for reproductive success. Ecology **70**:1061–1068.

Via, S. 1987. Genetic constraints on the evolution of phenotypic plasticity. *In* V. Loeschcke, ed., Genetic Constraints on Adaptive Evolution. Springer-Verlag, Berlin, pp. 47–71.

Via, S., and R. Lande. 1985. Genotype-environment interaction and the evolution of phenotypic plasticity. Evolution **39**:505–522.

Via, S., and R. Lande. 1987. Evolution of genetic variability in a spatially heterogeneous environment: Effects of genotype-environment interaction. Genet. Res. **49**:147–156.

Waser, N.M. 1984. Sex ratio variation in populations of a dioecious desert perennial, *Simmondsia chinensis*. Oikos **42**:343–348.

Young, H.J., and M.L. Stanton. 1990a. Influence of environmental quality on pollen competitive ability in wild radish. Science **248**:1631–1633.

Young, H.J., and M.L. Stanton. 1990b. Influences of floral variation on pollen removal and seed production in wild radish. Ecology **71**:536–547.

13

Development and the Evolution of Plant Reproductive Characters

Pamela K. Diggle
University of Colorado

The study of evolution is, in part, the study of variation and its consequences. Random mutation is generally accepted as the ultimate source of variation, which is, in turn, sorted by evolutionary forces including natural selection and genetic drift. Nevertheless, there is a growing recognition that random mutation, selection or drift, and environmental effects may not fully explain the kinds and patterns of phenotypic variation observed among organisms. Although random mutation provides the raw material of evolution, random changes at the level of the gene do not necessarily translate directly into random variation at the level of the phenotype. The lack of direct correspondence between genotypic and phenotpic variation originates, in part, because genetic changes are expressed during the ontogeny of an organism and development is inherently nonlinear; it is an interactive and contingent process. Thus, the effect of any particular genetic change depends on the genotype and the developmental program in which it occurs. Because of their key position as translators of genetic variation, developmental processes must play a critical role in determining the range of phenotypic variation among organisms. If evolution can be viewed as a process of generating and sorting phenotypic variation, then development, by presenting a potentially nonrandom subset of phenotypes to the environment, must play a fundamental role in the dynamics of evolutionary change. Thus, one of the challenges for evolutionary and developmental biologists is to recognize and understand the many ways that development influences the kinds and patterns of phenotypic variation observed among organisms.

During the past decade various authors have addressed this challenge by integrating an understanding of developmental processes into evolutionary theory (e.g., Gould, 1977; Bonner, 1982; Raff and Kaufman, 1983; Raff and Raff, 1987; Thomson, 1988; Wallace, 1988). Central concepts of development and evolution that have recently gained prominence include heterochrony, dissociation of mechanistically linked developmental processes, and the nature and role of developmental constraints (Wake et al., 1991). These concepts are reviewed below and then, in the following sections, they are considered in the context of plant reproductive biology.

Concepts of Development and Evolution

Heterochrony has come to be defined as a change in the relative timing of developmental processes (rate, initiation, or termination) in a descendant relative to its ancestor. Heterochrony has long been recognized as a major mode of evolutionary change (Garstang, 1922; Haldane, 1932; DeBeer, 1958; Gould, 1977; Raff and Kaufman, 1983; Raff and Wray, 1989). Recent publications by Gould (1977) and by Alberch et al. (1979) have reviewed and clarified the concepts and terminology associated with heterochrony and presented formal models of heterochrony. These models provide an important conceptual framework for the newly emerging field of development and evolution. They focus on discrete processes of initiation, termination, and rate of developmental processes and propose clear morphological outcomes for variation in these processes. The models also relate developmental characteristics to ecological circumstances and facilitate the formulation of testable hypotheses about morphological change. Most importantly, perhaps, the establishment of formal models has stimulated communication between evolutionary and developmental biologists.

Models of heterochrony have both developmental and evolutionary components. Development is studied via the comparative method: ontogenies of whole organisms or their individual characters can be compared, with the ultimate goal being the identification of particular developmental differences that result in divergent morphologies. The evolutionary component is added when the phylogenetic relationship of the organisms can be inferred. This requires an independent hypothesis of ancestor–descendant relationship. Only by merging developmental information with phylogenetic hypotheses can conclusions be drawn about the direction and type of change in development associated with the evolution of new morphologies.

Models of heterochrony also illustrate the importance of *dissociation of developmental processes* (e.g., Gould, 1977; Smith-Gill, 1983; Lord and Hill, 1987; Raff and Wray, 1989; Wake et al., 1991). For example, morphological change via heterochrony depends on dissociating the rate of a developmental process from the timing of initiation or termination of that process (Gould, 1977; Alberch et al., 1979). As originally conceived, heterochrony requires the dissociation of reproductive development from somatic development (Gould, 1977). More generally, and beyond the specific requirements of heterochrony, it is clear that for novel combinations of characters to occur, or for characters to evolve independently, the underlying developmental processes must be dissociated from one another. Morphological results of dissociation might include duplication, modification, or changes in position of characters. Dissociation may also juxtapose different developmental mechanisms or cues, providing opportunities for the evolution of novel morphological characters (Wake et al., 1991). Conversely, strong developmental association of characters or developmental processes might pose a significant constraint on the pattern of morphological evolution.

This potential for developmental processes to constrain morphological evolution has been widely discussed. There is general agreement that *developmental constraints* exist (Alberch, 1982; Kauffman, 1983; Cheverud, 1984; Maynard Smith et al., 1985; Levinton, 1986; Stearns, 1986; Gould, 1989), but there is no agreement on the definition of a developmental constraint. The likely reason for the lack of consensus is that the notion of constraint is ad hoc, defined by the organism and circumstances under study. Gould (1989) notes that "One identifies the canonical causes within an accepted theory; the directors of other kinds of change and the preventors of change by the canonical cause are constraint." Our accepted evolutionary theory concerns natural selection and adaptation; thus, to the extent that development might influence evolutionary change, it is a constraint. Development is an explanation imported from outside the local context to explain limits on the pattern observed or bias in the production of variation (Stearns, 1986; but see Wake et al., 1991 for a plea for a precise definition).

Constraint can have both positive and negative meanings. Constraint can connote a negative limitation imposed on the power of natural selection to change a character in certain directions. Alternatively, it can have a positive meaning, as in compelling or channeling phenotypic change in a particular direction (Gould, 1989). Constraints, both positive and negative, may be an important source of nonrandom or discontinuous morphological variation and, as a result, may limit or even direct the kinds of phenotypic variation on which natural selection or other evolutionary forces can act.

Thus, heterochrony, dissociation, and developmental constraints are all interrelated phenomena that affect the generation of phenotypic variation on which evolutionary forces can act. Because these concepts are interrelated, examples of each are not discussed separately. Instead, the following section includes an analysis of selected examples of work in various areas of plant reproductive biology from a developmental perspective. The examples are used to explore the possibility that phenomena such as heterochrony, developmental association and dissociation, and/or developmental constraints have played a role in shaping the system under study.

The examples are taken from studies of angiosperms, but the concepts may be applied equally to nonflowering plants.

Selected Topics

Timing of Reproduction

The age or size at first reproduction is an important variable in theories of life history evolution because it has a dramatic effect on total lifetime fecundity (e.g., Lewontin, 1965; MacArthur and Wilson, 1967). Shifts in timing of reproduction in plants are also important in reproductive isolation (Grant, 1971). Sympatric species or populations that flower during different parts of a season are effectively reproductively isolated from each other and, as a result, may have divergent evolutionary fates. Thus, timing of reproduction can have profound consequences for speciation and for individual success within a species.

It is possible to measure variation in timing of reproduction in many ways. For investigations of ecological factors or fitness consequences, it might be sufficient to examine variation in time (calendar date), size, or age of reproduction. To understand how this variation is generated, comparisons of both morphology and of developmental timing are important. There are at least two different modes of variation in timing of reproduction. First, variation in timing of reproduction might result from variation in developmental rate. Early-flowering individuals might go through all developmental stages and processes faster than late-flowering individuals, reaching a standard reproductive morphology at an earlier time (Fig. 13.1A). This represents a change in the *rate* of development. Alternatively, earlier flowering could result from altered *patterns* of development, in which flowers are produced at an earlier time during the ontogeny of the individual (i.e., at a new nodal position, Fig. 13.1B). The appearance of a structure, such as a flower, at a new nodal position can result from one of two developmental changes: (1) the character (the flower) becomes dissociated from other characters at a given nodal position (i.e., the flower appears earlier or later relative to other characters during the ontogeny of the individual); or (2) insertions or deletions of a portion of the individual's ontogeny could alter the position of the character. In the latter case, the absolute position of a character changes, but it does so in conjunction with other characters associated with that nodal position or stage of ontogeny (Jones, 1990).

Jones (1990) recognized these alternatives in her analysis of changes in reproductive timing associated with domestication of *Cucurbita argyrosperma*. *Cucurbita argyrosperma* subsp. *argyrosperma* is a cultivated subspecies, derived from the wild *Cucurbita argyrosperma* subsp. *sororia*. A

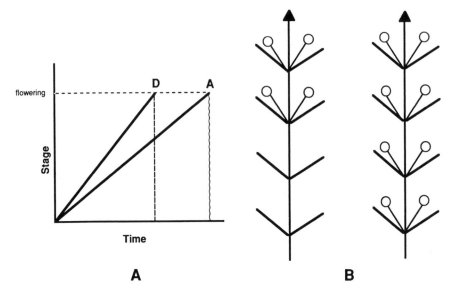

Figure 13.1. Two modes of variation in timing of reproduction. (A) Changes in developmental rate. Ontogenetic trajectories are shown for an ancestor (A) and descendant (D). The descendant passes through all of the same ontogenetic stages as the ancestor but does so at a greater rate, reaching the ancestral adult reproductive morphology more rapidly. (B) Changes in pattern of development. Flowers are produced at an earlier time during the ontogeny of the individual (i.e., at a new nodal position). Changes in pattern may occur by dissociation or by insertions and deletions.

change in the timing of reproduction has been associated with the domestication of this subspecies. The cultivar begins producing flowers earlier than the wild subspecies in terms of both real time (number of days) and in terms of morphology (Fig. 13.2). Rates of leaf production do not differ between the subspecies, indicating that early flowering was not achieved by increasing the growth rate. The change in timing of reproduction is morphological (i.e., a change in nodal position). The cultivar, on average, produces the first male flowers at node 12 and female flowers by node 30. In contrast, the wild progenitor does not produce male flowers until node 19 and female flowers until node 39. To distinguish between the alternative modes of morphological change, dissociation versus deletion, an assessment of the developmental and positional equivalence of organs or structures within the ontogeny of the whole plant (developmental homology sensu Guerrant, 1988) is required. As a criterion, Jones (1990) used a comparison of architectural ground plan and found that both subspecies share a basic ontogenetic pattern. They begin growth as orthotropic rosette

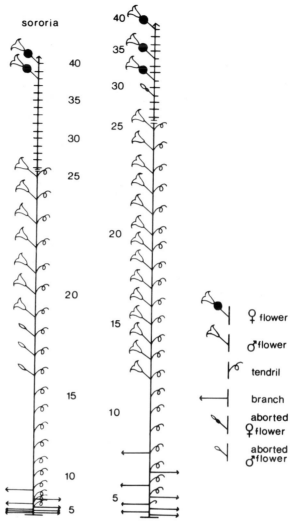

Figure 13.2. Architectural pattern of *Cucurbita argyrosperma* subsp. *argyrosperma*, a cultivar, and *Cucurbita argyrosperma* subsp. *sororia*, its wild progenitor. The numbers indicate nodal positions. The cultivar reproduces earlier than the wild species: the first male and female flowers are produced at earlier nodes and, thus, earlier in ontogeny. Based on longitudinal symmetry and fates of axillary meristems, the two subspecies share a common architectural ground plan. Thus, early reproduction results from the dissociation of flower development from the other developmental events occurring at each node.

shoots (upright shoots with short internodes) and, after similar numbers of internodes have been produced, both subspecies shift to plagiotropic (horizontal) growth with elongate internodes. Lateral branches and tendrils are formed similarly and at the same nodal positions in both subspecies. Thus, node for node, the fates of axillary meristems are equivalent except for those meristems producing flowers. Based on the similarity of architectural pattern and similarity of growth rate, there is a node-for-node correspondence between the two ontogenetic patterns. There is no evidence that a portion of the ontogeny of the cultivar has been deleted. Instead, the initiation of floral meristems has become dissociated from the fates of other meristems at the same node (those producing tendrils or branches) such that flowers are produced at ontogenetically earlier nodes. The onset of sexual reproduction has been accelerated in the life history of the cultivar compared to its ancestor by a dissociation of developmental processes. Moreover, dissociation of the developmental fates of meristems has been an important source of variation in life history.

Is there evidence among plants for the alternative pattern of morphological change (i.e., omission of nodes)? Herbert (1979) compared growth patterns of *Lupinus albus* cv. Ultra and *L. angustifolius* cv. Unicrop. Although the phylogenetic relationship of these two cultivars is not clear, the comparison is instructive. The cultivars differ in the number of days to first flowering, and this change in flowering time is due, in part, to changes in developmental pattern. Ultra produces 9–10 nodes before flowering, whereas Unicrop produces 20 vegetative nodes before flowering. Because inflorescences of these taxa are terminal, flowering occurs when the vegetative apical meristem converts to a floral meristem. This transition terminates the production of vegetative nodes. Because reproduction terminates vegetative growth, differences in timing of reproduction must occur by insertion or deletion of vegetative nodes during ontogeny. The architecture of *Lupinus* (terminal inflorescences) constrains the morphological options for altering the timing of reproduction: nodes (portions of the ontogeny) must be inserted or deleted. This architectural pattern has additional consequences for early reproduction: plants are smaller when flowering begins, and flowering terminates production of new leaves on that axis. To the extent that size and photosynthetic leaf area influence competitive ability or reproductive capacity, the capacity to respond to selection for early flowering is potentially decreased. If early-flowering individuals are much less competitive, then changes in reproductive timing may not evolve, despite possible demographic advantages or the potential for reproductive isolation. Architectural patterns, such as terminal inflorescences, can therefore be considered a developmental constraint; they determine both the type of variants available and may act to restrict the fitness of those variants.

Trade-offs between early reproduction and subsequent growth, survival, and late-life fecundity are well-known (Gadgil and Bossert, 1970; Schaffer, 1974; Stearns, 1977; Law, 1979; Charlesworth, 1980). In addition to architectural limitations, such as those that occur in *Lupinus*, reproduction entails substantial costs, and early reproduction may limit resources available for future growth and reproduction. Geber (1990) suggested that in addition to these costs, plants, due to their patterns of growth via meristems, may bear an additional cost of early reproduction. Both vegetative growth (proliferation of axes) and sexual reproduction require meristems, and commitment of a particular meristem to one of these functions precludes its commitment to the other. Early commitment of a meristem to flowering rather than to production of a vegetative branch has two direct effects: (1) it limits the production of new leaves, which supply the photosynthate necessary for current and future growth and reproduction, and (2) it precludes the production of new meristems (either vegetative or reproductive), which would have developed in the axils of those leaves (Fig. 13.3). Thus, plants reproducing early bear the additional cost of meristem limitation. Geber (1990) analyzed variation among individuals of *Polygonum arenastrum* for variation in timing and frequency of meristem commitment to inflorescences and branches. The results were consistent with the hypothesis that meristem availability and fate limits total reproduction and growth. There was a strong negative correlation between early flowering and total reproduction. Early-flowering individuals had high early fecundity, but reduced late fecundity and growth. Plants that delayed flowering grew much larger, had more branches, more meristems, and greater total reproductive output (Fig. 13.4). There was also genetic variation within the source population for position of first flowering and final total number of meristems, indicating the potential for evolutionary response in these traits. Geber (1990) noted that if resource limitation is stronger than meristem limitation in the field, the latter may not be an important constraining factor. There are as yet, however, no data on variation of meristem fates in natural populations. As empirical investigations of life-history trade-offs continue, meristem deployment is an important variable to consider.

The timing of reproduction in plants is a complex phenomenon and is determined by many factors. Developmental patterns and processes are potentially important both in defining options for variation in timing and in determining the consequences of this variation. Perhaps because of the open and notoriously plastic nature of plant growth, the potential for developmental patterns to constrain responses to selection on reproductive timing has rarely been considered. The comparisons cited above demonstrate that genetically determined patterns of growth (architecture) impose some limitation on the types of changes possible or on the ways that re-

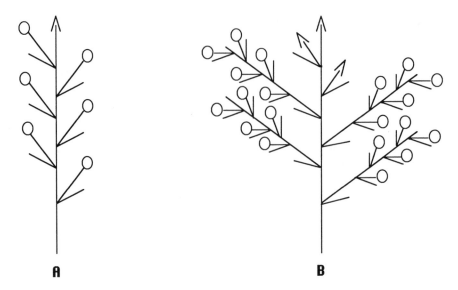

Figure 13.3. The cost of early reproduction. (A) Individual reproducing early in ontogeny. All axillary meristems are committed to the production of flowers or inflorescences. Production of photosynthetic area and new meristems is curtailed. (B) Individual delaying reproduction. Such plants have two advantages: increased vegetative proliferation yields (1) increased photosynthetic area and (2) increased number of new axillary meristems that can differentiate into flowers or inflorescences. By delaying reproduction, individual B has over three times the reproductive output of individual A.

production can vary. Patterns of plant architecture have been studied in a developmental context (e.g., Hallé et al., 1978) or ecological context (e.g., Bell, 1984; White, 1984; Watson, 1984; Watson and Casper, 1984; Waller and Steingraber, 1985), but rarely in the context of the evolution of plant reproductive characters (but see Waller, 1988). In addition to microevolutionary studies, such as the examples considered above, future research might include analyses of the range of successful architectural variation within a clade, given a certain architectural or developmental pattern, or of the extent of variation of architectural patterns within a lineage.

The timing of reproductive maturity is a common developmental marker used in studies of heterochrony. Are evolutionary changes in the timing of reproduction in plants heterochrony? Heterochrony is a change in the relative timing of developmental processes. If earlier reproduction is solely the result of a coordinated increase in the rate of all ontogenetic processes, all characters remain associated. Thus, there is no relative change, no resulting change in morphology, and no heterochrony. Heterochrony re-

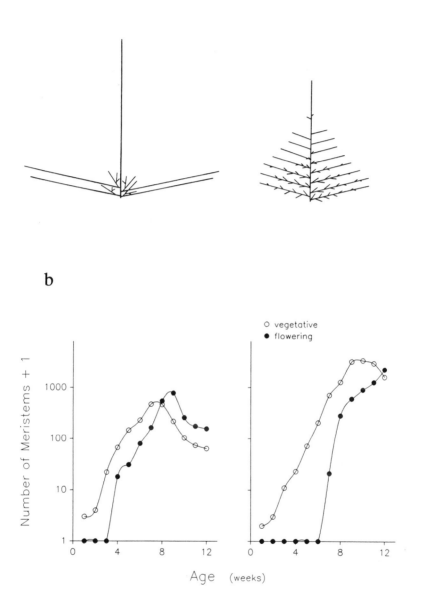

Figure 13.4. Meristem limitation in *Polygonum arenastrum*. (a) Branching patterns of two greenhouse-grown plants. Only vegetative branches are shown. (b) Semilog plot of age-specific growth and fecundity of the two individuals in A (from Geber, 1990).

quires the dissociation of developmental processes or events. In *Cucurbita argyrosperma*, the capacity of meristems to develop into flowers was dissociated from the overall architectural ground plan during the evolution of the cultivar. As a result, the descendant begins reproducing flowers while in a different morphology from its ancestor. This acceleration of reproduction can be recognized as a type of heterochrony. Insertions or deletions of a terminal portion of ontogeny are fundamental to heterochronic change. Changes that occur in earlier developmental stages, however, are not easily explained as heterochrony, which assumes no change in the sequence of developmental events (Alberch, 1985). The value of the formal models in comparing the examples discussed above lies less in labeling the changes as heterochronic or not than in focusing study on the developmental mechanism involved. Rather than stating that the species have undergone evolutionary changes in the timing of reproduction, the models focus attention on *how* the change has occurred.

Self-fertilization

The evolution of self-fertilizing species from outcrossing ancestors is one of the most common evolutionary transitions among flowering plants (Lloyd, 1965; Stebbins, 1970; Wyatt, 1983; Hill and Lord, 1990). Traditional hypotheses to explain the evolution of self-fertilization include reproductive assurance, the ability to set seed in the absence of pollinators; and reproductive isolation, a mechanism to restrict gene flow between sympatric species (Jain, 1976; summarized and reviewed by Wyatt, 1983). Underlying these hypotheses is the assumption that the ability to self-fertilize has been the character under selection. Developmental biologists have proposed an additional hypothesis in which self-fertilization, while adaptive, was not the selected character. Rather, rapid development was selected and small, self-fertilizing flowers were a result (Guerrant, 1984, 1988; Hufford, 1988b).

The flowers of self-fertilizing (autogamous) species are generally much smaller than those of their outcrossing ancestors (Ornduff, 1969; Wyatt, 1983), and it has been suggested that they have a juvenile morphology (Lord and Hill, 1987; Guerrant, 1988). In his discussion of heterochrony, Gould (1977) suggested that evolutionary juvenilization, or paedomorphosis, can result from two very different processes, both involving the dissociation of somatic development from reproductive development (Fig. 13.5). Paedomorphosis can result from acceleration of the development of reproductive tissues relative to the development of somatic tissues, resulting in a descendant that matures earlier in real time, while still in a juvenile morphology. This pattern of change is termed progenesis. Juvenilization can also result if the development of somatic tissues is retarded relative to development of reproductive structures. The result is again a

Paedomorphosis

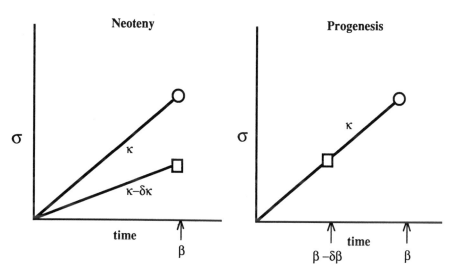

Figure 13.5. Two developmental patterns that result in paedomorphosis (evolutionary juvenilization). The graphs compare the developmental trajectories of an ancestor (circles) and descendant (squares). σ, morphology; β, reproductive maturity; k, rate of somatic (vegetative) development. Neoteny results when the rate of development of vegetative tissues or organs is retarded relative to the development of reproductive structures or tissues. The descendant reaches reproductive maturity while still in a morphology similar to the juvenile morphology of its ancestor. There is no change in the time required to reach maturity. Progenesis results when the rate of reproductive development is accelerated, while the rate of vegetative development is unchanged. Vegetative development is truncated as the descendant reaches reproductive maturity in less time but at a morphology characteristic of juvenile stages of the ancestor (modified from figures in Alberch et al., 1979).

juvenilized "adult," but the duration of development has not changed. This pattern of change is termed neoteny. Gould (1977) used the concepts of r- and K-strategies to formulate hypotheses predicting the occurrence of these two modes of evolutionary juvenilization. He suggested that progenesis might be expected in "r-selected" taxa of ephemeral but abundant resources where there is selection for rapid maturation and reproduction. In this case, selection would act directly on developmental rate, resulting in acceleration of reproductive development. The juvenile morphology of adults would, in such cases, be a consequence of selection on a developmental character. In contrast, neoteny might occur in stable, predictable environments that favor fine tuning of morphological adaptations. In this case there is no expectation about rates. Thus, if juvenile morphology occurs among adults, the expectation is that selection has acted on morphology itself and that this morphology is both adapted and adaptive (see also Gould, 1988).

Can Gould's hypotheses be applied to the evolution of small-flowered self-fertilizing species? The ability to flower and mature seeds earlier in the spring than their outcrossing ancestor is considered to have been important in the evolution of autogamy in species of at least three genera of annuals (Guerrant, 1984). Evolutionarily derived, autogamous species of *Leavenworthia* (Lloyd, 1965; Solbrig and Rollins, 1977), *Clarkia* (Moore and Lewis, 1965), and *Limnanthes* (Arroyo, 1973; Ritland and Jain, 1984) all live in habitats in which the summer drought arrives earlier in the spring than it does in the habitats of their outcrossing ancestors. Thus, rapid reproduction is likely to be advantageous to these species. Arroyo (1973) concluded that the main selective force leading to the evolution of self-fertilization in *L. floccosa* was for the ability to reproduce earlier than the putative ancestor, *L. alba*. The ability of *L. floccosa* to reproduce earlier is an important factor allowing this species to persist and flourish in the ephemeral habitats it now occupies. Guerrant (1984) used developmental analysis to test the hypothesis of a progenetic origin of *L. floccosa*. His comparative developmental study showed that flowers of *L. floccosa* and *L. alba* have considerable similarity in their size-shape growth trajectories (Fig. 13.6A,B) but that stages of reproductive development occur at an earlier time in *L. floccosa*, indicating an increased rate of reproductive development. In selfing flowers, microsporocyte meiosis, tetrad formation, and reproductive maturity occur earlier and, thus, at a smaller flower size than in flowers of *L. alba* (Fig. 13.6C). These results are consistent with a progenetic origin of *L. floccosa*. Guerrant (1984) suggested that small self-fertilizing flowers of *L. floccosa* are the result of a pattern of progenesis resulting from selection for rapid reproduction. Thus, selection has not been for the ability to self-fertilize, rather, selection has acted on development. Small flowers are a consequence of truncated development and self-fertilization is, in turn, a consequence of small flower size. Due to allometric relationships, smaller flowers have their stamens and gynoecium of more nearly equal length, providing the opportunity for self-fertilization. This scenario is in contrast to more common models of self-fertilization, in which small flower size is thought to be a consequence of selfing.

Guerrant's (1984) work stands as an innovative and important contribution to the study of development and mating system evolution. More recent phylogenetic analysis, however, suggests that *L. alba* does not represent the plesiomorphic morphology. Rather, *L. alba* and *L. floccosa* are each probably modified from a common ancestor (McNeil and Jain, 1983). This does not substantially change the interpretation of the data: the de-

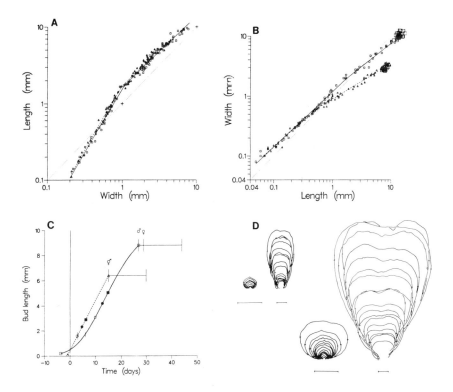

Figure 13.6. Flower growth of *Limnanthes alba* (squares) and *L. floccosa* (triangles). (A) Allometric plot of bud growth in length and width; the species cannot be distinguished statistically in these features. (B) Allometric plot of petal growth in length and width; slopes and intercepts are statistically indistinguishable during the initial portion of growth, but both slopes and intercepts differ after the slopes change. (C) Growth in length over time and timing of ontogenetic events during flower growth. (D) Tracings of some of the petals whose dimensions are indicated in (B), showing the progression of growth in size and shape. The two figures on the left are *L. floccosa* before and after the slope changed, and the two on the right are *L. alba*. All scale bars are 1 mm, but note that the scale is different before and after the slope changed. The smaller illustration in each pair fits into the open space of the larger (from Guerrant, 1988).

velopmental relationship between the two taxa is such that relatively few developmental differences distinguish them, and these are associated with ecological and life-history differences (Guerrant, 1988).

The studies of Lord and co-workers on development and evolution of cleistogamous species have also shown that a hypothesis of progenesis can explain many of the developmental differences between chasmogamous and cleistogamous flowers (Lord, 1981, 1984; Lord and Hill, 1987). Cleistogamous flowers are bud-like in morphology and self-pollinate without undergoing anthesis. For a thorough review of heterochrony and development in the evolution of cleistogamy, see Lord and Hill (1987).

The scenario proposed for the evolution of self-fertilization in *Limnanthes* can be contrasted to that proposed for *Arenaria uniflora*, a winter annual that occurs on granite outcrops in the southeastern United States. Populations of *A. uniflora* vary in flower size, degree of protandry, and levels of self-fertilization (Wyatt, 1984; Hill, 1989). Wyatt (1983) suggested that self-fertilization may have arisen in response to competition for pollinators between *A. uniflora* and a second winter annual of granite outcrops, *A. glabra*, which has larger, showier flowers and shares the same pollinators. Wyatt (1986) found no difference in flowering time or presence of pollinators between outcrossing and autogamous populations of *A. uniflora*, and it appeared that the autogamous populations occurred in more mesic habitats. However, whenever *A. uniflora* occurs with *A. glabra*, *A. uniflora* is autogamous. Presumably, outcrossed individuals either receive no pollen when growing with *A. glabra* or receive enough *A. glabra* pollen to depress their reproductive success sufficiently for selfing to spread in these populations (Wyatt, 1983).

The life history of these autogamous populations of *A. uniflora* contrasts with that of *Limnanthes*. The *A. uniflora* populations occur in stable, moist habitats, and there is no evidence of early maturation. These observations suggest that neoteny might be an appropriate model for a comparison of floral development in these two populations. In fact, there is no difference in the duration of floral development in these two morphs (Hill, 1989; Hill and Lord, 1990). The rate of development of the "somatic" tissues of the flower is decreased in the selfer, but the timing of differentiation events in the reproductive organs remains unchanged. The result is a smaller flower size at reproductive maturity. This general pattern of floral development is consistent with models of neoteny and, thus, is consistent with the suggestion that it is small flower size itself that was selected in these populations. Small flower size might have been selected because it facilitated selfing. Alternatively, reduced size might have been selected after selfing was established due to reallocation of resources.

These examples demonstrate that there are multiple developmental mechanisms leading to the same morphological and functional result (small, self-fertilizing flowers). Although there are too few studies to draw from, the consistency of the observed developmental patterns with theoretical predictions based on ecology is intriguing. Particularly significant is the notion that development itself can be a character subject to selection and that selection on development can have profound effects on such characters as the ability to self-fertilize. These examples also demonstrate the use of developmental models to test or corroborate evolutionary (adaptive) hypotheses and to generate new hypotheses.

Floral Form and Function

Among angiosperms, flowers show tremendous diversity of size, shape, symmetry, number of parts, herkogamy, dichogamy, and gender. Because floral traits have important consequences for the reproductive success of individuals and for the mating systems of populations, much work in plant reproductive biology has focused on demonstrating the function and/or adaptive nature of floral morphology. Can examination of patterns and sources of variation in floral morphology provide additional information about the function and evolution of floral form?

Flowers are considered to provide more reliable taxonomic characters than do vegetative parts (Stebbins, 1950). Numbers of floral parts, and their size and shape are traits that serve to distinguish taxa. Yet variation of floral form within a species is prevalent at many levels: among organs within flowers, among flowers within individuals, and among individuals within a population. Because floral morphology has such complex and significant consequences for reproduction, the pattern and range of floral variation is critical to the study of a species' reproductive system.

Floral variation within individuals of some species can be attributed to flower position (e.g., Ellstrand et al., 1984; Thomson, 1989; Seburn et al., 1990; Diggle, 1991a). Continuous, quantitative change in flower size with position on an inflorescence is common, with later flowers typically being smaller (e.g., Holtsford, 1985; Thomson, 1989; Diggle, 1991a). In addition, flower buds at the extreme distal portions of most indeterminate inflorescences rarely complete development and reach anthesis (Weberling, 1983; D. R. Kaplan, personal communication). Variation in flower size can be associated with differences in function: smaller, distal flowers have fewer ovules and, hence, set fewer seeds. Smaller flowers with decreased seed-set may be a facultative response to depletion of resources by earlier developing flowers and fruits (e.g., Lloyd, 1980; Stephenson, 1981; Bertin, 1982; Thomson, 1989; Byrne and Mazer, 1990). Alternatively, smaller flowers may be the consequence of a developmental constraint, such as a pattern imposed by properties of the inflorescence meristem.

Examination of the resource depletion hypothesis requires the establishment of a standard pattern of size variation in all floral organs, occurring in the absence of resource depletion, and the demonstration of a change in the pattern of variation in response to resource depletion. Data from andromonoecious species can be used to explore this type of floral variation. *Solanum hirtum* is an andromonoecious species with positional variation in floral morphology and function. Within an inflorescence, floral buds in basal positions are always hermaphroditic, whereas development of buds in distal positions is labile: buds may develop as either hermaphroditic or staminate flowers (Diggle, 1988). If there are no fruits already developing on a plant, distal buds develop as hermaphroditic flowers. If there are already numerous fruits developing and utilizing resources, subsequently developing distal buds will have a reduced gynoecium and be functionally staminate. These observations would seem to support the hypothesis that resource limitation is an underlying source of variation in floral form and function. In *S. hirtum*, however, even nonfruiting individuals show variation in flower size with position in the inflorescence. These individuals have hermaphroditic flowers at all positions, but all floral organs of distal hermaphroditic flowers are smaller than those of the basal hermaphroditic flowers. Distal staminate flowers, on fruit-bearing (resource-depleted) individuals, show the same variation in size associated with position, plus additional variation in the gynoecium, which results in the absence of female function (Diggle, 1991b). The resource limitation hypothesis is supported, but not as the source of variation in general floral size. Alternative explanations of positional variation in floral morphology should be sought. For example, changes in leaf size and shape are common during vegetative growth, and it has been suggested that this change is due, in part, to fluctuations in meristem size during shoot ontogeny (Troll and Rauh, 1950; Chazdon, 1991). Fluctuation of meristem size during inflorescence development has not been investigated, but might be associated with variation in the size of flowers produced. Alternatively, size variation in preformed inflorescences may result from mechanical effects of the packing of floral primordia in dormant buds. Consideration of such developmental correlates of positional variation does not provide an ultimate explanation, but it does provide alternatives to strictly functional and extrinsic arguments.

The data from *Solanum hirtum* also raise the issue of intrafloral variation (i.e., covariation vs. dissociation among floral organs). In fruit-bearing individuals of *S. hirtum*, gynoecial development in distal floral buds varies independently of the development of the other organs within the same flower (Diggle, 1991a,b). Such developmental dissociation of organs within individual flowers suggests that independent response to selection could occur, given genetic variation for these developmental traits. Independence of gynoecial and androecial development, in particular, is relevant to the subject of phenotypic gender. Variation of this trait among flowers and among individuals must require the dissociation of gynoecial and androecial development. The extensive variation of phenotypic gender documented in natural populations (reviewed by Lloyd and Bawa, 1984; Goldman and Willson, 1986) indicates that independence of gynoecial and androecial development is a common characteristic of developing flowers.

Independent evolutionary changes in gynoecial and androecial development can also be documented among species within a clade. Hufford (1988a,b) examined diversification of floral morphology within the genus *Eucnide*. If only mature morphology of these species is examined, there appears to be an invariant interspecific allometry of style and stamen lengths. Although the absolute lengths of these organs vary among the species, they covary such that the stigma and anthers of an individual flower are always at approximately the same level. This invariant allometry might be interpreted as a constraint on floral evolution within this clade. Yet there is a diversity of patterns of postanthesis development among these species that results in different patterns of dichogamy and probably in different levels of self- and cross-fertilization. Figure 13.7 shows four species, all appearing very similar in bud. Based on cladistic analysis, *Eucnide bartonioides* is most similar to the ancestral morphology (and development) of this group. In this species the style elongates first and becomes receptive. Then the stamens elongate and dehisce. There is always some separation of stigma and anthers; the stamens are below the level of the stigma during dehiscence. In *E. lobata*, the style also elongates and becomes receptive before the stamens elongate, but the stamens reach the height of the stigma before dehiscence, providing an opportunity for self-fertilization. In *E. cordata*, protogyny is well developed: style elongation and receptivity precede stamen elongation and dehiscence. In addition, stamens begin dehiscing while still well below the level of the stigma. Male and female function are well separated in both time and space. In *E. hirta*, elongation of style and stamens is much more synchronous, and dehiscence follows receptivity very closely, making self-fertilization highly likely. Thus, changes in the timing of elongation of the style and stamen filaments and in the timing of receptivity and dehiscence are diverse within this clade. There is no evidence of constraint on independent variation of male and female function during phylogenesis within *Eucnide*. The functional independence, however, results as much from dissociation of processes occurring within each organ or organ type as from dissociation of gynoecial and androecial development per se. Within the gynoecium, the process of stylar elongation has become dissociated from stigma receptivity, and, within the androecium, the process of filament elongation has been uncoupled from dehiscence of the anthers. Thus, inherited clade-specific patterns of developmental association or covariation, such as the invariant allometry of *Eucnide*, may not prevent a particular functional change in response to selection. It may, however, determine the particular route taken to arrive at that function.

Although it appears that the gynoecium and androecium may vary independently at many levels of organization, from intrafloral to interspecific, other floral organs do show a variety of patterns of covariation (e.g.,

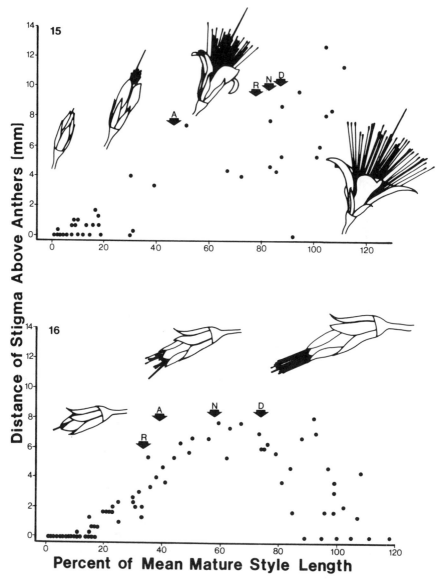

Figure 13.7. Plots of the vertical distance of the stigma from the uppermost anthers versus a percentage of the mean style length, including diagrams showing morphological changes associated with postanthesis development in *Eucnide*. (A) *E. bartonioides*, (B) *E. cordata*, (C) *E. hirta*, (D) *E. lobata*. A, anthesis (exsertion of the stigma beyond the corolla); D, onset of anther dehiscence; N, onset of nectar secretion; R, onset of stigma receptivity. Each data point represents a single dissected flower (from Hufford, 1988a).

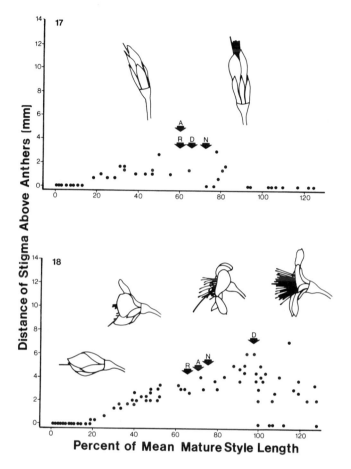

Figure 13.7. (*Continued*).

Erickson, 1948; Berg, 1960; Richards and Barrett, 1987; Campbell, 1989; Galen, 1989; Herrera, 1990b). Two hypotheses have been suggested to explain patterns of covariation among phenotypic characters. Covariation might be due to common genetic control (pleiotropy) or common developmental origin of those characters. Alternatively, covariation might result from natural selection for integration of functionally related characters (Stebbins, 1950; Olson and Miller, 1958; Cheverud, 1982; Kingsolver and Wiernasz, 1987; Zelditch and Carmichael, 1989).

As part of an investigation of the nature of genetic correlations, Conner and Via (1992) have measured phenotypic, genetic, and environmental correlations among floral and vegetative characters in wild radish (*Raphanus raphanistrum*). They found significant phenotypic correlations among all floral characters, and positive genetic correlations among several floral characters. Particularly high correlations exist between stamen filament lengths and floral tube length. They concluded that some of the positive correlations among floral traits are probably due to genes affecting overall flower size (i.e., pleiotropy). In addition, they noted that the strong correlation between filament length and corolla tube length might be due to either a close developmental relationship or selection for functional integration. Elongation of both corolla and stamen filaments probably results from the action of growth regulators such as gibberellic acid (reviewed by Lord, 1981). Thus, genetic variation for production of this underlying developmental signal might explain covariation. Alternatively, they suggested that strong correlation might be maintained by selection for proper placement of pollen on pollinators. Filament length must covary with tube length in order to place the anthers at the proper height above the tube for pollen dispersal. Current research is aimed at distinguishing among these alternative hypotheses by analysis of genetic structure and of selection by pollinators for covariation of floral structures.

Herrera (1990b) has also examined patterns of intrafloral covariation. He studied floral variation among disjunct, and putatively genetically distinct, populations of *Viola cazorlensis* that are each pollinated exclusively by the same species of day-flying hawkmoth. Despite large differences among populations for character means, Herrera (1990b) found that patterns of covariation among those characters were uniform across all populations. Assuming that phenotypic associations are indicative of genetic associations, he concluded that the genetic determinants of floral phenotype have not undergone significant differentiation in local populations, despite long-term isolation and restricted gene flow. It is not clear, however, whether uniform selection by the common pollinator or an inherent stability of the developmental–genetic associations among floral characters is maintaining constant patterns of covariation. Herrera (1990a) measured the effect of individual floral traits on maternal fitness (seed production) and concluded that there was no evidence of selection on particular floral traits. As Conner and Via (1992) suggested, however, selection for covariation among floral traits is the critical issue, and this is yet to be addressed for *V. cazorlensis*. The source of uniform covariation among these isolated populations remains unknown.

Further analysis of floral variation in *V. cazorlensis* provides evidence for functionally integrated subsets of characters (Herrera, 1990b). Factor analysis of the data pooled across all populations showed that there were four subsets of floral characters that covaried with one another but that were orthogonal to the other three subsets of (internally covarying) floral characters. These characters may represent functional subsets: characters that function together in pollinator attraction or pollen deposition or dispersal. Alternatively, these characters may share a common developmental origin, perhaps being initiated at the same time in response to the same cue or they may develop from a common primordium.

Many features of flowers have been related to the selective power of insects over the course of evolutionary time (Crepet, 1983). Many plant taxa with specialized pollinators have complex floral morphology with considerable connation and adnation of parts. The fusion of stamens to the corolla is a particularly common form of adnation, and these structures often develop from a common primordium. Such flowers are likely to show a stronger covariation of stamens and corolla than flowers with separate petal and stamen primordia. As the studies cited above demonstrate, however, separating "internal" (developmental) from "external" (selective) sources of floral covariation will be difficult. The strong association between insect pollination and fusion of floral organs does raise the possibility that selection for covariation has resulted in the evolution of fused organs because they are more likely to covary and in this way enhance floral integration and function. Direct selection for functional integration might indirectly select for developmental modifications that result in fused floral organs.

Covariation of developmental processes themselves is undoubtedly common and can result in unexpected changes in form and function. Genetic data from a study of flower size variation in *Mimulus* (Fenster, Diggle, Barrett, and Ritland, unpublished results) showed that developmental rate and duration of development, both processes that affect floral size, have a strong negative genetic correlation: flowers that elongate for the shortest period of time also elongate at the greatest rate, and vice versa. The negative correlation could be due to linkage disequilibrium or could imply that genes (or genetic factors) that act to increase the rate of bud elongation also act to decrease duration. If the latter is true, growth rate cannot evolve independently of duration, and selection on either developmental character will always affect flower size in *Mimulus*. A negative association of growth rate and duration is also apparent in phenotypic data reported for *Limnanthes floccosa* (Fig. 13.6) and cleistogamous flowers of *Lamium amplexicaule* (Lord, 1982, Fig. 29), indicating that it may be a common constraint among rapidly developing (progenetic) self-fertilizing flowers. If a negative genetic correlation of rate and duration of floral growth is common, then small flowers may be the inevitable outcome of selection for rapid floral development. The mating system of rapidly developing species might thus be constrained by the developmental correlation of rate and duration of floral bud elongation.

Changes of the mating system among species in the *Scutellaria angustifolia* complex may also be due to developmental covariation. Within this species complex, large-flowered (outcrossing) species have evolved from small-flowered (self-fertilizing) ancestors in three separate lineages (Olmstead, 1989). As discussed above, the pattern of evolutionary change from large- to small-flowered taxa is exceedingly common, but the change from small to large flowers has been documented only rarely (Kress, 1983; Olmstead, 1989). Olmstead (personal communication) suggests that, in *Scutellaria*, this transition may be the result of strong genetic correlations between vegetative and floral growth or size. The small-flowered species also have small leaves and grow in dry sagebrush habitats. The large-flowered species have much larger leaves and grow in more mesic woodland habitats. If strong genetic correlations between leaf and flower size exist, then selection for increased leaf area in low-light, mesic habitats would indirectly select for larger flowers. The change in breeding system would, in turn, be an indirect effect of larger flower size. Genetic correlations in *Scutellaria* would therefore not be a limitation; rather, they may have enabled an otherwise rare morphological transition to occur repeatedly within this species complex and resulted, in turn, in a shift in mating system.

In summary, patterns of variation and covariation occur at all levels of plant structure from intrafloral to interspecific. These patterns most certainly arise during development, but the ultimate explanation for the patterns observed at each level remains unclear. Patterns of genetic covariation are an important component of theories of multivariate morphological evolution (Lande, 1979) and are a recognized source of constraints on evolutionary change (Lande, 1979, 1982; Cheverud 1982, 1984; Maynard Smith et al., 1985; Riska, 1989). The studies cited above confirm that covariation can affect the evolution of reproductive characters in plants. Genetic associations can cause traits to evolve in concert (e.g., flower and leaf size in *Scutellaria*, stamen and corolla lengths in many species) and can cause certain combinations of traits to be more easily achieved than others (e.g., small, rapidly developing flowers in *Mimulus*). Covariation of traits may also cause individual traits to evolve to less adapted states and/or retard progress in the direction of selection (Lande, 1979).

Conclusions

The examples discussed in the preceding sections show that patterns or rules of growth and patterns of association or dissociation can have important consequences for the generation of morphological variation among individuals. Developmental processes may determine the type of variation possible and may limit the fitness of the variant individual or the function of a variant structure. Formal models of development, such as models of heterochrony, focus attention on mechanisms of morphological change. By focusing attention on development, these models can suggest new questions or hypotheses and identify circumstances in which developmental processes themselves might be targets of selection.

The examples also demonstrate the complex relationship between development and other factors of evolution. The diversity of form and function observed among organisms must reflect the evolutionary outcome of an interaction of internal genetic and developmental processes and of external ecological forces. Genetics and development may well define what is possible, whereas the environment ultimately defines what is successful. Thus, developmental patterns and processes are an integral and fundamental component of evolutionary dynamics. The interaction of genetics, development, and processes of natural selection and genetic drift in determining patterns of morphological diversity is complex, and these factors are intricately linked. Rather than studying processes occurring at any one of these levels (genetic, developmental, and phenotypic), a more powerful approach for future research will be aimed at understanding the interactions between the levels and between processes. This type of integrative approach must certainly contribute an additional dimension to the explanatory power of theories of plant evolution.

Acknowledgments

I thank Jeffrey Conner, Charles Fenster, Cynthia Jones, Richard Olmstead, and Sara Via for generously sharing unpublished data and ideas. I also thank Jeffrey Conner, Charles Fenster, William Friedman, Jeffrey Hill, Larry Hufford, Cynthia Jones, Stephanie Mayer, Jennifer Richards, and Garland Upchurch for comments that have substantially improved this manuscript.

Literature Cited

Alberch, P. 1982. Developmental constraints in evolutionary processes, *In* J.T. Bonner, ed., Evolution and Development. Springer-Verlag, New York, pp. 313–332.

Alberch, P. 1985. Problems with the interpretation of developmental sequences. Syst. Zool. **34**:46–58.

Alberch, P., S.J. Gould, G.F. Oster, and D.B. Wake. 1979. Size and shape in ontogeny and phylogeny. Paleobiology **5**:296–317.

Arroyo, M.T.K. 1973. Chiasma frequency evidence of the evolution of autogamy in *Limnanthes floccosa* (Limnanthaceae). Evolution **27**:679–688.

Bell, A.D. 1984. Dynamic morphology: A contribution to plant population ecology. *In* R. Dirzo and J. Sarukhán, eds., Perspectives on Plant Population Biology. Sinauer, Sunderland, MA, pp. 48–65.

Berg, R.L. 1960. The ecological significance of correlation pleiades. Evolution **14**:171–180.

Bertin, R.I. 1982. The evolution and maintenance of andromonoecy. Evol. Theory **6**:25–32.

Bonner, J.T. 1982. Evolution and Development: Report of the Dahlem Workshop on Evolution and Development. Springer-Verlag, New York.

Byrne, M., and S.J. Mazer. 1990. The effect of position on fruit characteristics, and relationships among components of yield in *Phytolacca rivinoides* (Phytolaccaceae). Biotropica **22**:353–365.

Campbell, D.R. 1989. Measurement of selection in a hermaphroditic plant: Variation in male and female pollination success. Evolution **43**:318–334.

Charlesworth, B. 1980. Evolution in Age-Structured Populations. Cambridge University Press, Cambridge.

Chazdon, R.L. 1991. Plant size and form in the understory palm genus *Geonoma*: Are species variations on a theme? Am. J. Bot. **78**:680–694.

Cheverud, J.M. 1982. Phenotypic, genetic, and environmental morphological integration in the cranium. Evolution **36**:499–516.

Cheverud, J.M. 1984. Quantitative genetics and developmental constraints on evolution by selection. J. Theor. Biol. **110**:155–172.

Conner, J., and S. Via. 1992. Patterns of phenotypic and genetic correlations in wild radish: Possible evidence for the effects of selection on floral morphology. Evolution, in press.

Crepet, W.L. 1983. The role of insect pollination in the evolution of the angiosperms. *In* L. Real, ed., Pollination Biology. Academic Press, New York, pp. 31–50.

DeBeer, G.R. 1958. Embryos and Ancestors, 3rd ed. Clarendon Press, Oxford.

Diggle, P.K. 1988. Labile Sex Expression in the Andromonoecious *Solanum hirtum*. Ph.D. Dissertation, University of California, Berkeley.

Diggle, P.K. 1991a. Labile sex expression in the andromonoecious *Solanum hirtum*: Pattern of variation in floral structure. Can. J. Bot. **69**:2033–2043.

Diggle, P.K. 1991b. Labile sex expression in the andromonoecious *Solanum hirtum*: Floral morphogenesis and sex determination. Am. J. Bot. **78**:377–393.

Ellstrand, N.C., E.M. Lord, and K.J. Eckard. 1984. The inflorescence as a metapopulation of flowers: Position-dependent differences in function and form in the cleistogamous species *Collomia grandiflora* Dougl. *ex* Lindl. (Polemoniaceae). Bot. Gaz. **145**:329–333.

Erickson, R.O. 1948. Cytological and growth correlations in the flower bud and anther of *Lilium longiflorum*. Am. J. Bot. **35**:729–739.

Gadgil, M., and W.H. Bossert. 1970. Life historical consequences of natural selection. Am. Nat. **102**:52–64.

Galen, C. 1989. Measuring pollinator-mediated selection on morphometric floral traits: Bumblebees and the alpine sky pilot, *Polemonium viscosum*. Evolution **43**:882–890.

Garstang, W.J. 1922. The theory of recapitulation: A critical restatement of the biogenetic law. Zool. J. Linn. Soc. **35**:81–101.

Geber, M.A. 1990. The cost of meristem limitation in *Polygonum arenastrum*: Negative genetic correlations between fecundity and growth. Evolution **44**:799–819.

Goldman, D.A., and M.F. Willson. 1986. Sex allocation in functionally hermaphroditic plants: A review and critique. Bot. Rev. **52**:157–194.

Gould, S.J. 1977. Ontogeny and Phylogeny. Belknap Press of Harvard University Press, Cambridge, MA.

Gould, S.J. 1988. The uses of heterochrony. *In* M.L. McKinney, ed., Heterochrony in Evolution. Plenum Press, New York, pp. 1–13.

Gould, S.J. 1989. Developmental constraints in *Cerion*, with comments on the definition and interpretation of constraint in evolution. Evolution **43**:516–539.

Grant, V. 1971. Plant Speciation. Columbia University Press, New York.

Guerrant, E.O. 1984. The role of ontogeny in the evolution and ecology of selected species of *Delphinium* and *Limnanthes*. Ph.D. Dissertation, University of California, Berkeley.

Guerrant, E.O. 1988. Heterochrony in plants: The intersection of evolution, ecology, and ontogeny. *In* M.L. McKinney, ed., Heterochrony in Evolution. Plenum Press, New York, pp. 111–133.

Guerrant, E.O. 1989. Early maturity, small flowers and autogamy: A developmental connection? *In* J.H. Bock and Y.B. Linhart, eds., The Evolutionary Biology of Plants. Westview Press, Boulder, CO, pp. 61–84.

Haldane, J.B.S. 1932. The time of action of genes and its bearing on some evolutionary problems. Am. Nat. **66**:5–24.

Hallé, F., R.A.A. Oldeman, and P.B. Tomlinson. 1978. Tropical Trees and Forests: An Architectural Approach. Springer-Verlag, New York.

Herbert, S.J. 1979. Density studies on lupins. I. Flower development. Ann. Bot. **43**:55–63.

Herrera, C.M. 1990a. The adaptedness of the floral phenotype in a relict endemic, hawkmoth-pollinated violet. 1. Reproductive correlates of floral variation. Biol. J. Linn. Soc. **40**:263–274.

Herrera, C.M. 1990b. The adaptedness of the floral phenotype in a relict endemic, hawkmoth-pollinated violet. 2. Patterns of variation among disjunct populations. Biol. J. Linn. Soc. **40**:275–291.

Hill, J.P. 1989. Homeosis, heterochrony, and the evolution of floral form. Ph.D. Dissertation, University of California, Riverside.

Hill, J.P., and E.M. Lord. 1990. The role of developmental timing in the evolution of floral form. Dev. Biol. **1**:281–287.

Holtsford, T.P. 1985. Non-fruiting hermaphroditic flowers of *Calochortus leichtlinii:* Potential reproductive functions. Am. J. Bot. **72**:1687–1694.

Hufford, L.D. 1988a. The evolution of floral morphological diversity in *Eucnide* (Loasaceae): The implications of modes and timing of ontogenetic changes on phylogenetic diversification. *In* P. Leins, S.C. Tucker and P.K. Endress, eds., Aspects of Floral Development. Schweizerbartsche Verlagbuchhandlung, Stuttgart, pp. 103–119.

Hufford, L.D. 1988b. Potential roles of scaling and post-anthesis developmental changes in the evolution of floral forms of *Eucnide* (Loasaceae). Nord. J. Bot. **8**:147–157.

Jain, S.K. 1976. The evolution of inbreeding in plants. Annu. Rev. Ecol. Syst. **7**:469–495.

Jones, C.S. 1990. The developmental basis of leaf shape variation in a wild and a cultivated subspecies of *Cucurbita*. Ph.D. Dissertation, University of California, Berkeley.

Jones, C.S. 1992. Comparative ontogeny of a wild cucurbit and its derived cultivar. Evolution, in press.

Kauffman, S.A. 1983. Developmental constraints: Internal factors in evolution. *In* B.C. Goodwin, N. Holder, and C.C. Wylie, eds., Development and Evolution. Cambridge University Press, Cambridge, pp. 195–225.

Kingsolver J.G., and D.C. Wiernasz. 1987. Dissecting correlated characters: Adaptive aspects of phenotypic covariation in melanization patterns of *Pieris* butterflies. Evolution **43**:485–503.

Kress, W.J. 1983. Self-incompatibility in Central American *Heliconia*. Evolution **37**:735–744.

Lande, R. 1979. Quantitative genetic analysis of multivariate evolution, applied to brain: Body size allometry. Evolution **33**:402–416.

Lande, R. 1982. A quantitative genetic theory of life history evolution. Ecology **63**:607–615.

Law, R. 1979. The costs of reproduction in an annual meadow grass. Am. Nat. **113**:3–16.

Levinton, J.S. 1986. Developmental constraints and evolutionary saltations: A discussion and critique. *In* J.P. Gustafson, G.L. Stebbins, and F.J. Ayala, eds., Genetics, Development and Evolution, Seventeenth Stadler Genetics Symposium, pp. 253–288.

Lewontin, R.C. 1965. Selection for colonizing ability. *In* H.G. Baker and G.L. Stebbins, eds., The Genetics of Colonizing Species. Academic Press, New York, pp. 77–91.

Lloyd, D.G. 1965. Evolution of self compatibility and racial differentiation in *Leavenworthia* (Crucifereae). Contrib. Gray Herb. **195**:3–195.

Lloyd, D.G. 1980. Sexual strategies in plants. I. An hypothesis of serial adjustment of maternal investment during one reproductive session. New Phytol. **86**:69–79.

Lloyd, D.G., and K.S. Bawa. 1984. Modification of the gender of seed plants in varying conditions. Evol. Biol. **17**:255–338.

Lord, E.M. 1981. Cleistogamy: A tool for the study of floral morphogenesis, function and evolution. Bot. Rev. **47**:421–449.

Lord, E.M. 1982. Floral morphogenesis in *Lamium amplexicaule* L. (Labiatae) with a model for the evolution of the cleistogamous flower. Bot. Gaz. **143**:63–72.

Lord, E.M. 1984. Cleistogamy: A comparative study of intraspecific floral variation. *In* R.A. White and W.C. Dickison, eds., Contemporary Problems in Plant Anatomy. Academic Press, New York, pp. 451–494.

Lord, E.M., and J.P. Hill. 1987. Evidence for heterochrony in the evolution of plant form. *In* R.A. Raff and E.C. Raff, eds., Development as an Evolutionary Process. Alan R. Liss, New York, pp. 47–70.

MacArthur, R.H., and E.O. Wilson. 1967. The Theory of Island Biogeography. Princeton University Press, Princeton, NJ.

Maynard Smith, J., R. Burian, S. Kaufman, P. Alberch, J. Campbell, B. Goodwin, R. Lande, D. Raup, and L. Wolpert. 1985. Developmental constraints and evolution. Quart. Rev. Biol. **60**:265–287.

McKinney, M.L. 1988. Heterochrony in Evolution. Plenum Press, New York.

McNeill, C.I., and S.K. Jain. 1983. Genetic differentiation studies and phylogenetic inference in the plant genus *Limnanthes* (section *Inflexae*). Theor. Appl. Genet. **66**:257–269.

Moore, D.M., and H. Lewis. 1965. The evolution of self-pollination in *Clarkia xantiana*. Evolution **19**:104–114.

Olmstead, R. 1989. Phylogeny, phenotypic evolution, and biogeography of the *Scutellaria angustifolia* complex (Lamiaceae): Inference from morphological and molecular data. Syst. Bot. **14**:320–338.

Olson, E., and R. Miller. 1958. Morphological Integration. University of Chicago Press, Chicago.

Ornduff, R. 1969. Reproductive biology in relation to systematics. Taxon **18**:121–133.

Plack, A. 1957. Sexual dimorphism in Labiatae. Nature (London) **180**:1218–1219.

Raff, R.A., and T.C. Kaufman. 1983. Embryos, Genes and Evolution. Macmillan, New York.

Raff, R.A., and E.C. Raff. 1987. Development as an Evolutionary Process. Alan R. Liss, New York.

Raff, R.A., and G.A. Wray. 1989. Heterochrony: Developmental mechanisms and evolutionary results. J. Evol. Biol. **2**:409–434.

Richards, J.H., and S.C.H. Barrett. 1987. Development of tristyly in *Pontederia cordata* (Pontederiaceae). I. Mature floral structure and patterns of relative growth of reproductive organs. Am. J. Bot. **74**:1831–1841.

Riska, B. 1989. Composite traits, selection response, and evolution. Evolution **43**:1172–1191.

Ritland, K., and S. Jain. 1984. The comparative life histories of two annual *Limnanthes* species in a temporally variable environment. Am. Nat. **124**:656–679.

Schaffer, W.M. 1974. Selection for optimal life histories: The effect of age structure. Ecology **55**:291–303.

Seburn, C.N.L., T.A. Dickinson, and S.C.H. Barrett. 1990. Floral variation in *Eichhornia paniculata* (Spreng.) Solms (Pontederiaceae): I. Instability of stamen position in genotypes from northeast Brazil. J. Evol. Biol. 3:103–123.

Smith-Gill, S.J. 1983. Developmental plasticity: Developmental conversion versus phenotypic modulation. Am. Zool. 23:47–55.

Solbrig, O.T., and R.C. Rollins. 1977. The evolution of autogamy in species of the mustard genus *Leavenworthia*. Evolution 31:265–281.

Stearns, S.C. 1977. Life history tactics: A critique of the theory and a review of the data. Annu. Rev. Ecol. Syst. 8:145–171.

Stearns, S.C. 1986. Natural selection and fitness, adaptation and constraint, *In* D.M. Raup and J. Jablonski, eds., Patterns and Processes in the History of Life. Springer-Verlag, New York, pp. 23–44.

Stebbins, G.L. 1950. Variation and Evolution in Plants. Columbia University Press, New York.

Stebbins, G.L. 1970. Adaptive radiation in angiosperms. I. Pollination mechanisms. Annu. Rev. Ecol. Syst. 1:307–326.

Stephenson, A.G. 1981. Flower and fruit abortion: Proximate causes and ultimate functions. Annu. Rev. Ecol. Syst. 12:253–279.

Thomson, J.D. 1989. Deployment of ovules and pollen among flowers within inflorescences. Evol. Trends Plants 3:65–68.

Thomson, K.S. 1988. Morphogenesis and Evolution. Oxford University Press, New York.

Troll, W., and W. Rauh. 1950. Das Erstarkungswachstum krautiger Dikotylen, mit besonderer Berücksichtigung der primären Verdickungsvorgänge. Sitzungsverichte der Heidelberger Akademie der Wissenschaften, Mathematisch-naturwissenschaftliche Klasse. Springer-Verlag, Heidelberg.

Wake, D.B., P. Maybee, J.H. Hanken, and G. Wagner. 1991. Development and evolution: The emergence of a new field. *In* T.R. Dudley, ed., Proceedings of the IV International Congress of Systematics and Evolutionary Biology. Dioscorides Press, Portland, OR.

Wallace, A. 1988. A Theory of Evolution and Development. John Wiley, New York.

Waller, D.M. 1988. Plant morphology and reproduction. *In* J. Lovett Doust and L. Lovett Doust, eds., Plant Reproductive Ecology: Patterns and Strategies. Oxford University Press, New York, pp. 203–227.

Waller, D.M., and D.A. Steingraber. 1985. Branching and modular growth: Theoretical models and empirical patterns. *In* J.B.C. Jackson, L.W. Buss and R.E. Cook, eds., Population Biology and Evolution of Clonal Organisms. Yale University Press, New Haven, CT, pp. 225–257.

Watson, M.A. 1984. Developmental constraints: Effect on population growth and patterns of resource allocation in a clonal plant. Am. Nat. 123:411–426.

Watson, M.A., and B.B. Casper. 1984. Morphological constraints on the patterns of carbon distribution in plants. Annu. Rev. Ecol. Syst. 15:133–258.

Weberling, F. 1983. Fundamental features of modern inflorescence morphology. Bothalia **14**:917–922.

White, J. 1984. Plant Metamerism, *In* R. Dirzo and J. Sarukhán, eds., Perspectives on Plant Population Biology. Sinauer, Sunderland, MA, pp. 15–47.

Wyatt, R. 1983. Pollinator-plant interactions and the evolution of breeding systems. *In* L. Real, ed., Pollination Biology. Academic Press, New York, pp. 51–95.

Wyatt, R. 1984. The evolution of self-pollination in granite outcrop species of *Arenaria* (Caryophyllaceae). I. Morphological correlates. Evolution **38**:804–816.

Wyatt, R. 1986. Ecology and evolution of self-pollination in *Arenaria uniflora* (Caryophyllaceae). J. Ecol. **74**:403–418.

Zelditch, M.L., and A.C. Carmichael. 1989. Ontogenetic variation in patterns of developmental and functional integration in skulls of *Sigmodon fulviventer*. Evolution **43**:814–824.

14

The Evolution of Endosperm: A Phylogenetic Account

Michael J. Donoghue and Samuel M. Scheiner
University of Arizona
Northern Illinois University

The seeds of flowering plants differ in many ways from those of other seed plants, and any explanation of the origin of angiosperms must account for the evolution of these differences. Prominent among the characteristic features of the angiosperm seed is the development of endosperm tissue from the product of a second fertilization. A variety of explanations have been proposed for the evolution of this unique tissue. It has been suggested, for example, that the greater genetic relatedness of the endosperm to the embryo (relative to the female gametophyte) provides a better environment for embryo growth (Sargant, 1900; Sporne, 1975). The most detailed treatment of the problem was provided by Brink and Cooper (1940, 1947), who supported the widespread view that increased vigor resulting from the union of genetically distinct nuclei results in a superior nurse tissue for the developing embryo. Stebbins (1976) proposed that endosperm is "doubly equipped for rapid growth" in that heterozygosity yields hybrid vigor and triploidy provides a larger number of RNA templates, making possible an increased rate of protein synthesis.

Recent interest in this problem was sparked by Charnov's (1979) suggestion that intersexual and/or kin selection might account for the evolution of double fertilization and endosperm. Explanations involving kin selection have now been elaborated (or extended to other aspects of seed biology) by many authors, including Cook (1981), Westoby and Rice (1982), Queller (1983, 1984, 1989), Willson and Burley (1983), Law and Cannings (1984), Bulmer (1986), Mazer (1987), Haig and Westoby (1988, 1989a,b), Uma Shaanker et al. (1988), and Haig (1990). As applied to seeds, such arguments are meant to explain or predict the way(s) in which selection will act to alter the resource allocation of tissues based on their genetic relatedness. In general, it is argued that a typical triploid endosperm will be selected to garner fewer resources (i.e., be less aggressive) than its associated embryo, but should be selected to garner more resources than its female gametophyte.

In an outstanding review of the subject, Queller (1989) noted a significant shortcoming of arguments based on inclusive fitness: they do not actually account for the origin of new tissues, nor for their maintenance for a long enough period of time that kin selection could have a significant impact on patterns of resource allocation (also see Bulmer, 1986). Our aim is to focus directly on explanations for the origin and persistence of events in the development of the angiosperm seed. Accordingly, our arguments can be viewed as complementary to kin selection hypotheses. We attempt to explain the origin and maintenance of seed tissues, whereas kin selection arguments properly apply to the subsequent modification of these tissues with respect to allocation. Nevertheless, we also will advance the stronger claim that arguments based on intersexual or kin conflict are not necessary to explain what we know about the behavior of seed tissues. In arguing this point, our intention is not to take issue with kin selection in general, nor with kin selection as applied to plants or even to seed tissues. Instead, we contend that it is possible to identify explanations that are at least as simple, and that it is therefore unnecessary to invoke kin selection to explain either the origin *or* the subsequent modification of seed tissues.

Our analysis is consciously historical in that we begin by inferring a chronicle of evolutionary events based on recent phylogenetic analyses and then provide a narrative account for this sequence of changes (O'Hara, 1988; Donoghue, 1989; Brooks and McLennan, 1991). Consideration of factors involved in the origin of angiosperms suggests that selection for rapid reproduction and increased rate of development played a critical role (either directly or indirectly) in the evolution of seed tissues. We will argue that double fertilization was an incidental outcome of changes related directly to more rapid reproduction, and that polyploidy was selected to increase the growth rate and storage capacity of the resulting tissue. Similar arguments are also applied to the evolution of the female gametophyte within angiosperms. Tests of these hypotheses depend critically on obtaining better data on the relative and absolute timing of developmental events, and we hope to encourage more detailed analyses along these lines.

A Chronicle of Events

The analysis below relies directly on the phylogenetic results summarized in Figure 14.1, which are derived primarily from the cladistic studies of extant and extinct lines of seed plants conducted by Crane (1985, 1988) and Doyle and Donoghue (1986, 1992; Donoghue and Doyle, 1989a). Based mainly on morphological features, these studies concluded that an-

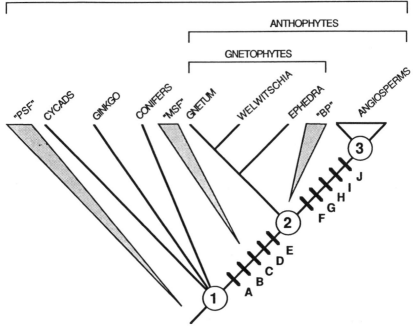

Figure 14.1. A phylogeny of relevant seed plants based on morphological and molecular evidence (references in text). Parsimony optimizations of relevant features at the internal nodes labeled 1, 2, and 3 are based upon the seven recent groups, for which the most complete data are available. The relative positions of various fossil taxa (presumably paraphyletic groups) are shown by shaded triangles: "PSF," various Paleozoic "seed ferns"; "MSF," various Mesozoic "seed ferns," including *Caytonia*; "BP," Bennettitales and *Pentoxylon*. Selected evolutionary changes between nodes 1 and 2: (A) more rapid reproduction; (B) smaller seeds/gametophytes; (C) siphonogamy/nonmotile sperm; (D) retention of the ventral canal nucleus; and (E) double fertilization. Selected evolutionary changes between nodes 2 and 3: (F) further reduction of the female gametophyte; (G) partial alveolarization; (H) loss of archegonia; (I) addition of a second female gametophyte nucleus/triploidy; and (J) other specializations of male and female gametophytes. See text for discussion.

giosperms are monophyletic and belong to a clade including the extant Gnetales and the extinct groups Bennettitales and *Pentoxylon*. The relatives of this anthophyte clade are various "seed ferns" of Mesozoic age. The monophyly of Gnetales is very well supported, as is the conclusion that *Welwitschia* and *Gnetum* are more closely related to each other than either is to *Ephedra*. Relationships among the other major lines of extant seed plants (ginkgos, cycads, conifers) are still uncertain (Doyle and Donoghue, 1992).

These basic phylogenetic results are consistent with morphological analyses of extant seed plant groups alone (Loconte and Stevenson, 1990), as well as with analyses of ribosomal RNA (Zimmer et al., 1989; Hamby and Zimmer, 1991) and nucleotide sequences of the chloroplast gene *rbc*L (M. Chase and R. Olmstead, personal communication). A combined analysis of morphological and ribosomal data yielded the same results (Doyle, Donoghue, and Zimmer, unpublished).

The sequence of character changes leading to endosperm can be determined by parsimoniously reconstructing (optimizing) the states at the internal nodes (presumed common ancestors) labeled 1, 2, and 3 in Figure 14.1. This approach was highlighted by Donoghue (1989) in reference to the evolution of endosperm and other characters, and was also employed recently by Friedman (1990a,b). Here we present a more detailed account of the basic conclusions. Before proceeding, however, it must be noted that most embryological features are unknown for the relevant fossil groups, and for this reason these characters were generally not included in the phylogenetic studies involving fossils. Unfortunately, missing data on fossils also prevent us from obtaining a more complete assessment of the states at the several nodes of interest, or the exact sequence of changes along particular branches. Instead, our reconstructions are based only on extant groups, for which more or less complete data are available. Except as noted below, basic information on seed development was obtained from Maheshwari and Singh (1967), Singh (1978), Foster and Gifford (1974), Sporne (1974, 1975), and Johri (1984). In *Ephedra*, we have relied primarily on the studies of Friedman (1990a,b), who also provided a thorough summary of previous literature.

Changes between Nodes 1 and 2

Cycads and ginkgos (represented by the extant species *Ginkgo biloba*) are critical in reconstructing the condition of the seed at node 1 in Figure 14.1 (representing the ancestor of modern seed plants). In both groups the megaspore mother cell (megasporocyte) divides by meiosis to produce a linear tetrad of spores. Three of the haploid products abort, leaving only the cell at the chalazal end of the ovule (monosporic development). This cell enlarges and undergoes a series of mitotic divisions without the formation of cell walls between nuclei, which results in a large number of free nuclei (e.g., more than 3000 nuclei in some cycads and over 8000 in some ginkgo plants). This phase is followed by "alveolarization," wherein cell walls are laid down between nuclei in a characteristic centripetal pattern.

The differentiation of (1–)2–6 archegonia at the micropylar end of the gametophyte begins with the division of an archegonial initial, resulting in a primary neck cell and a central cell. The primary neck cell gives rise to a four-celled neck, and the central cell divides to form the ventral canal nucleus and the egg nucleus. In ginkgo and most cycads a wall usually forms between the egg and the ventral canal nucleus, and, except under very rare circumstances (Bryan and Evans, 1957), the ventral canal nucleus soon disintegrates.

In both cycads and ginkgo two large multiflagellate sperms are differentiated after the last mitotic division in the male gametophyte. The sperm are released into the archegonial chamber, where usually only one of the two enters an archegonium and fertilizes an egg. Rarely, the second sperm enters the same archegonium, but it soon degenerates. Simple polyembryony (i.e., the development of embryos from more than one archegonium) is common in both groups, but typically only one embryo survives to seed maturity. Although it is possible for the two sperm from one male gametophyte to fertilize eggs in different archegonia, multiple embryos probably more often involve sperm from different male gametophytes. The two cotyledons remain inside the seed on germination and obtain nutrients for the developing seedling. A rather long time period is required to complete sexual reproduction in cycads and ginkgos, and we assume that was also true of the first seed plants. In ginkgo, for example, there may be as many as 5 months between pollination and fertilization, and maturation of the embryo may require another 8 months.

The phylogenetic relationship of conifers to other extant groups is still not well resolved (Doyle and Donoghue, 1992), and their bearing on the condition at node 1 is therefore uncertain. We have based our assessment at node 1 primarily on cycads and ginkgo rather than conifers. In most respects conifer seed development is similar to cycads and ginkgo, and where there are differences it appears that the condition in conifers is derived. Thus, the sperms of conifers lack flagellae and are delivered by the pollen tube directly to the archegonium (siphonogamy). In comparison with cycads and ginkgo, the female gametophyte is generally reduced. Molecular data sets, and those morphological analyses including fossil groups, favor the view that these derived traits of conifers evolved independently in anthophytes (everything from node 2 up in Fig. 14.1; see references above). On the other hand, the analysis of extant seed plant groups by Loconte and Stevenson (1990) concluded that conifers are directly united with anthophytes, implying that reduction of the gametophyte and siphonogamy are homologous in the two groups. However, even if this phylogenetic conclusion were correct, these features still appear to have evolved independently when fossils (e.g., early conifers) are taken into account (Donoghue et al., 1989; Doyle and Donoghue, 1992). It should also be noted that a connection between conifers and anthophytes would not change our basic conclusions on character evolution. If Loconte and Stevenson (1990) are correct, an additional node would be inserted between the base of the seed plants and the anthophytes, and the shift to siphonogamy and reduction of the gametophyte would simply be interpreted as antedating the origin of anthophytes.

The states present at node 2 in Figure 14.1 (the common ancestor of anthophytes) are probably best inferred by reference to extant Gnetales. Based on relationships within Gnetales, and on comparison with cycads, ginkgo, and conifers, the basal condition in Gnetales is best estimated by the condition in *Ephedra* (Donoghue, 1989; Friedman, 1990a,b). Many embryological features of *Welwitschia* and *Gnetum* (e.g., tetrasporic development, partial alveolarization) apparently evolved within Gnetales (Doyle and Donoghue, 1986). Although there is little direct information on seed development in the relevant fossil groups (*Caytonia*, Bennettitales, *Pentoxylon*), the seeds of these plants were smaller than those found in modern cycads or ginkgo, and we presume that their female gametophytes were correspondingly reduced (Harris, 1951, 1964; Stewart, 1983; Tiffney, 1986).

Initial stages in the development of the female gametophyte of *Ephedra* are similar to those in cycads, ginkgo, and conifers (Land, 1904; Maheshwari, 1935; Lehmann-Baerts, 1967). Meiosis results in a linear tetrad of megaspores, of which three degenerate, leaving the chalazal cell to undergo a series of free-nuclear divisions. Alveolarization results in a cellular gametophyte, with generally from 250 to 512 nuclei present at the time of wall formation. From 1–3(–12) archegonia are differentiated at the micropylar end of the gametophyte. Each archegonial initial divides to produce a primary neck cell and a central cell; in turn, the neck cell divides to produce a multicellular, multitiered neck, while the central cell (following pollination) divides to form the ventral canal nucleus and the egg nucleus. Most importantly, a wall is not formed between these two nuclei and the ventral canal nucleus is generally present at the time of fertilization (Friedman, 1990a,b).

In *Ephedra* a binucleate sperm cell is produced by each male gametophyte. In contrast to cycads and ginkgo, the sperm are nonmotile and are delivered directly to the archegonium, where both nuclei are discharged into the egg cell. One of the sperm nuclei fuses with the egg nucleus to form the zygote, and it appears that the second sperm often fuses with the persistent ventral canal nucleus (Land, 1907; Herzfeld, 1922; Maheshwari, 1935; Khan, 1940, 1943; Mulay, 1941; Narang, 1955; Moussel, 1978; Friedman, 1990a,b). The earlier reports are reviewed by Friedman (1990a,b), who has provided the most complete documentation of the process. In addition to demonstrating that this second fertilization is a regular event in *Ephedra*, Friedman (1990a,b) confirmed that the additional fusion occurs when the ventral canal nucleus is displaced away from the vacuolate apex of the egg cytoplasm toward the chalazal end of the archegonium, where it enters a distinctive column of cytoplasm rich in organelles (also see Moussel, 1977). Similar displacement of the ventral canal nucleus also seems to be correlated with anomalous egg-like behavior in cycads (Bryan and Evans, 1957) and conifers (Murrill, 1900; Hutchinson, 1915).

The fate of the diploid product of the second fertilization is not yet clear (Friedman, 1990a,b, and personal communication). It may undergo a limited number of mitotic divisions (Kahn, 1940, 1943), but there is no evidence of development of a second embryo or of a nutritive tissue, nor for the fusion of additional nuclei. Double fertilization has also been reported in *Gnetum* (Waterkeyn, 1954; Vasil, 1959; Martens, 1971) and *Welwitschia* (Pearson, 1909, 1929). In these plants additional free nuclei are present in the vicinity of the egg at the time of fertilization owing to incomplete alveolarization.

Embryo development in *Ephedra* begins with a series of free-nuclear divisions, with around eight nuclei produced before cell walls are laid down. Precocious cleavage polyembryony, wherein any or all of the first tier of eight cells independently develop into an embryo, occurs frequently (as in conifers). Simple polyembryony may also occur, but in either case only one embryo usually survives in each mature seed. As in cycads and ginkgo, two cotyledons are produced, and the female gametophyte serves as the nutritive tissue. The overall length of time required to complete the sexual cycle is considerably reduced in Gnetales as compared to cycads and ginkgo. For example, the time from pollination to fertilization may be as little as 10–36 hr in *Ephedra* (Land, 1970; Moussel, 1977).

Assessment of the states present at nodes 1 and 2 in Figure 14.1 allows us to infer a set of changes along the intervening branch, although it is not possible to determine the exact order in which these occurred. It appears that there was a shift to a more rapid sexual cycle and a reduction in overall size of the seed and female gametophyte. There also appears to have been a switch from haustorial male gametophyte development and production of swimming sperm to siphonogamous delivery of nonmotile sperm directly to the archegonium. Finally, we support previous arguments that double fertilization evolved between nodes 1 and 2, associated with the delivery of both sperm nuclei to the egg and the persistence of the ventral canal nucleus until the time of fertilization (Donoghue, 1989; Friedman, 1990a,b).

Changes between Nodes 2 and 3

Node 3 in Figure 14.1 represents the common ancestor of angiosperms. Within angiosperms there are a number of different patterns of female gametophyte development (for recent reviews see Battaglia, 1989; Haig,

1990). For the reasons given below, we agree with most other authors (e.g., Maheshwari, 1950; Davis, 1966; Takhtajan, 1969; Stebbins, 1974; Palser, 1975; Cronquist, 1988; Battaglia, 1989; Haig, 1990) that the ancestral form of female gametophyte development in angiosperms was probably most like the *Polygonum* type.

The development of the *Polygonum*-type embryo sac begins as it does in other seed plants (except *Gnetum* and *Welwitschia*): the megaspore mother cell divides by meiosis to produce a linear tetrad, of which three products abort and the remaining chalazal spore undergoes free-nuclear mitotic divisions. The mature female gametophyte consists of only eight nuclei in seven cells. Cell walls are formed so as to delimit three antipodal cells at the chalazal end of the gametophyte, two synergids and an egg cell at the micropylar end (the egg apparatus), and a large central cell containing two free polar nuclei (one the sister nucleus of the egg and the other the sister of an antipodal: Brink and Cooper, 1947). The egg in angiosperms is a free cell (i.e., an archegonium is not differentiated).

Two sperms are produced by the angiosperm male gametophyte and delivered directly to the female gametophyte by the pollen tube. One sperm nucleus fuses with the egg to form the zygote (which develops directly into the embryo), while the second sperm nucleus fuses with the two polar nuclei to form a triploid primary endosperm nucleus. In most cases the polar nuclei fuse prior to fertilization, but in others the three nuclei fuse more or less simultaneously (Davis, 1966). In either case, the primary endosperm nucleus then undergoes a series of mitotic divisions to become triploid endosperm (Vijayaraghavan and Prabhakar, 1984). In some cases this involves a series of free-nuclear divisions before cell walls are formed (nuclear development), whereas in other cases cell walls are laid down from the start (cellular development); "helobial" development, found among monocots, is an intermediate condition. Phylogenies of angiosperms support the view that cellular development is ancestral (Donoghue and Doyle, 1989b).

The view that the *Polygonum* type is ancestral has mainly been based on the observation that it is the most common and widespread type within angiosperms. Our own assessment is based first of all on outgroup comparison (Maddison et al., 1984). Although the *Polygonum* type is unknown outside of angiosperms, there are elements of this developmental program that are found in other seed plants, suggesting that other embryo sac types were derived within angiosperms. In particular, outgroup comparison implies that monosporic development, with mitotic divisions of the chalazal nucleus, is the primitive condition in angiosperms. Within angiosperms only the *Polygonum* type fits this description. The *Oenothera* type is monosporic, but its development is derived in that it is the micropylar nucleus that undergoes divisions. Other gametophyte types are bi- or tetrasporic.

Comparison with other seed plants also favors the *Polygonum* type on the grounds that it entails the largest number of mitotic divisions. The *Polygonum* type is said to be three-phasic, because a series of three divisions result in the mature eight-nucleate condition. By contrast, with only very rare exceptions, other types are two- or one-phasic (Battaglia, 1989; Haig, 1990).

Phylogenetic analyses within angiosperms also support the hypothesis that the *Polygonum* type is primitive. Optimization of an embryo sac development character on the most parsimonious trees of Donoghue and Doyle (1989b) unequivocally assigns the *Polygonum* type to the basal node. Indeed, among primary lines of angiosperms (e.g., Magnoliales, Laurales, Winterales, tricolpates, and paleoherbs, all sensu Donoghue and Doyle, 1989b), other types of development are very rare, so that almost any rooting would find the *Polygonum* type to be ancestral. Although the *Fritillaria* and *Peperomia* types (both tetrasporic and two-phasic) are found in Piperaceae, these appear to have been derived within Piperales (because the *Polygonum* type is found in Saururaceae). Therefore, even if the root were placed among paleoherbs (as suggested especially by ribosomal RNA sequences: Zimmer et al., 1989; Hamby and Zimmer, 1991), it would still be most parsimonious to conclude that the *Polygonum* type was ancestral in angiosperms.

We conclude from the foregoing considerations that between nodes 2 and 3 in Figure 14.1 there must have been a substantial reduction in size of the female gametophyte, a switch from complete to partial alveolarization (such that two free nuclei remain in the central cell at maturity), and the complete loss of archegonia. Most importantly, there was a change in the source of the nutritive tissue, from the haploid female gametophyte to triploid endosperm, which required the fusion of a second female gametophyte nucleus (the second polar nucleus). There were also significant changes in the structure and function of the angiosperm female gametophyte, such as specialization of the synergids for their role in fertilization (Willemse and van Went, 1984; van Went and Willemse, 1984). Related changes appear to have occurred in the angiosperm male gametophyte, including reduction of the number of divisions and especially derivation of the "male germ unit," which entailed the physical connection of the two sperm cells and their association with the vegetative nucleus (Dumas et al., 1984; Russell, 1991).

A Narrative Account

Having inferred a series of evolutionary events leading to endosperm, we now consider the possible causes of these changes. We recognize that this involves considerable speculation, perhaps more than many readers would condone. Our aim is to try to formulate a model that is consistent with what we know about phylogeny, and to provide an alternative to kin selection models (which are themselves highly speculative). In doing so, we hope to identify and encourage critical tests of these alternative explanations.

Changes between Nodes 1 and 2

We suggest that the changes between nodes 1 and 2 (more rapid reproduction, decreased seed size, siphonogamy and nonmotile sperm, double fertilization) were directly or indirectly related to the spread of seasonally drier, less equable, and generally less predictable climates during the Mesozoic (e.g., Parrish et al., 1986), and to the shift in anthophytes to growth in disturbed (e.g., riparian) habitats. These factors may have selected directly for increased speed of reproduction and possibly smaller seeds, and perhaps indirectly for an earlier onset of alveolarization and differentiation of the female gametophyte.

The origin of siphonogamy and nonmotile sperms might also have been related to life in drier habitats and selection for more rapid reproduction. Haustorial male gametophyte growth in cycads and ginkgo is a slow process (generally requiring months), and fertilization is dependent on the presence of liquid in the archegonial chamber. It is noteworthy that siphonogamy and nonmotile sperm probably evolved independently within conifers, also perhaps as an adaptation to drier climates (Doyle and Donoghue, 1986, 1992).

We agree with several previous suggestions that double fertilization, when it first originated, was an accidental consequence of the delivery of both male gametes directly to the archegonium and the presence of an additional female gametophyte nucleus at that time (see Meeuse, 1963, 1986; Donoghue, 1989; Friedman, 1990a,b). The persistence of the second nucleus, the sister nucleus of the egg, may have been an effect of selection for more rapid reproduction (i.e., the ventral canal nucleus simply had not disintegrated by the time of fertilization). This was probably largely a function of a shortening of the life cycle, especially the time between pollination and fertilization. Fusion of the second sperm with the ventral canal nucleus may simply have been a consequence of the tendency for compatible nuclei to undergo fusion, given the opportunity to do so (Khan, 1943).

This interpretation is consistent with the observation that fusion evidently occurs whenever sperm are delivered directly to the archegonium and there is an additional female nucleus present at the time of fertilization. Double fertilization is also found, for example, in *Welwitschia* and *Gnetum*, where both conditions hold. Moreover, when both of these conditions are not met, double fertilization has not been reported. Although conifers are siphonogamous, free female nuclei are usually not present at the time of fertilization. The few instances of double fertilization reported in conifers seem always to be associated with abnormal persistence of the ventral canal nucleus (Land, 1902; Nichols, 1910; Hutchinson, 1915; Allen, 1946).

Changes between Nodes 2 and 3

Many of the changes between nodes 2 and 3 in Figure 14.1 (further reduction of the female gametophyte, partial alveolarization, loss of archegonia) probably resulted from continued selection for rapid growth and reproduction. Paleoecological data suggest that the first angiosperms evolved in marginal, unpredictable habitats, where they were probably selected for an accelerated life cycle (Doyle and Hickey, 1976; Doyle, 1978, 1984; Hickey and Doyle, 1977; Crane, 1987; also see Cornet, 1986). Although it could be argued that each characteristic of the angiosperm seed and gametophyte was independently selected, other traits of angiosperms (e.g., the carpel) appear to be paedomorphic and may have originated through progenesis (Takhtajan, 1969; Doyle, 1978). This entire suite of characters might best be accounted for by continuation of a general progenetic shift that began earlier in the anthophyte line, in which case the angiosperm ovule can be interpreted as "underdeveloped" at the time of fertilization.

Why did a new tissue largely replace the nutritive function of the female gametophyte, and why was another female gametophyte nucleus added? In some nonangiospermous seed plants (cycads, ginkgo) the female gametophyte is already fully developed at the time of fertilization; in others (conifers, gnetophytes), the gametophyte is usually well developed (generally having reached mature size), although some provisioning occurs after fertilization (Haig and Westoby, 1989b). Because of the presumed shift in the timing of fertilization in the first angiosperms, the gametophyte would have been considerably reduced in size during early embryo development. This would have set the stage for the evolution of a nutritive tissue other than the gametophyte (Queller, 1983). We suggest that the diploid product of the second fertilization was selected over the haploid gametophyte, owing both to heterotic effects (Brink and Cooper, 1940, 1947) and to the effects of increased ploidy on cell size, tissue growth rate, and nutrient storage capacity (see below). The fusion of an additional female nucleus may have further enhanced the rate of growth and storage capacity of the nutritive tissue.

The switch from gametophyte to endosperm set the stage for further modifications. First, double fertilization became a highly dependable event. This presumably meant ensuring a favorable position for the second female nucleus within the egg cytoplasm (cf. *Ephedra*: Friedman, 1990a,b), which would also facilitate the subsequent growth of the product of the second fertilization. Second, with the evolution of a new nutritive tissue, the female gametophyte was selected for specialized functions during the earliest stages of development. Thus, the egg apparatus (the synergids in particular) appears to have evolved to facilitate the movement of sperms during fertilization. In addition, the central cell became specialized for nutrient storage and transfer very early in embryo development (Willemse and van Went, 1984), and the timing of fusion of the polar nuclei shifted. These observations contradict the common assumption that the angiosperm female gametophyte is a functionless vestige. Finally, there were significant changes in the structure of the male gametophyte, especially the evolution of the male germ unit, which presumably increased the effectiveness of sperm delivery and ensured simultaneous transmission of the two gametes (Russell, 1991).

Possible Objections

The new and probably most controversial element in this explanation concerns the impact of increasing the amount of DNA. An increase in DNA content is generally correlated with an increase in total cell volume, presumably due to mechanisms that maintain a more or less fixed ratio of nucleus to cytoplasm volume (e.g., Price, 1976; D'Amato, 1977; Cavalier-Smith, 1978, 1985a,b,c). In turn, cellular demand for metabolites and metabolic rate might rise through an increase in the rate of transcription and translation owing to the availability of additional DNA templates. Although an increase in the volume of DNA also tends to slow down the rate of cell division (Nagl, 1974; Nagl and Ehrendorfer, 1975; Price and Bachmann, 1976; Cavalier-Smith, 1978), this is not necessarily the case when DNA content is increased by polyploidy. Indeed, polyploids apparently do not show an increase in the DNA synthesis period (Troy and Wimber, 1968), and there are cases in which somatic cell cycle time is significantly shorter in polyploids in comparison to related diploid species or to diploid tissues in the same species (Bennett, 1973, and references therein). Cavalier-Smith (1985b) suggested a mechanism to explain such observations: multiplication rate in polyploids may not be greatly impacted by DNA amount because cell volume is increased without a corresponding decrease in gene concentration.

Based on these considerations, we propose that increasing the amount of DNA by polyploidy would have had the effect of increasing the rate at which a resulting tissue could fill a given volume in comparison to the haploid gametophyte tissue (also see Grime and Mowforth, 1982; Grime et al., 1985). This follows if polyploid cells are larger but divide at about the same or at a greater rate. Furthermore, we reason that the maximum rate of volume growth might be achieved with some intermediate amount of DNA per cell. Haploid cells would be smaller, so that the overall growth of the tissue might be retarded, even if cell division rates were somewhat higher. In contrast, much larger amounts of DNA may slow cell division rate, such that the rate of growth would be reduced.

In view of the evident relationship between genome size and cell division rate, Cavalier-Smith (1978) actually made the opposite suggestion, namely that selection for rapid seed development should favor haploidy of endosperm cells. However, we believe that this expectation fails to take into account the contribution of increased cell size to filling rate, and the observation that polyploidy may not have a significant impact on cell division rate. In any case, this difference of opinion highlights that the critical factor in our account is the net effect of the interaction of cell size and division rate on volume increase. Although the evidently rapid development of endosperm (e.g., Brink and Cooper, 1940, 1947) is certainly consistent with our scenario, it is clear that detailed quantitative comparisons are needed to evaluate critically the interaction of cell size and division rate on the growth of seed tissues.

The argument just given concerns the rate of growth of the nutritive tissue, but there are perhaps more important factors that must be considered, especially the function of endosperm as a storehouse of energy and nutrients to be utilized during later stages of embryo growth and on seed germination. We suggest that an increase in the amount of DNA and in cell size would also have resulted in a more efficient storage tissue (cf. Cavalier-Smith, 1978). In the early stages of endosperm development, the need for rapid growth may have imposed limits on the increase in cell size, as suggested above. During later stages of development, however, as cell division rate declines, there appear to be distinct advantages to large cell size and increased metabolic activity. For lipid storage, the same vacuole volume can· be accommodated in a smaller number of larger cells, with fewer resources tied up in such things as cell membranes, cell walls, and organelles. Likewise, in the case of carbohydrates stored in thick cell walls, larger cells are less costly than an equivalent volume of smaller cells. Increase in cell volume and activity seems to have been achieved primarily through endoreduplication, wherein chromosome replication occurs without mitotic reduction. It is well known that mature endosperm cells generate very large volumes of DNA through this process (D'Amato, 1984) and obtain very large sizes (Vijayaraghavan and Prabhakar, 1984; Jacobsen, 1984). We also note that other plant tissues specialized for rapid nutrient uptake and transfer (e.g., the tapetum in anthers) are characterized by high DNA content because of endopolyploidy (Kapil and Tawari, 1978).

According to our account, the dual requirements of rapid growth and nutrient storage are accomplished by two separate mechanisms affecting the amount of DNA: fusion of nuclei early in development and endopolyploidy later on. Greatly increasing the amount of DNA by endopolyploidy at an early stage might constrain growth rate by slowing the rate of cell division, whereas fusion of a very large number of nuclei would be necessary to increase DNA content to the levels achieved through endopolyploidy later in development.

Aside from concerns about the role of nucleotypic effects, our account may be seen as incomplete for several reasons. One might ask, for example, why there should have been fusion of a male with a female nucleus (rather than two female nuclei, for example), if increased vigor is simply a function of the amount of DNA? In response, it should first be noted that we have not rejected the arguments of Brink and Cooper (1940, 1947) regarding the benefits of heterozygosity. We have simply added another argument based on nucleotypic effects. We think that both played a role in the selection of polyploid endosperm over the female gametophyte. Second, based on the phylogenetic arguments above, double fertilization probably originated in anthophytes prior to the divergence of the angiosperm line. If not, it would have been very likely to occur early in angiosperm evolution owing to siphonogamy and the probable presence of a free female nucleus at the time of fertilization. Thus, we would argue that retention and modification of the product of the second fertilization would have been the path of least resistance, simply because this product was already available.

Another problem concerns the identity of the added nucleus: why was this a second female gametophyte nucleus rather than some other nucleus? For that matter, why not modify a portion of the developing embryo, the nucellus, or the integumentary tissue? Although available evidence cannot rule out the possibility that double fertilization in the angiosperm line yielded a triploid product directly, there is no precedent for an initial fusion of two female nuclei in related plants (except perhaps in *Welwitschia* and *Gnetum*, which represent a highly specialized line within Gnetales). If we therefore assume an initially diploid product, we would argue that the fusion of a second female nucleus would have been the most likely outcome. Fusion of an additional male nucleus would require the presence in the embryo sac of additional male gametes. This would entail the production of additional sperm by increasing the number of mitotic divisions in the development of the male gametophyte. Production of additional sperm is known to occur in some seed plants (especially some conifers: Willson and Burley, 1983), but not in Gnetales or among potentially basal angiosperms (Maheshwari, 1950). Fusion of a diploid nucleus of sporophyte origin seems unlikely in view of the fact that any such nuclei would already have been walled off. Elaboration of other seed tissues as a nutritive tissue would entail a change in the developmental program of the relevant cells, which would already be committed to different functions. Of course, in some cases with angiosperms, nucellar tissue has been modified for a nutritive function (perisperm), which indicates that this is possible and perhaps an option that was easier than reinstating the nutritive role of the gametophyte.

A third problem arises from concerns that underlie kin conflict arguments (discussed below). Rapid early growth and increased storage/transfer capacity could be viewed as increasing the "aggressiveness" of the endosperm. This is seen as conflicting with the interests of the sporophyte, which is viewed as partitioning a fixed (or significantly limited) quantity of nutrients among its offspring. We believe that the strength of this conflict may have been exaggerated, and that evidence presented in favor of conflict is consistent with alternative explanations. The development of invasive haustoria (Masand and Kapil, 1966) may not reflect conflict; instead, such structures may simply have been selected for their role in garnering nutrients for the rapidly developing embryo. The observation that such haustoria are most often formed by embryos and endosperms (Queller, 1983) may simply reflect the fact that these tissues are undergoing growth at the appropriate time and are not already committed to other functions (as opposed to the gametophyte or adjacent sporophyte tissues). Likewise, the role of hypostase tissue, found in only some angiosperm seeds, is still disputed (see Masand and Kapil, 1966; Haig and Westoby, 1988, and references therein).

In this regard, it is important to consider relevant aspects of source-sink relationships in plants, in particular the finding that there is not a fixed pool of photosynthate resource. Seeds, particularly endosperm tissues, tend to be rich in lipids relative to other plant tissues, providing a high-energy source of carbon for embryo and seedling development (Mooney, 1972; Levin, 1974; Vijayaraghavan and Prabhakar, 1984; Jacobsen, 1984). Thus, the developing endosperm is a sink for carbon and energy. The increased demand for photosynthates by the endosperm sink can result in an increased rate of photosynthesis in source tissues (Sweet and Wareing, 1966; Evans and Rawson, 1970; Geiger, 1976; Watson and Casper, 1984). Embryos with different photosynthate sources, such as those in different fruits, would not be affected. Furthermore, if the sporophyte is limited by some other resource, such as nitrogen, a more "aggressive" endosperm can increase carbon and energy resources for its associated embryo without decreasing the amount available to the sporophyte. Indeed, it is often nitrogen, not photosynthate, that limits sporophyte growth (see Tilman, 1984, and references therein).

Even though the pool of photosynthate may not be limited on a whole plant basis, for a single sink (e.g., a seed or fruit) there is still a limit to the amount of photosynthate that can be supplied per unit time (Garrish and Lee, 1989). In cycads, ginkgo, and conifers, seed development may take many months. Even with the lower photosynthetic rates characteristic of conifers (Mooney, 1972), there is ample time to supply the developing seed with sufficient photosynthate. If the amount of time for seed development were significantly limited, however, then the rate of photosynthesis might become a limiting factor. Under these circumstances a polyploid tissue might even have been selected so as to increase the carbon and energy sink and, thereby, increase photosynthetic rate.

Explanations Involving Kin Selection

The application of kin selection arguments to the evolution of endosperm has been carefully reviewed elsewhere (e.g., Queller, 1989), and it is not our intention to provide a detailed critique. Instead, we wish to highlight several general features that are relevant in evaluating the role of kin selection and then to contrast aspects of previous scenarios with our own. In making comparisons with other explanations, our aim is not to give an exhaustive account of previous arguments, but rather to begin to identify important differences and critical tests. More generally, we hope to convince the reader that explanations such as ours, which do not invoke kin selection, warrant renewed attention.

General Considerations

Kin selection explanations are largely elaborations of two models presented by Charnov (1979). In the first model, double fertilization makes the endosperm more similar genetically to its associated embryo than to embryos in other seeds borne by the same sporophyte and, therefore, a stronger competitor for nutrients for its embryo. In Charnov's second model, double fertilization permits the male gametophyte to play a role in garnering resources for its offspring. The increased competition for nutrients is viewed in this case as a conflict of interest between the male and female parents as to how the seed should be provisioned. Both models assume resource trade-offs leading to conflict over the distribution of investment (Trivers, 1974): when more resources are directed to one embryo, other embryos on the parent sporophyte suffer a cost, either by a direct reduction in resources or indirectly through effects on the survival of the sporophyte. In general, provisioning by tissues within the seed is selected to increase at a cost to other seeds on the sporophyte to the extent that the cost–benefit ratio is less than the coefficient of relatedness (r) with other seeds (Hamilton, 1964). When embryos borne on a given sporophyte are sired by different fathers and are therefore half-sibs ($r = \frac{1}{4}$), the embryo should favor its seed more than the endosperm will, the endosperm more than the female gametophyte, and the female gametophyte more than the parent sporophyte.

Several assumptions of published kin selection models may limit their relevance in accounting for the evolution of endosperm. Kin selection models generally have assumed (for simplicity) the existence of a single locus that controls the nutrient-garnering ability ("aggressiveness") of the embryo, endosperm, gametophyte, and parent sporophyte (Queller, 1984; Law and Cannings, 1984; Bulmer, 1986). There is as yet no direct evidence for the existence of such a locus, and we interpret several studies as casting doubt on this assumption (Nakamura and Stanton, 1989; Schwaegerle and Levin, 1990). Haig and Westoby (1989b) reviewed evidence for parent-specific gene expression in maize and postulated the existence of a second set of loci (epistatic with the first) that would yield these genotypic effects. Multilocus models may give the same qualitative outcome as single-locus models, but this will depend on details of the genetic system.

In addition, genetic models are not entirely robust to changes in assumptions regarding the degree of relatedness and the timing of the effects of selfishness. Models that result in strong kin selection effects are based on outcrossing plants that store significant amounts of energy, so that the immediate effects of selfishness are not felt by contemporary embryos. The outcome predicted on the basis of simple calculations of relatedness may not be obtained when selfishness affects contemporary seeds that may be more closely related than half-sibs (compare Queller, 1984; Law and Cannings, 1984). Queller (1989) provided a thorough analysis of this point and concluded that the predicted rank-order of aggressiveness (embryo>endosperm>gametophyte>mother) is generally upheld. However, he also identified several conditions under which it is not; in general, when overconsumption affects siblings nonrandomly. Even if the rank order held, however, the predicted strength of selection would differ depending on the level of outcrossing and the life form of the plants under consideration.

The fossil record of angiosperms suggests that they were colonizing plants at first, living primarily in disturbed environments, where they were strongly selected for rapid maturation and reproduction (Stebbins, 1974; Doyle and Hickey, 1976). The first angiosperms were probably insect-pollinated (most likely by beetles and/or flies: Crepet and Friis, 1987) and may have produced flowers with multiple ovules. Under these circumstances, inbreeding might have been common and seeds in a single fruit might frequently have been full-sibs. Moreover, it may well have been contemporary siblings that would have felt the effects of an overconsuming embryo. If so, the likelihood that kin selection played an important role in the evolution of endosperm is decreased.

It is also important to note that most such models assume that the fitness of a seed is a concave monotonically increasing function of the amount of resource that it garners. The optimal resource allocation for a given seed vis-à-vis the parent sporophyte exists at the inflection point of the fitness function, where more benefit to the sporophyte is gained by allocating additional resources to a different seed. These assumptions may be inappropriate for several reasons. First, if increased demand by the seed increases the available pool of resources, as described above, then an optimal allocation point may not exist. Second, even if an optimal resource allocation exists, seeds may be constrained below the optimal level, perhaps by a limit on the rate of development (e.g., Benner and Bazzaz, 1985). Finally, kin selection models (as standardly formulated) assume that the critical determinant of fitness is the amount of resource, but if a particular size or shape of seed enhances dispersal or germination (or some other critical attribute), then increasing the amount of resource obtained could actually decrease fitness. Our own model, in contrast, focuses directly on rate of development and storage efficiency of the nutritive tissue as critical factors. In this connection it is noteworthy that weedy angiosperms in ephemeral environments typically produce small seeds (presumably as an adaptation for rapid development and/or dispersal: Salisbury, 1942; Harper et al., 1970; Levin, 1974), and that the early angiosperms also produced small seeds (Tiffney, 1986).

Scenarios, Predictions, and Tests

Kin selection arguments have sometimes been accompanied by specific scenarios for the evolution of double fertilization and endosperm. In this section we briefly consider several of these (Westoby and Rice, 1982; Queller, 1983; Willson and Burley, 1983) in order to highlight ways in which our own account differs. Predictions based on these alternatives suggest a number of tests.

Westoby and Rice (1982) proposed that selection for deferment of maternal investment in offspring, which would allow a better assessment of offspring quality, was a driving force in the evolution of endosperm. The first evolutionary step was a switch in the timing of fertilization, such that it occurred prior to extensive development of the female gametophyte. This allowed maternal investment after fertilization and placed the embryo in a position to garner resources aggressively. The evolution of endosperm was a response on the part of the sporophyte to interpose a less aggressive tissue between it and the embryo.

Westoby and Rice (1982) implied that precocious fertilization arose as an adaptation for deferment of investment. According to our explanation (and others: e.g., Takhtajan, 1976; Queller, 1983), this switch was an effect of selection for rapid maturation and was therefore not itself an adaptation (sensu Gould and Vrba, 1982; Greene, 1986; Coddington, 1988). We agree that precocious fertilization set the stage for the evolution of endosperm, but we believe, with Queller (1983), that it did so by placing the product of double fertilization on nearly equal footing with the much-reduced gametophyte. Under these circumstances, the polyploid tissue would have been favored by selection because of its increased growth rate and ability to garner and store nutrients. Whereas Westoby and Rice (1982) would predict a decrease in "aggressiveness" with the addition of a second maternal nucleus, we argue that the triploid tissue would have been more aggressive.

According to Westoby and Rice (1982), alternative hypotheses do not explain the observation that endosperm is found only where maternal investment is deferred until after fertilization or the fact that endosperm contains more doses of maternal than paternal genes. Regarding the correlation, reduction of the female gametophyte (and deferment of maternal investment) created the need for a rapidly growing storage tissue, setting the stage for the evolution of endosperm. According to our hypothesis, triploidy was selected for its effects on vigor and an extra maternal genome was added because it was readily available, whereas other possibilities were not (see above).

Queller (1983) recognized that reduction of the female gametophyte, brought about by selection to reproduce quickly, might permit double fertilization to occur and would put the gametophyte and the product of the second fertilization on nearly equal footing. Kin selection enters his argument in explaining the presumably gradual switch in the source of the nutritive tissue. The role of the (diploid) "endosperm" would increase because it would be selected to garner more resources for its embryo. This increase would be countered by a concomitant reduction in the gametophyte, because it would presumably be advantageous to the sporophyte to maintain a fixed level of resource allocation. Triploidy is seen as a possible maternal strategy to reduce the aggressiveness of the endosperm, but its appearance was not integrated by Queller (1983) into the sequence of events.

This scenario is generally consistent with our own except for one important point. In our view, the switch to endosperm came about because it grew more rapidly and yielded a better storage tissue. Any further reduction in the size of the gametophyte occurred because there were limits on seed size and only the most efficient storage tissue was retained. Furthermore, the presence of a new nutritive tissue meant that the gametophyte could become increasingly specialized for its role in fertilization. We hypothesize that the addition of a second female gametophyte nucleus was favored because it further increased the endosperm's growth rate and storage capacity.

Queller (1983) made several predictions based on kin selection, including (1) metabolic activity should be highest in endosperm, which should predominate over the female gametophyte in nourishing the embryo; (2) when endosperm is lost, its function should be taken over by the embryo; and (3) there should be reduced conflict in self-fertilizing or apomictic species. The first prediction is consistent with our hypothesis, but we assume that increased metabolic activity resulted from the nucleotypic effects of polyploidy, not from single locus genotypic effects. Our argument also implies that triploidy should increase metabolic activity, not decrease it, as predicted by most kin selection models.

In support of the second prediction, Queller (1983) noted that most angiosperms store seed reserves either in the endosperm or in the embryo, rather than in the gametophyte or maternal tissue. However, as he also noted, the evidence is equivocal. Thus, loss of endosperm function in the Caryophyllidae was accompanied by the evolution of perisperm, which originates from the nucellus (parent sropopophyte), while in grasses, antipodal cells (of gametophytic origin) may play a nutritional role early in development (cf. Willemse and van Went, 1984). Queller (1983) observed that endosperm has been lost in Orchidaceae, Trapaceae, and Podostemaceae, and that in these cases there has been compensation in the form of absorptive suspensors (proembryo tissue). As we suggested above, haustoria do not provide unambiguous evidence of conflict. In any case, Queller (1983) also indicated that there may be compensation by the parental sporophyte in these plants.

Regarding the third prediction, Queller (1983) cited a comparison of an apomictic and a sexual species of *Taraxacum* (Cooper and Brink, 1949). In support of decreased conflict, he noted that apomicts averaged fewer endosperm cells at a given embryo size, but we find these data difficult to interpret. Cooper and Brink (1949) showed that in the absence of normal fertilization the development of the embryo and endosperm are uncoupled, but it is not clear that this resulted from selection for diminished conflict. The number of endosperm cells varies enormously during early stages of embryo growth, but apomictic endosperms apparently grow more rapidly at later stages and reach larger sizes. This is consistent with our account, as the apomicts also seem to have a larger genome size in this case.

Willson and Burley (1983) summarized a variety of kin selection arguments, but they reemphasized Charnov's (1979) idea that the evolution of double fertilization reflected male–female conflict. They argued that selection for polyploidy is due to "gene level" effects (not "individual level" effects), because polyploidy changes the number of loci but does not affect relatedness. The possible effects of polyploidy on nucleus and cell volume, growth rate, and RNA synthesis were discussed by Willson and Burley (1983), but these factors were not explicitly incorporated into a model for the origin of double fertilization and endosperm. Finally, based on the observation that in many modern angiosperms fusion of maternal nuclei occurs before arrival of the sperm, they suggested that the evolution of the fusion of polar nuclei might have preceded double fertilization.

In our model, double fertilization is seen as an incidental by-product of selection for rapid reproduction, continued reduction of the female gametophyte, and siphonogamy, not the outcome of male–female conflict. In contrast to Willson and Burley (1983) we argue that polyploidy was favored because of its nucleotypic, not genotypic, effects. Seed plant phylogenies, and the evidence from *Ephedra* reviewed above, indicate that double fertilization evolved before the addition of a second female nucleus, and we therefore interpret the precocious fusion of polar nuclei as having been derived within the angiosperm line.

From the foregoing comparisons it should be clear that even when kin selection is invoked, it is envisioned to have played a rather limited role. Many of the critical events in these scenarios, such as the reduction of the female gametophyte, the origin of double fertilization, and the initial persistence of the product of double fertilization, are simply not explained by kin selection. Because much of our account deals with these events, many of our suggestions are not at odds with kin selection arguments. In other words, if there are disagreements about the mechanisms we have proposed for these events, these should not be interpreted as disagreements over the role of kin selection.

There do appear to be a few critical differences between our account and kin selection accounts—differences that actually concern the role of kin selection. In several of the scenarios presented above, kin selection is thought to have been critical in the transition from dominance of the female gametophyte to dominance of polyploid endosperm. A diploid tissue derived from a second fertilization should be selected to garner more resources, and the gametophyte responds by garnering less, thereby maintaining the same overall level of support on the part of the sporophyte. Our alternative is that the polyploid cell line was selected because it grew faster than the haploid cell line and made a more efficient storage tissue. The gametophyte was then reduced in favor of the more efficient tissue and became increasingly specialized for fertilization. Our account identifies an immediate outcome on which selection could act, whereas kin selection arguments do not (Queller, 1989).

A second difference concerns the consequences of adding a second female gametophyte nucleus. Some kin selection arguments interpret this as the sporophyte's way of making a less aggressive nutritive tissue, which implies that the triploid endosperm should show a decrease in metabolic activity and growth rate. This has been a major difficulty, because the endosperm contains qualitatively the same alleles as the embryo. To explain how a decrease might come about, Haig and Westoby (1989b) invoked parent-specific gene expression coupled with competition for limiting factors (e.g., nucleotides) in polyploid cells (see Queller, 1989). In contrast, we propose that the triploid tissue would have had a higher metabolic level, growth rate, and storage capacity (also see Bulmer, 1986). Here, it will be critical to make quantitative comparisons of the growth rates of diploid versus triploid endosperms, making use of naturally occurring variants or experimentally manipulated tissues. Broad phylogenetic comparisons (e.g., comparing Onagraceae, which have diploid endosperms, to related plants with triploid endosperm) might not be very revealing, as a variety of factors may have subsequently influenced endosperm growth. Because our suggestions concern the immediate consequences of an increase in ploidy, comparisons of tissues within species or among close relatives are more appropriate. Additional comparisons of sexual and apomictic plants, along the lines reported by Cooper and Brink (1949), might be especially useful.

These considerations highlight a more fundamental difference between our account and kin selection models. Kin selection arguments assume the existence of one or more genes that control the nutrient garnering ability of the endosperm. While we assume that there are genes that are expressed in the endosperm, which influence endosperm structure and function, it remains to be seen whether there are genes that have the particular effects required by kin selection models. It would obviously be of great significance in evaluating the alternatives if genes of this sort were discovered. In our view the amount of DNA can have overriding and more immediate effects upon which selection might act. As outlined above, evidence already exists for such effects, but quantitative studies of tissue growth are needed to evaluate our proposal further.

Finally, a point of departure between our hypothesis and the male–female conflict models (Charnov, 1979; Willson and Burley, 1983) concerns whether or not double fertilization arose as an adaptive strategy. An assumption of male–female conflict is that some immediate advantage existed for those organisms in which double fertilization first occurred. In contrast, we favor the view that double fertilization was an accident and did not confer any immediate advantage. We predict that double fertilization will occur automatically whenever one or more free female nuclei are present in the vicinity of the egg at the time that sperm are delivered (if both sperm are viable). Friedman's (1990a,b) studies of double fertilization in *Ephedra* provide support for this position, but the movement of the ventral canal nucleus to a favorable position within the egg cytoplasm requires further attention. In particular, we suppose that this movement occurred (at least initially) as a consequence of other changes during the development of the cell (e.g., movement of the egg nucleus/zygote). If there were significant variation in the occurrence of double fertilization in *Ephedra* and if it were possible to determine whether it had taken place, this might provide a means of testing whether there are any immediate advantages associated with double fertilization. Clearly, we need to know more about the fate of the product of double fertilization in *Ephedra*. In any case, Friedman's (1990a,b) observations suggest that double fertilization may be largely under maternal control, which conflicts with the view that it is a male strategy.

Diversity within Angiosperms

The account presented above points to the importance of selection for increased rate of development, the role of chance in relation to fusion of nuclei, and the benefits of increased DNA content in connection with tissue growth rate and storage capacity. These same factors could also explain the origin and maintenance of much of the embryological variation seen within angiosperms, some of which has been interpreted as the outcome of kin selection (e.g., Haig, 1990).

Based on our best understanding of angiosperm phylogeny, modifications of *Polygonum*-type development are known to have originated many times independently and to occur sporadically throughout angiosperms, but the regular occurrence of derived types is phylogenetically quite limited and evidently nonrandom in distribution. In particular, bisporic and tetrasporic types appear to us to have evolved most often in aquatic plants (e.g., Podostemaceae), in parasitic plants (e.g., Loranthaceae), and in plants that are otherwise highly reduced or grow in extreme environments (e.g., *Adoxa*). Better phylogenetic tests of the sequence of evolutionary events are needed, but these correlations suggest that selection to speed up sexual reproduction might have driven the reduction and elimination of steps in the development of the female gametophyte and endosperm.

The occurrence of bisporic and tetrasporic types in Liliaceae (sensu lato) appears to be an exception to the idea that modifications of female gametophyte development are related to selection for rapid reproduction. In some geophytic groups, however, the time available to complete early reproductive events may be highly constrained, especially where these take place just before the onset of cold weather and dormancy. There may also be other factors involved, which foster selection for derived types. For example, it has been suggested (Geeta Bharathan, personal communication) that the large genome sizes characteristic of many of these monocots (Grime and Mowforth, 1982) may slow the rate of cell division, and that this might have selected for the elimination of developmental events (e.g., cell wall formation).

The fusion of additional female nuclei in some angiosperms (resulting in endosperm of higher ploidy) and changes from a bipolar to tetrapolar arrangement of nuclei in the embryo sac (e.g., the *Peperomia* type in Piperaceae) might have originated as an outcome of shifts in the timing of mitotic divisions in relation to cell size, cell shape, and the timing of vacuole formation. Such shifts may have altered the proximity of female gametophyte nuclei (and hence their propensity to undergo fusion) and the orientation of divisions. Although the fusion of additional nuclei may have originated as an accidental consequence of other developmental changes, endosperms of higher ploidy may have been selected for their effects on growth rate and storage capacity.

The arguments just presented to account for changes within angiosperms obviously closely parallel our arguments for the origin of endosperm. Selection to speed up reproduction may have resulted in reduction of the female gametophyte in some lines of angiosperms and elimination of developmental steps. For example, elimination of wall formation after meiosis and maintenance (and subsequent division) of all four products of meiosis would result in the tetrasporic pattern of development. These changes may have brought about the fusion of additional female gametophyte nuclei, especially when accompanied by reduction of the ovule and changes in embryo sac shape. As noted above, the evolution of much higher ploidy levels may have been constrained by the negative effects of very large amounts of DNA on cell division rate, with further changes in cell size and storage capacity being brought about instead by endoreduplication later in development.

Our reactions to kin selection arguments that have been applied to the evolution of the angiosperm female gametophyte also parallel the arguments given above. First, it is important to note that kin selection explanations have not been given for the basic changes in embryo sac development described above (Haig, 1990). For example, no explanation has been provided for the shift from monosporic to bisporic or tetrasporic embryo sacs or for the shift occurring in particular habitats or plant groups. Instead, kin selection has been viewed as effecting subsequent changes in derived embryo sacs, where there might have been conflict among genetically distinct nuclei. Haig (1990) suggested, for example, that the elimination in some tetrasporic embryo sacs of some or all of the chalazal divisions leading to the antipodals (a phenomenon known as "strike") may have been the result of selection to minimize competition for fertilization between these nuclei and the egg. He postulated that if the antipodals were retained this would have allowed the production of "antipodal eggs" and that suppression of divisions was selected as a mechanism to avoid this.

As Haig (1990) himself noted, there is no evidence for this scenario. In particular, antipodal eggs are not found in tetrasporic groups, so that the supposed problem to which "strike" is the solution might never have existed. Furthermore, we can provide a simpler explanation that does not involve kin selection, which is consistent with our observations on the distribution and possible causes of tetrasporic development. Given that antipodals are expendable (as they seem to be in most groups), and given that selection for rapid reproduction was intense, it stands to reason that the divisions giving rise to these cells might have been omitted. Again, our explanation highlights timing and efficiency.

Conclusions

Previous discussions of the evolution of endosperm have focused on mechanisms such as heterosis or kin selection and have argued from supposed selection pressures to a sequence of evolutionary events (e.g., Haig and Westoby, 1989b). In contrast, we inferred a chronicle of evolutionary events by reference to a hypothesis of seed plant relationships and then considered the possible causes of the hypothesized character state changes (O'Hara, 1988). This approach helps to focus attention on how changes inside the seed might have been related to environmental and morphological changes associated with the origin of angiosperms and to ecological shifts within angiosperms. Moreover, analysis of evolutionary changes in a phylogenetic context focuses attention on interrelationships among changes and possible common causes. For example, a single cause (viz. selection for more rapid reproduction) may have been responsible for siphonogamy, reduction of the female gametophyte, and, incidentally, double fertilization. Likewise, selection within angiosperms for rapid reproduction may have driven the evolution of bisporic and tetrasporic embryo sacs.

A phylogenetic perspective also suggests how one change may have set the stage for another (Donoghue, 1989). Thus, with double fertilization already established, further reduction in the size of the gametophyte apparently set the stage for the elaboration of a polyploid nutritive tissue. In this light, double fertilization can be seen as an exaptation (Gould and Vrba, 1982; Greene, 1986; Baum and Larson, 1991), its regular occurrence having preceded the function with which it later became associated (Donoghue, 1989). Even changes that are spatially or historically separated from one another may have had a common underlying cause. For example, it may be that a polyploid nutritive tissue, endoreduplication, and fusion of additional nuclei in the development of some angiosperm female gametophytes were all selected as a function of the beneficial effects of increasing the amount of DNA per cell on tissue growth rate and/or storage capacity.

We have argued that the available evidence is consistent with an explanation for the origin of endosperm that does not require intersexual or kin selection. Indeed, conflict arguments are unable to explain some of the changes that appear to have been critical in the evolution of endosperm (e.g., selection for rapid reproduction), and they fail to account for the origin and persistence of the tissues on which kin selection is supposed to have acted. The mechanisms we have identified (e.g., incidental fusion of nuclei; benefits associated with increased DNA) are meant to explain the origin and maintenance of seed tissues. Kin selection might have operated to shift the allocation of resources in these tissues, but, as we have suggested, there may be even simpler explanations. The same sorts of arguments also apply to changes within angiosperms. Kin conflict arguments have not been advanced for the origin of derived embryo sac types, whereas we can provide explanations for these shifts in terms of development time, as well as for the minor changes to which kin selection arguments have been applied.

Most importantly, our analysis focuses attention on the need for more concrete information on the timing of developmental events. In the past, emphasis has been placed primarily on describing the fusion of various nuclei and on the theoretical consequences of the existence of different genotypes within the seed. Our account points to the possibly overriding effects of the volume of DNA present in cells and its impact on growth rates. The limited information available on rates of development is consistent with our arguments, but much more data of this sort are needed in order to choose among competing hypotheses.

Additional Evidence on Post-Fertilization in Ephedra

Friedman (1992) recently confirmed earlier suggestions (Kahn, 1943) that the product of the second fertilization event in *Ephedra* can undergo several mitotic divisions. The four nuclei produced by two sets of free-nuclear divisions are diploid, and there is no evidence that they function in nutrition of the embryo. Therefore, it appears that triploidy and nutritive function are derived features of angiosperms.

According to Friedman (1992) the development of the second fertilization product in *Ephedra* is "fundamentally similar" to free-nuclear endosperm development in angiosperms, which he presumed to be the primitive condition (following Stebbins, 1974). However, as he noted, free-nuclear development in angiosperms is far more extensive than it is in *Ephedra*. Furthermore, in the case of free-nuclear development in angiosperms cell walls are usually not formed until the embryo is differentiated. Finally, Friedman's (1992) assumption that free-nuclear endosperm development is primitive in angiosperms is questionable: it is not supported by phylogenetic analyses of angiosperms (Donoghue and Doyle, 1989a), which imply that cellular development is ancestral. If cellular endosperm development is primitive in angiosperms, then the transition to endosperm must have involved a change in the timing of cell wall formation.

Acknowledgments

We thank Larry Venable and his Plant Population Ecology seminar of 1985 for stimulating our interest in this subject, and L. Aviles, B. Baldwin, G. Bharathan, J. Doyle, M. Folsom, W. Friedman, B. Knox, D. Lloyd, S. Mazer, D. Queller, and R. Wyatt for thoughtful comments on the ideas presented here. David Queller provided a very helpful review of the manuscript.

Literature Cited

Allen, G.S. 1946. Embryogeny and development of the apical meristems of *Pseudotsuga*. I. Fertilization and early embryogeny. Am. J. Bot. **33**:666–677.

Battaglia, E. 1989. The evolution of the female gametophyte of angiosperms: An interpretative key. (Embryological questions: 14). Ann. Bot. (Rome) **47**:7–144.

Baum D.A., and A. Larson. 1991. Adaptation reviewed: A phylogenetic methodology for studying character macroevolution. Syst. Zool. **40**:1–18.

Benner, B.L., and F.A. Bazzaz. 1985. Response of the annual *Abutilon theophrasti* Medic. (Malvaceae) to timing of nutrient availability. Am. J. Bot. **72**:320–323.

Bennett, M.D. 1973. The duration of meiosis. *In* M. Balls and F.S. Billett, eds., The Cell Cycle in Development and Differentiation. Cambridge University Press, Cambridge, pp. 111–131.

Brink, R.A., and D.C. Cooper. 1940. Double fertilization and development of the seed in angiosperms. Bot. Gaz. **102**:1–25.

Brink, R.A., and D.C. Cooper. 1947. The endosperm in seed development. Bot. Rev. **13**:423–541.

Brooks, D.R., and D.A. McLennan. 1991. Phylogeny, Ecology, and Behavior. University of Chicago Press, Chicago.

Bryan, G.S., and R.I. Evans. 1957. Types of development from the central nucleus of *Zamia umbrosa*. Am. J. Bot. **44**:404–415.

Bulmer, M.G. 1986. Genetic models of endosperm evolution in higher plants. *In* S. Karlin and E. Nevo, eds., Evolutionary Process and Theory. Academic Press, New York, pp. 743–763.

Cavalier-Smith, T. 1978. Nuclear volume control by nucleoskeletal DNA, selection for cell volume and cell growth rate, and the solution of the DNA C-value paradox. J. Cell Sci. **34**:247–278.

Cavalier-Smith, T. 1985a. Introduction: The evolutionary significance of genome size. *In* T. Cavalier-Smith, ed., The Evolution of Genome Size. John Wiley, New York, pp. 1–36.

Cavalier-Smith, T. 1985b. Cell volume and the evolution of eukaryotic genome size. *In* T. Cavalier-Smith, ed., The Evolution of Genome Size. John Wiley, New York, pp. 105–184.

Cavalier-Smith, T. 1985c. DNA replication and the evolution of genome size. *In* T. Cavalier-Smith, ed., The Evolution of Genome Size. John Wiley, New York, pp. 211–251.

Charnov, E.L. 1979. Simultaneous hermaphroditism and sexual selection. Proc. Natl. Acad. Sci. U.S.A. 76:2480–2484.

Coddington, J. 1988. Cladistic tests of adaptational hypotheses. Cladistics **4**:3–22.

Cook, R.E. 1981. Plant parenthood. Nat. Hist. **90**:30–35.

Cooper, D.C., and R.A. Brink. 1949. The endosperm-embryo relationship in an autonomous apomict, *Taraxacum officinale*. Bot. Gaz. **111**:139–153.

Cornet, B. 1986. The leaf venation and reproductive structures of a Late Triassic angiosperm, *Sanmiguelia lewisii*. Evol. Theory 7:231–309.

Crane, P.R. 1985. Phylogenetic analysis of seed plants and the origin of angiosperms. Ann. Missouri Bot. Gard. **72**:716–793.

Crane, P.R. 1987. Vegetational consequences of the angiosperm diversification. *In* E.M. Friis, W.G. Chaloner, and P.R. Crane, eds., The Origins of Angiosperms and Their Biological Consequences. Cambridge University Press, Cambridge, pp. 107–144.

Crane, P.R. 1988. Major clades and relationships in the "higher" gymnosperms. *In* C.B. Beck, ed., Origin and Evolution of Gymnosperms. Columbia University Press, New York, pp. 218–272.

Crepet, W.L., and E.M. Friis. 1987. The evolution of insect pollination in angiosperms. *In* E.M. Friis, W.G. Chaloner, and P.R. Crane, eds., The Origins of Angiosperms and Their Biological Consequences. Cambridge University Press, Cambridge, pp. 181–201.

Cronquist, A. 1988. The Evolution and Classification of Flowering Plants, 2nd ed. New York Botanical Garden, Bronx, NY.

D'Amato, F. 1977. Nuclear Cytology in Relation to Development. Cambridge University Press, Cambridge.

D'Amato, F. 1984. Role of polyploidy in reproductive organs and tissues. *In* B.M. Johri, ed., Embryology of Angiosperms. Springer-Verlag, Berlin, pp. 519–566.

Davis, G.L. 1966. Systematic Embryology of Angiosperms. John Wiley, New York.

Donoghue, M.J. 1989. Phylogenies and the analysis of evolutionary sequences, with examples from seed plants. Evolution 43:1137–1156.

Donoghue, M.J., and J.A. Doyle. 1989a. Phylogenetic studies of seed plants and angiosperms based on morphological characters. *In* B. Fernholm, K. Bremer, and H. Jornvall, eds., The Hierarchy of Life. Elsevier, Amsterdam, pp. 181–193.

Donoghue, M.J., and J.A. Doyle. 1989b. Phylogenetic analysis of angiosperms and the relationships of Hamamelidae. *In* P. Crane and S. Blackmore, eds., Evolution, Systematics and Fossil History of the Hamamelidae, Vol. 1. Clarendon Press, Oxford, pp. 17–45.

Donoghue, M.J., J.A. Doyle, J. Gauthier, A.G. Kluge, and T. Rowe. 1989. The importance of fossils in phylogeny reconstruction. Annu. Rev. Ecol. Syst. **20**:431– 460.

Doyle, J.A. 1978. Origin of angiosperms. Annu. Rev. Ecol. Syst. **9**:365–392.

Doyle, J.A. 1984. Evolutionary, geographic, and ecological aspects of the rise of angiosperms. *In* Proceedings of the 27th International Geological Congress, 2 (Moscow, 1984). VNU Science Press, Utrecht, pp. 23–33.

Doyle, J.A., and M.J. Donoghue. 1986. Seed plant phylogeny and the origin of angiosperms: An experimental cladistic approach. Bot. Rev. **52**:321–431.

Doyle, J.A., and M.J. Donoghue. 1992. Fossils and seed plants revisited. Brittonia 44:in press.

Doyle, J.A., and L.J. Hickey. 1976. Pollen and leaves from the mid-Cretaceous Potomac group and their bearing on early angiosperm evolution. *In* C.B. Beck, ed., Origin and Early Evolution of Angiosperms. Columbia University Press, New York, pp. 139–206.

Dumas, C., R.B. Knox, C.A. McConchie, and S.D. Russell. 1984. Emerging physiological concepts in fertilization. What's New Plant Phsiol. **15**:17–20.

Evans, L.T., and H.M. Rawson. 1970. Photosynthesis and respiration by the flag leaf and components of the ear during grain development in wheat. Aust. J. Biol. Sci. **23**:245–254.

Foster, A.S., and E.M. Gifford. 1974. Comparative Morphology of Vascular Plants, 2nd ed. W. H. Freeman, San Francisco.

Friedman, W.E. 1990a. Double fertilization in *Ephedra*, a nonflowering seed plant: Its bearing on the origin of angiosperms. Science **247**:951–954.

Friedman, W.E. 1990b. Sexual reproduction in *Ephedra nevadensis* (Ephedraceae): Further evidence of double fertilization in a nonflowering seed plant. Am. J. Bot. **77**:1582–1598.

Friedman, W.E. 1992. Evidence of a pre-angiosperm origin of endosperm: Implications for the evolution of flowering plants. Science **225**:336–339.

Garrish, R.S., and T.D. Lee. 1989. Physiological integration in *Cassia fasciculata* Michx.: Inflorescence removal and defoliation experiments. Oecologia **81**:279– 284.

Geiger, D.R. 1976. Effects of translocation and assimilate demand on photosynthesis. Can. J. Bot. **54**:2337–2345.

Gould, S.J., and E.S. Vrba. 1982. Exaptation—a missing term in the science of form. Paleobiology **8**:4–15.

Greene, H.W. 1986. Diet and arboreality in the monitor, *Varanus prasinus*, with comments on the study of adaptation. Fieldiana Zool. **31**:1–12.

Grime, J.P., and M.A. Mowforth. 1982. Variation in genome size—an ecological interpretation. Nature (London) **299**:151–153.

Grime, J.B., J.M.L. Shacklock, and S.R. Band. 1985. Nuclear DNA contents, shoot phenology and species co-existence in a limestone grassland community. New Phytol. **100**:435–445.

Haig, D. 1990. New perspectives on the angiosperm female gametophyte. Bot. Rev. **56**:236–274.

Haig, D., and M. Westoby. 1988. Inclusive fitness, seed resources, and maternal care. *In* J. Lovett Doust and L. Lovett Doust, ed., Plant Reproductive Ecology Patterns and Strategies. Oxford University Press, Oxford, pp. 60–79.

Haig, D., and M. Westoby. 1989a. Selective forces in the emergence of the seed habit. Biol. J. Linn. Soc. **38**:215–238.

Haig, D., and M. Westoby. 1989b. Parent-specific gene expression and the triploid endosperm. Am. Nat. **134**:147–155.

Hamby, R.K., and E.A. Zimmer. 1991. Ribosomal RNA as a phylogenetic tool in plant systematics. *In* P.S. Soltis, D.E. Soltis, and J.J. Doyle, eds., Molecular Systematics in Plants. Chapman & Hall, New York, pp. 50–91.

Hamilton, W.D. 1964. The genetical evolution of social behavior. J. Theor. Biol. **7**:1–52.

Harper, J.L., P.H. Lovell, and K.G. Moore. 1970. The shapes and sizes of seeds. Annu. Rev. Ecol. Syst. **1**:327–356.

Harris, T.M. 1951. The relationships of the Caytoniales. Phytomorphology **1**:29–39.

Harris, T.M. 1964. The Yorkshire Jurassic flora. II. Caytoniales, Cycadales & Pteridosperms. British Museum (Natural History), London.

Herzfeld, S. 1922. *Ephedra campylopoda* Mey. Morphologie der weiblichen Blute und Befruchtungvorgang. Denkschriften Acad. Wiss. Wein. **98**:243–268.

Hickey, L.J., and J.A. Doyle. 1977. Early Cretaceous fossil evidence for angiosperm evolution. Bot. Rev. **43**:3–104.

Hutchinson, A.H. 1915. Fertilization in *Abies balsamea*. Bot. Gaz. **60**:457–472.

Jacobsen, J.V. 1984. The seed: Germination. *In* B.M. Johri, ed., Embryology of Angiosperms. Springer-Verlag, Berlin, pp. 611–646.

Johri, B.M., ed. 1984. Embryology of Angiosperms. Springer-Verlag, Berlin.

Johri, B.M., and K.B. Ambegaokar. 1984. Embryology: Then and now. *In* B.M. Johri, ed., Embryology of Angiosperms. Springer-Verlag, Berlin, pp. 1–52.

Kapil, R.N., and S.C. Tawari. 1978. The integumentary tapetum. Bot. Rev. **44**:457–490.

Kahn, R. 1940. A note on "double fertilization" in *Ephedra foliata*. Curr. Sci. **9**:323–334.

Khan, R. 1943. Contributions to the morphology of *Ephedra foliata* Boiss. II. Fertilization and embryogeny. Proc. Natl. Acad. Sci. India **13**:357–375.

Land, W.J.G. 1902. A morphological study of *Thuja*. Bot. Gaz. **34**:249–259.

Land, W.J.G. 1904. Spermatogenesis and oogenesis in *Ephedra trifurca*. Bot. Gaz. **38**:1–18.

Land, W.J.G. 1907. Fertilization and embryogeny in *Ephedra trifurca*. Bot. Gaz. **44**:273–292.

Law, R., and C. Cannings. 1984. Genetic analysis of conflicts arising during development of seeds in the Angiospermophyta. Proc. R. Soc. London Ser. B **221**:53–70.

Lehmann-Baerts, M. 1967. Etude sur les Gnetales—XII. Ovule, gametophyte femelle et embryogenese chez *Ephedra distachya* L. Cellule **67**:53–87.

Levin, D.A. 1974. The oil content of seeds: An ecological perspective. Am. Nat. **108**:193–206.

Loconte, H., and D.W. Stevenson. 1990. Cladistics of the Spermatophyta. Brittonia **42**:197–211.

Maddison, W.P., M.J. Donoghue, and D.R. Maddison. 1984. Outgroup analysis and parsimony. Syst. Zool. **33**:83–103.

Maheshwari, P. 1935. Contributions to the morphology of *Ephedra foliata* Boiss. I: The development of the male and female gametophytes. Proc. Indian Acad. Sci. **1**:586–601.

Maheshwari, P. 1950. An Introduction to the Embryology of Angiosperms. McGraw-Hill, New York.

Maheshwari, P., and H. Singh. 1967. The female gametophyte of gymnosperms. Biol. Rev. **42**:88–130.

Martens, P. 1971. Les Gnetophytes (Handbuch der Pflanzenanatomie, Band 12, Teil 2). Gebrueder Borntraeger, Berlin.

Masand, P., and R.N. Kapil. 1966. Nutrition of the embryo sac and embryo—a morphological approach. Phytomorphology **16**:158–175.

Mazer, S.J. 1987. Maternal investment and male reproductive success in angiosperms: Parent-offspring conflict or sexual selection. Biol. J. Linn. Soc. **30**:115–133.

Meeuse, A.D.J. 1963. Some phylogenetic aspects of the process of double fertilization. Phytomorphology **13**:136–144.

Meeuse, A.D.H. 1986. Again: Double fertilization and the mono- versus pleiophyletic evolution of angiosperms. Phytomorphology **36**:17–21.

Mooney, H.A. 1972. The carbon balance of plants. Annu. Rev. Ecol. Syst. **3**:315–346.

Moussel, B. 1977. Structures protoplasmiques au sein de la cavite archegoniale et modalites de la proembryogenese chez l'*Ephedra distachya* L. Rev. Cytol. Biol. Veg. **40**:73–123.

Moussel, B. 1978. Double fertilization in the genus *Ephedra*. Phytomorphology **28**:336–345.

Mulay, B.N. 1941. The study of the female gametophyte and a trophophyte in *Ephedra foliata* Boiss. J. Univ. Bombay **10**:56–69.

Murrill, W.A. 1900. The development of the archegonium and fertilization in the hemlock spruce *Tsuga canadensis*. Ann. Bot. **14**:583–607.

Nagl, W. 1974. Role of heterochromatin in the control of cell cycle duration. Nature (London) **249**:53–54.

Nagl, W., and F. Ehrendorfer. 1975. DNA content, heterochromatin, mitotic index, and growth in perennial and annual Anthemidae (Asteraceae). Plant Syst. Evol. **123**:35–54.

Nakamura, R.R., and M.L. Stanton. 1989. Embryo growth and seed size in *Raphanus sativus*: Maternal and paternal effects *in vivo* and *in vitro*. Evolution **43**:1435–1443.

Narang, N. 1955. Contributions to the life history of *Ephedra campylopoda*. II. Fertilization and embryogeny. Proc. Indian Sci. Cong. **3**:224.

Nichols, G.E. 1910. A morphological study of *Juniperus communis* var. *depressa*. Beiheft Bot. Zentralb. **25**:201–241.

O'Hara, R.J. 1988. Homage to Clio, or, toward an historical philosophy for evolutionary biology. Syst. Zool. **37**:142–155.

Palser, B.F. 1975. The bases of angiosperm phylogeny: Embryology. Ann. Missouri Bot. Gard. **62**:621–646.

Parrish, J.M., J.T. Parrish, and A.M. Ziegler. 1986. Permian-Triassic paleogeography and paleoclimatology and implication for therapsid distribution. *In* N. Hotton, P.D. MacLean, J.J. Roth, and E.C. Roth, eds., The Ecology and Biology of Mammal-like Reptiles. Smithsonian Institution Press, Washington, D. C., pp. 109–131.

Pearson, H.H.W. 1909. Further observations on *Welwitschia*. Phil. Trans. R. Soc. London Ser. B **200**:331–402.

Pearson, H.H.W. 1929. *Gnetales*. Cambridge University Press, Cambridge.

Price, J.H. 1976. Evolution of DNA content in higher plants. Bot. Rev. **42**:27–52.

Price, H.J., and K. Bachmann. 1976. Mitotic cycle time and DNA in annual and perennial Microseridinae (Compositae, Cichoriodeae). Plant Syst. Evol. **126**:323–330.

Queller, D.C. 1983. Kin selection and conflict in seed maturation. J. Theor. Biol. **100**:153–172.

Queller, D.C. 1984. Models of kin selection on seed provisioning. Heredity **53**:151–165.

Queller, D.C. 1989. Inclusive fitness in a nutshell. *In* P.H. Harvey and L. Partridge, eds., Oxford Surveys in Evolutionary Biology, Vol. 6. Oxford University Press, pp. 73–109.

Russell, S.D. 1991. Isolation and characterization of sperm cells in flowering plants. Annu. Rev. Plant Physiol. Plant Mol. Biol. **42**:189–204.

Salisbury, E.J. 1942. The Reproductive Capacity of Plants. G. Bell and Sons, London.

Sargant, E. 1900. Recent work on the results of fertilization in angiosperms. Ann. Bot. **14**:689–712.

Schwaegerle, K.E., and D.A. Levin. 1990. Quantitative genetics of seed size variation in *Phlox*. Evol. Ecol. **4**:143–148.

Singh, H. 1978. Embryology of Gymnosperms (Handbuch der Pflanzenanatomie 10). Gebrueder Borntraeger, Berlin.

Sporne, K.R. 1974. The Morphology of Gymnosperms. Hutchinson, London.

Sporne, K.R. 1975. The Morphology of Angiosperms. St. Martin's Press, New York.

Stebbins, G.L. 1974. Flowering Plants: Evolution Above the Species Level. Harvard University Press, Cambridge, MA.

Stebbins, G.L. 1976. Seeds, seedlings, and the origin of angiosperms. *In* C.B. Beck, ed., Origin and Early Evolution of Angiosperms. Columbia University Press, New York, pp. 300–311.

Stewart, W.N. 1983. Paleobotany and the Evolution of Plants. Cambridge University Press, Cambridge.

Sweet, G.B., and P.E. Wareing. 1966. Role of plant growth in regulating photosynthesis. Nature (London) **210**:77–79.

Takhtajan, A. 1969. Flowering Plants—Origin and Dispersal. Edinburgh University Press, Edinburgh.

Takhtajan, A. 1976. Neoteny and the origin of flowering plants. *In* C.B. Beck, ed., Origin and Early Evolution of Angiosperms. Columbia University Press, New York, pp. 207–219.

Tiffney, B.H. 1986. Evolution of seed dispersal syndromes according to the fossil record. *In* D.R. Murray, ed., Seed Dispersal. Academic Press, Sydney, pp. 273–305.

Tilman, G.D. 1984. Plant dominance along an experimental nutrient gradient. Ecology **65**:1445–1453.

Trivers, R.L. 1974. Parent-offspring conflict. Am. Zool. **14**:249–264.

Troy, M.R., and D.E. Wimber. 1968. Evidence for a constancy of the DNA synthetic period between diploid-polyploid groups in plants. Exp. Cell Res. **53**:145–154.

Uma Shaanker, R., K.N. Ganeshaiah, and K.S. Bawa. 1988. Parent offspring conflict, sibling rivalry and brood size patterns in plants. Annu. Rev. Ecol. Syst. **19**:177–205.

van Went, J.L., and M.T.M. Willemse. 1984. Fertilization. *In* B.M. Johri, ed., Embryology of Angiosperms. Springer-Verlag, Berlin, pp. 273–318.

Vasil, V. 1959. Morphology and embryology of *Gnetum ula* Brongns. Phytomorphology **9**:167–215.

Vijayaraghavan, M.R., and K. Prabhakar. 1984. The endosperm. *In* B.M. Johri, ed., Embryology of Angiosperms. Springer-Verlag, Berlin, pp. 319–376.

Waterkeyn, L. 1954. Etude sur les Gnetales—I. Le strobile femelle, l'ovule et de grains de *Gnetum africanum* Welw. Cellule **56**:103–146.

Watson, M.A., and B.B. Casper. 1984. Morphogenetic constraints on patterns of carbon distribution in plants. Annu. Rev. Ecol. Syst. **15**:233–258.

Westoby, M., and B. Rice. 1982. Evolution of the seed plants and inclusive fitness of plant tissues. Evolution **36**:713–724.

Willemse, M.T.M., and J.L. van Went. 1984. The female gametophyte. *In* B.M. Johri, ed., Embryology of Angiosperms. Springer-Verlag, Berlin, pp. 159–196.

Willson, M.F., and N. Burley. 1983. Mate Choice in Plants. Princeton University Press, Princeton.

Zimmer, E.A., R.K. Hamby, M.L. Arnold, D.A. LeBlanc, and E.C. Theriot. 1989. Ribosomal RNA phylogenies and flowering plant evolution. *In* B. Fernholm, K. Bremer, and H. Jornvall, eds., The Hierarchy of Life. Elsevier, Amsterdam, pp. 205–214.

Index